Starch Nanomaterials and Food Applications

Germán Ayala Valencia
Editor

Starch Nanomaterials and Food Applications

Springer

Editor
Germán Ayala Valencia
Department of Chemical and Food Engineering
Federal University of Santa Catarina
Florianópolis, SC, Brazil

ISBN 978-3-031-60085-2   ISBN 978-3-031-60086-9 (eBook)
https://doi.org/10.1007/978-3-031-60086-9

This Springer imprint is published by the registered company Springer Nature Switzerland AG
The registered company address is: Gewerbestrasse 11, 6330 Cham, Switzerland

This book is dedicated especially to my daughter and wife Liz and Talita, respectively.

Mi pequeña Liz, tu llegada ilumino mi vida.
Tu haz hecho los días más felices, divertidos y cálidos.
A tu lado siendo que todo lo he encontrado.
Talita, las palabras se quedan cortas para expresar el amor que siento por ti.
Tu eres tinta indeleble y parte de mí.
Simplemente, eres mi alma gemela

I love you so much!

Germán Ayala Valencia, Ph.D.
Editor

Preface

Starch nanomaterials are a promising alternative for obtaining advanced structures with improved physicochemical properties. These nanomaterials can be used to stabilize hydrophilic and hydrophobic biomolecules, and to modify food texture and the physicochemical properties of packaging materials. Starch nanomaterials can be obtained by "top-down" and "bottom-up" approaches and are classified as starch nanocrystals, nanoparticles, nanofibers, nanomicelles, and nanovesicles. This book reviews concepts related to starch nanomaterials and their food and packaging applications. Thus, Chap. 1 introduces and classifies different starch nanoarchitectures. The methods used to produce starch nanocrystals, nanoparticles, nanofibers, nanomicelles, and nanovesicles have been reviewed in Chaps. 2, 3, 4, and 5. The potential applications of starch nanomaterials as adsorptive and encapsulating materials are reviewed in Chap. 6. Chapters 7, 8, and 9 analyze the use of starch nanomaterials as food emulsifiers and ingredients for the manufacture of functional foods and packaging, respectively. The digestibility of starch nanomaterials has been revised in Chap. 10. Finally, Chap. 11 reviews the degradation and environmental impacts of starch nanomaterials.

I would like to sincerely thank the contributors for their insights into *Starch Nanomaterials and Food Applications*.

Florianópolis, SC, Brazil — Germán Ayala Valencia

Contents

About the Editor

Germán Ayala Valencia received his Ph.D. in food engineering in 2017 from the University of São Paulo, Brazil. Dr. Valencia is a professor at the Department of Chemical and Food Engineering in the Federal University of Santa Catarina, Florianópolis, Brazil. Prof. Valencia was the guest editor of a special issue published by Starch—Stärke (11–12/2021). Nowadays, Prof. Valencia is a Board Member of the *Journal of Packaging Technology and Science* and editor of the book *Natural Additives in Foods* published by Springer Nature. Prof. Valencia mainly works in the food science and technology area with an emphasis on packaging, pigments, nanotechnology, agro-industrial waste, and encapsulation of bioactive compounds.

Contributors

Edith Agama-Acevedo Instituto Politécnico Nacional, CEPROBI, Yautepec, Mexico

Carla Ivonne La Fuente Arias Research Institute of Food Engineering for Development (IIAD), Universitat Politècnica de València (UPV), Valencia, Spain

Germán Ayala Valencia Department of Chemical and Food Engineering, Federal University of Santa Catarina, Florianópolis, SC, Brazil

Luis A. Bello-Perez Instituto Politécnico Nacional, CEPROBI, Yautepec, Mexico

Valeria Cedeño-Fierro Faculty of Engineering and Administration, Universidad Nacional de Colombia, Palmira, Valle del Cauca, Colombia

Wilson Daniel Caicedo Chanon Department of Chemical and Food Engineering, Federal University of Santa Catarina, Florianópolis, SC, Brazil

Raul Remor Dalsasso Department of Chemical and Food Engineering, Federal University of Santa Catarina, Florianópolis, SC, Brazil

Samile Bezerra de Aguiar São Carlos Institute of Chemistry, University of São Paulo – USP, São Carlos, SP, Brazil

Shirley Duarte Faculty of Chemistry, National University of Asunción, San Lorenzo, Paraguay

Omayra Ferreiro Faculty of Chemistry, National University of Asunción, San Lorenzo, Paraguay

Juan F. Grisales-Mejia Faculty of Engineering and Administration, Universidad Nacional de Colombia, Palmira, Valle del Cauca, Colombia

Maria A. Hernandez-Aguirre Instituto Politécnico Nacional, CEPROBI, Yautepec, Mexico

Marco A. Maldonado-Campos Instituto Politécnico Nacional, CEPROBI, Yautepec, Mexico

Bianca Chieregato Maniglia São Carlos Institute of Chemistry, University of São Paulo – USP, São Carlos, SP, Brazil

Alcilene Rodrigues Monteiro Department of Chemical and Food Engineering, Federal University of Santa Catarina, Florianópolis, SC, Brazil

Sarah L. Paz-Arteaga Faculty of Sciences, Universidad Nacional de Colombia, Medellín, Antioquia, Colombia

Milena Dutra Pierezan Department of Food Science and Technology, Agricultural Sciences Center, Federal University of Santa Catarina, Florianópolis, SC, Brazil

Vania Zanella Pinto Food Engineering, Graduate Program of Food Science and Technology, Universidade Federal da Fronteira Sul, UFFS, Laranjeiras do Sul, PR, Brazil

Anna María Polanía Faculty of Engineering, School of Food Engineering, Universidad del Valle, Tuluá, Valle del Cauca, Colombia

Nathiely Ramírez-Gúzman Cuerpo Académico Agro-biotecnología, Facultad de Ciencias Biológicas, Universidad Autonoma de Coahuila, Unidad Torreón, Torreón, Coahuila, Mexico

Lina Maria Rayo-Mendez Food, Bioprocessing and Nutrition Department, North Carolina State University, Raleigh, NC, USA

Jaiber Humberto Rodriguez-Llanos Engineering Department, Anhanguera University, São Paulo, Brazil

Pedro Augusto Invernizzi Sponchiado Departamento de Química, Faculdade de Filosofia, Ciências e Letras de Ribeirão Preto – FFCLRP, Universidade de São Paulo – USP, Ribeirão Preto, SP, Brazil

Cristian Torres-León Cuerpo Académico Agro-biotecnología, Centro de Investigación y Jardín Etnobiológico (CIJE), Universidad Autonoma de Coahuila, Unidad Torreón, Viesca, Coahuila, Mexico

Natalia S. Vargas-Marín Agroindustrial Engineering. Faculty of Engineering, Corporación Universitaria Minuto de Dios - UNIMINUTO, Valle del Cauca, Colombia

José Vega-Baudrit Centro Nacional de Alta Tecnología "Dr. Franklin Chang Díaz", Laboratorio Nacional de Nanotecnología (LANOTEC), San José, Costa Rica
POLIUNA, Escuela de Química, Universidad Nacional, Heredia, Costa Rica

Silvani Verruck Department of Food Science and Technology, Agricultural Sciences Center, Federal University of Santa Catarina, Florianópolis, SC, Brazil

Abbreviations

AgNPs	Silver nanoparticles
AH	Acid hydrolysis
ALNs	Amylolipid nano-vesicles
AM	Amylose
ANN	Annealing
AP	Amylopectin
AS	Acetylated starch
ASSR	Anti-solvent/solvent ratio
ASP	Anti-solvent precipitation
ASTM	American Society for Testing and Materials
BSN	Banana starch nanoparticle
BIO	Biodegradation
BOD	Biological oxygen demand
C	Control
CABSN	Citric-acid banana starch nanoparticles
CI	Complexity index
CM	Chlorpheniramine maleate
CMC	Critical micelle concentration
CNC	Cellulose nanocrystal
CNFs	Cellulose nanofibers
DDSA	Dodecenylsuccinic acid
DIC	Disseminated intravascular coagulation
DIS	Disintegration
DMSO	Dimethyl sulfoxide
DNA	Deoxyribonucleic acid
DPG	Dipalmitoylglycerol
DS	Degree of substitution
DSC	Differential scanning calorimetry
EAI	Emulsifying activity index

EE	Encapsulation efficiency
EH	Enzymatic hydrolysis
EN	European Standards
ESI	Emulsifying stability index
FAO	Food and agriculture organization
FDA	Food and drug administration
FTIR	Fourier-transform infrared spectroscopy
GHG	Greenhouse gas
GI	Gamma irradiation
GP	Gastric phase
GRAS	Generally recognized as safe
GRS	Glutinous rice starch
HMT	Heat-moisture-treatment
HDPE	High-density polyethylene
HPH	High-pressure homogenization
HS	Hydroxypropylated starch
ICM	Isolated crystalline material
IP	Intestinal phase
ISO	International organisation for standardisation
ISNPs	Insoluble starch nanoparticles
L	Large
LA	Lauric acid
LC	Loading capacity
LCA	Life cycle assessments
LDPE	Low-density polyethylene
M	Medium
MntMD	Montmorillonite modified with dimethyl dihydrogenated tallow ammonium
MPG	Monopalmitoylglycerol
MVBs	Multivesicular bodies
Ni	Not informed
NMnt	Natural montmorillonite
mRNAs	Messenger RNA
NREU	Non-renewable energy use
OP	Oral phase
OS	Octenyl-succinate starch
OSA	Octenyl succinic anhydride
O/W	Oil-in-water
PA	Palmitic acid
PBAT	Poly(butylene adipate-co-terephthalate)
PBS	Buffered solution
PCL	Polycaprolactone
PDI	Polydispersity index
PEG	Polyethylene glycol

PEGME	Poly(ethylene glycol) methyl ether
PHU	Polyhydroxyurethane
PLA	Poly (lactic acid)
PPC	Poly(propylene carbonate)
PU	Polyurethane
PVA	Polyvinyl alcohol
RC	Relative crystallinity
RDS	Rapidly digestible starch
RNA	Ribonucleic acid
RS	Resistant starch
RS1	Resistant starch (type 1)
RS2	Resistant starch (type 2)
RS3	Resistant starch (type 3)
RS4	Resistant starch (type 4)
RS5	Resistant starch (type 5)
RVA	Rapid visco analyser
S	Small
SDS	Slowly digestible starch
SEM	Scanning electron microscopy
SGC	Short glucan chains
SGF	gastrointestinal fluid
SIF	simulated intestinal fluid
SLR	Solid-liquid ratio
SMPs	Starch microparticles
SNCs	Starch nanocrystals
SNFs	Starch nanofibers
SNPs	Starch nanoparticles
SSNPs	Soluble starch nanoparticles
SNMs	Starch nanomicelles
SNMts	Starch nanomaterials
SNVs	Starch nanovesicles
SPI	Soy protein isolate
SSPS	Soluble soybean polysaccharide
TEM	Transmission electron microscopy
TG	Thermogravimetry
TPS	Thermoplastic starch
TPU	Thermoplastic polyurethane
WHO	World health organization
WPI	Whey protein isolate
W/O	Water-in-oil
XRD	X-ray diffraction
βLG	β-lactoglobulin
ΔH	Enthalpy change

Chapter 1
A Brief Introduction to Starch-Based Nanomaterials

Wilson Daniel Caicedo Chacon, Raul Remor Dalsasso, Vania Zanella Pinto, and Germán Ayala Valencia

1.1 Introduction

Starch is a natural semicrystalline polymer composed of amylose, a linear biopolymer of α-D-glucose units linked by α-1,4-glycosidic bonds, and amylopectin, a highly branched biopolymer in which the α-D-glucose units are linked via α-1,4- and α-1,6-glycosidic bonds [1]. Starch is considered the second most common biomass on earth. This macromolecule is produced by green plants as an energy reserve and stored in tubers/roots, fruits, and grains/seeds in the form of granules with different shapes and particle sizes varying between 2 and 100 μm [2, 3].

Starch is used as a thickener, binding agent, emulsifier, clouding agent, or gelling agent in the food sector. However, this macromolecule has many disadvantages associated with poor water solubility and shear and thermal stability, resulting in insufficient physical and functional properties [4, 5]. Physical and chemical treatments have been used to improve the physicochemical properties of starch [6].

Modifying the starch architecture from the micro to the nanoscale is a recent alternative to alter the physicochemical properties of the resulting starch nanomaterials. Furthermore, these starch nanoarchitectures can be functionalized with bioactive compounds to acquire antimicrobial and antioxidant properties [2]. Starch nanostructures are defined as materials having at least one dimension

Authors W. D. C. Chacon and R. R. Dalsasso have equally contributed to this chapter.

W. D. C. Chacon · R. R. Dalsasso · G. Ayala Valencia (✉)
Department of Chemical and Food Engineering, Federal University of Santa Catarina, Florianópolis, SC, Brazil
e-mail: g.ayala.valencia@ufsc.br

V. Z. Pinto
Engenharia de Alimentos, Programa de Pós-Graduação em Ciência e Tecnologia de Alimentos, Universidade Federal da Fronteira Sul, Laranjeiras do Sul, Paraná, Brazil
e-mail: vania.pinto@uffs.edu.br

G. Ayala Valencia (ed.), *Starch Nanomaterials and Food Applications*,
https://doi.org/10.1007/978-3-031-60086-9_1

between 1 and 500 nm [7]. Dimensions lower than 1000 nm have also been used to classify this type of nanomaterial [2]. Several starch architectures can be produced at the nanoscale, and the most common starch nanomaterials (SNMts) are nanocrystals (SNCs), nanoparticles (SNPs), nanofibers (SNFs), nanomicelles (SNMs), and nanovesicles (SNVs) [2, 3, 8].

SNMts can be obtained using "top-down" approaches such as acid and enzymatic hydrolysis (EH), anti-solvent precipitation (ASP), high-pressure homogenization (HPH), and ultrasonication, as well as by means of "bottom-up" approaches using electrospinning and electrospraying, self-assembly, and emulsification [2, 3, 8, 9].

The present chapter aims to define and classify the different starch nanoarchitectures, as well as to discuss briefly the methodologies used to produce these nanomaterials. More information about the production of SNMts is discussed in Chaps. 2, 3, 4 and 5.

1.2 Preparation of Starch Nanomaterials

The building of nanostructures is based on the "top-down" and "bottom-up" approaches (Fig. 1.1). The "bottom-up" approach is based on building nanoparticles from smaller units or precursor molecules, whereas the "top-down" method involves reducing larger materials, such as starch blocks, to the nanoscale through decomposition or milling processes, resembling acidic, alkaline, and enzymatic hydrolysis [10].

Acid and enzymatic hydrolysis, ball milling, anti-solvent precipitation, high-pressure homogenization, microfluidization, sonication, electrospun fibers or electrospray, self-assembly, and emulsification can result in SNMts. Starch granule fragmentation can result in SNCs or SNPs, described as top-down processing, whereas starch solutions electrospun or precipitation or emulsification, described as bottom-up processing can result in SNFs, SNPs, SNMs, or SNVs.

The fragmentation of starch granules into a nano-sized structure is the main challenge for "top-down" materials building up. There are a few strategies that result in the formation of starch nanocrystals [11, 12]. However, most starch processing results in nanoparticles from starch granules. This is due to the structure crystallinity lost during the nanosizing.

The starch granules are insoluble in water and some other solvents at room temperature. To allow their use to build up SNPs or SNFs, the starch granules must be gelatinized by heating or chemically [13–15]. In addition, to obtain nanomicelles and nanovesicles, the starch granules must be in solution or gelatinized [16].

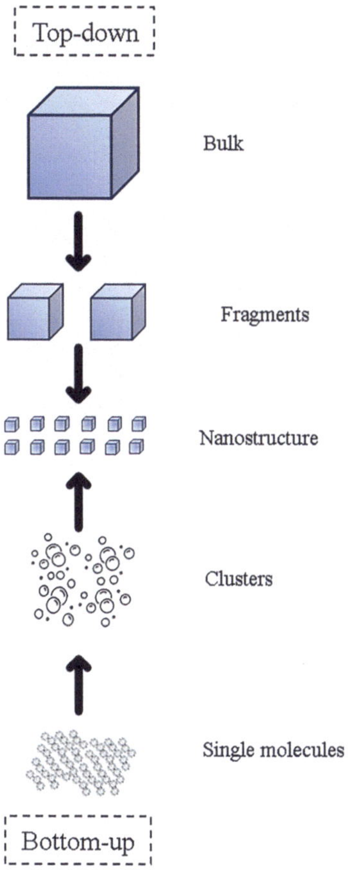

Fig. 1.1 Schematic representation of the top-down and bottom-up building up of nanostructures

1.3 Classification of Starch Nanomaterials

It is essential to specify the types of SNMts (Fig. 1.2) because this significantly influences the stabilization capacity of the compounds and/or their use as a reinforcing material. In general, SNMts can be classified as nanocrystals (SNCs), nanoparticles (SNPs), nanofibers (SNFs), nanomicelles (SNMs), and nanovesicles (SNVs).

1.3.1 Starch Nanocrystals

Starch granules have a highly organized structure, with amorphous and crystalline regions. The amorphous regions are the most susceptible to acid or enzymatic hydrolysis, as the crystallinity of starch is preserved under conditions below the

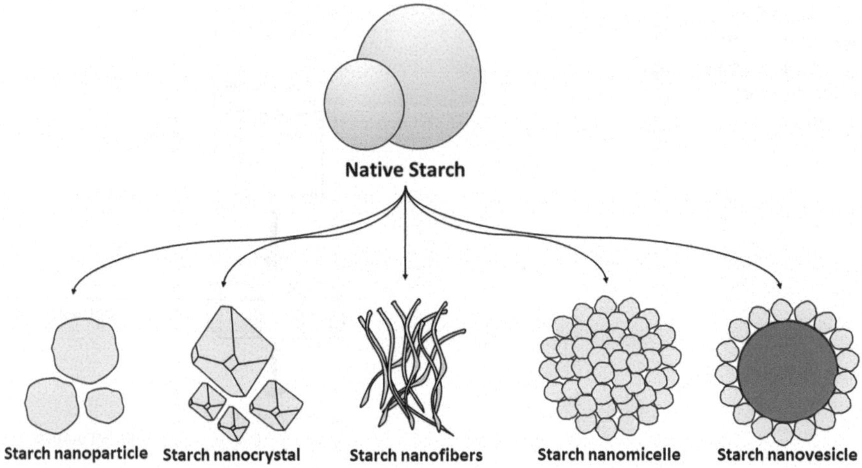

Fig. 1.2 Nanostructures derived from starch

gelatinization temperature [17, 18]. Therefore, by removing all or part of this amorphous region from the granule, platelets with high crystallinity are released, which occur between 30 and 80 nm, and are known as nanocrystals (SNCs) [8, 19–21]. Acid hydrolysis processes can also result in fragment starch [22], or even other crystalline structures [23], or amorphous starch nanoparticles (SNPs) [24]. This hydrolysis is a top-down approach using concentrated acids to isolate the SNC.

SNCs have often been prepared using waxy starch via sulfuric acid hydrolysis [25]. However, some other starches and hydrolysis processes have also been studied to produce SNCs. Corn or maize starch is the main source used to produce SNCs, with several reports about rice, potato, wheat, and bean starches, as well as cassava, quinoa, amaranth, and some fewer common sources such as starchy seeds from fruits and pines [2, 8, 11, 12, 26, 27]. Enzymatic hydrolysis, mixed acids, ball milling, and sonication-assisted isolation are also efficient for producing this crystalline nanoscale material [28–31].

The low SNCs yield after the acid hydrolysis isolation encouraged the alternative development to improve it. Therefore, starch pretreatments, such as physical, chemical, or enzymatic modifications, can affect not only the recovery yield of SNCs, but also their properties. Therefore, the new challenge is to understand how different starch sources and pretreatments, and the isolation processing reflect in the properties and applications of SNCs.

1.3.2 Starch Nanoparticles

SNPs are produced through controlled fragmentation of starch using the "top-down" and "bottom-up" techniques [10, 32–35]. The antisolvent or nanoprecipitation method is frequently used to produce SNPs [32]. This process involves dissolving starch in a miscible solvent such as water, typically hot water, and then adding it to another non-miscible solvent or antisolvent in water, such as ethanol (Fig. 1.3). The insolubility in the solvent triggers the precipitation of starch into nanoparticles. Solvents such as ethanol act as precipitation agents and help control the size and morphology of SNPs. This method is widely employed in starch nanotechnology because of its versatility and capability to generate nanoscale particles.

The synthesis of SNPs represents a significant advancement with substantial improvements in conventional starch properties. These SNMts, when reduced to the nanoscale, undergo a significant increase in their specific surface area. This not only enhances their capacity for compound absorption and release, as documented in previous studies [36–38], but also provides a versatile platform for multiple innovative applications. In addition to their impact on compound absorption and release, the synthesis of SNPs has proven particularly valuable in emulsion stabilization within the food industry [39]. These nanoparticles have also excelled in increasing the water adsorption capacity [40]. Furthermore, their versatility is evident in their ability to function as cross-linking agents and fillers with thermo-sensitive properties

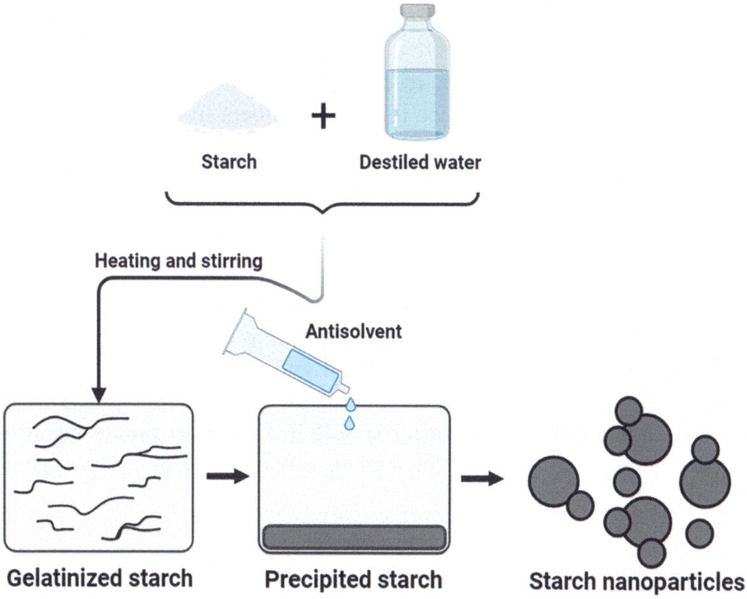

Fig. 1.3 General process for the production of starch nanoparticles by nanoprecipitation or antisolvent precipitation

[41]. However, their versatility is not limited solely to these applications. SNPs have served as precursors for the synthesis of carbon nanodots, expanding their utility in fields such as nanotechnology and electronics [42]. Beyond their physical properties and practical applications, it is essential to highlight that these SNPs can be functionalized with bioactive compounds, as documented in previous research [43–45]. This functionalization allows SNPs to acquire antimicrobial and antioxidant properties, further enhancing their utility in sectors that require food preservation and the improvement of pharmaceutical and cosmetic products. Collectively, these advancements illustrate the potential and versatility of SNPs in enhancing various areas of science and industry [46-48].

1.3.3 Starch Nanofibers

SNFs formation is based on a bottom-up approach by electrospinning. It involves an electrodynamic process that was first patented in 1934 by the invention "method and apparatus for preparing artificial threads", describing the production system of polymer-based fibers using high-voltage static electricity [49]. Electrospun fibers are synthesized by applying a high-voltage electric field on the surface of polymer solution droplets, causing a liquid jet to be ejected through a spinneret [50]. By controlling the polymer concentration, molecular weight and viscosity, appropriate solvent, low surface tension, the solution electrical conductivity and feeding rate, and applied voltage, it is possible to produce fibers with different sizes and morphologies [51, 52]. Starch granules are not easy solubilized to produce an appropriate electrospun solution. Therefore, recently electrospinning setups and different solvents were investigated to produce this type of fibers from the starch polymers [13, 14, 53–55].

The two main approaches are using an electro-wet-spinning setup, which involves the formation of the fibers on a grounded metal mesh immersed in a pure ethanol bath [14, 55]. The starch granules were dissolved in an aqueous dimethyl sulfoxide solution of 95% (v/v). The other approach uses a traditional electrospinning setup and a starch dispersion in aqueous formic acid of 90% (v/v) [56]. Starch undergoes rapid esterification, generating esters in the starch glucose units. This process is also known as chemical gelatinization at room temperature [13]. In addition, starch modification and/or blending with other polymers were used to produce an appropriate dispersion for spin [57]. Each fiber processing is driven under particular conditions and reflects various properties, sizes, and morphologies [14, 55, 56].

1.3.4 Starch Nanomicelles

SNMs can be used to increase the solubility of compounds lower polarity compounds and have been studied as delivery systems for medical [58–60] and food applications, such as the delivery of curcumin [16, 61, 62] and β-carotene [63]. They are conventionally obtained using a down-top approach, as modified amphiphilic starch molecules emulsify an active hydrophobic compound. Starch is modified by hydrolysis esterification to form smaller chains with a hydrophilic and hydrophobic regions [60–62]. In this form, starch-based micelles act as the emulsifiers of hydrophobic compounds or solutions.

Therefore, the methods to form nanomicelles are based on using amphiphilic modified starch as the emulsifier of a hydrophobic phase, which is located in the core of the micelles. Usually, pre-emulsions are formed by self-assembly by mixing the hydrophobic solution, the amphiphilic starch, in an aqueous phase under high-shear homogenization [63]. If better homogenization or droplet size reduction is required, a high-energy nanoemulsification method can be further used on the pre-emulsion, such as ultrasonic treatment [16, 58–60]. Other methods, such as microfluidization, high pressure, and membrane homogenization, could also be effective for this purpose, but no studies using these techniques for obtaining starch nanomicelles have been informed. A solid-separation step such as filtration or centrifugation can be performed to use undissolved particles [59, 62].

Some specific procedures to obtain SNMs systems use organic solvents such as ethanol [64], chloroform [61], and acetone [58] to compose the hydrophobic solution. However, organic solvents are frequently dangerous and environmentally harmful; therefore, it is interesting to avoid their use when possible. If necessary, the use of solvents generally recognized as safe or with GRAS status, such as ethanol, should be preferable, especially for food and health applications.

Focusing on food applications of SNMs and micelles, it is noticeable that starch is a versatile, accessible, and environmentally friendly raw material for the formation of nanostructured systems aimed at the protection and delivery of bioactive compounds. The consequent protection and prolongation of the activity of sensible natural active compounds make the substitution of synthetic and unsustainable food additives more feasible in the future. This exchange can fulfill the expectations of health-worried consumers. It also contributes to the establishment of a greener chemistry and industry. However, there is a broad field to be explored, which includes the development systems for different encapsulated bioactive compounds, such as other vegetal and animal-derived extracts, such as essential oils and propolis, respectively. Studying how the nanostructures interact with different types of food media, such as muscle-based foods, water-based juices, milk-fat-based foods, vegetable cells, and bakery products, is also highly relevant.

1.3.5 Starch Nanovesicles

SNVs are nano-dimensioned structures that encapsulate hydrophobic materials. Nanovesicles do not have a specified structure; however, they are usually charged with lipophilic bioactive substances aimed at their protection, controlled release, and increased stability in aqueous media [43, 65, 66]. Starches have gained attention for their applications because they are highly available, non-toxic, safe, biocompatible, and biodegradable, and their properties can be easily changed by chain modification and hydrolysis [65, 67, 68]. This characteristic is interesting for tailoring nanoparticle properties, such as size, active release rate, etc. The most studied applications of SNVs are in the medical field, for the delivery of drugs in medical applications [65, 66]. However, agriculture can also benefit from nanocarriers of pesticides [68, 69], nutrients, and biocontrol agents [70]. Food products have great potential to be improved as well, with the development of nanocarriers of natural compounds, including preservers and bioactives that benefit the consumer's health, such as curcumin [43], but, this is a field of application that is yet to be explored.

The conventional way to obtain nanovesicles is by the dilution of an oil-in-water nanoemulsion of the desired hydrophobic molecule into a gelatinized starch solution, followed by its drying, forming oil-charged microcapsules [43, 65]. The conventional emulsification method uses high-shear mixing, but alternative methods can be used. An alternative method was reported by Li et al. [68], the premix glass membrane emulsification method. For this, an oil-in-water or water-oil-water pre-emulsion was obtained by mechanical stirring or sonication (top-down approach), followed by a membrane homogenization stage to reduce the particle's sizes by shear forces. This method has the advantage of easily controlling the particles size, being reproducible, and having a high production rate. To obtain nanovesicles, some systems may require organic solvents such as ethanol. When this is the case, the nanoemulsion loading stage must enable sufficient time and temperature to allow the evaporation of the organic solvent [43].

Native starches are sensitive to enzymatic degradation and water and tend to release the encapsulated material quickly because of the reduction of intermolecular forces and swelling when applied in water-based systems [71]. Like other polysaccharides, starch is hydrophilic and relatively chemically stable, and its easy modification or functionalization can be applied to solve these issues [72]. Several studies have applied modified starches with amphiphilic properties to form nanomicelles, which are starch-based polymers modified to have hydrophobic and hydrophilic regions [60–62]. SNVs have shown great stability and controlled release of active compounds, besides the advantages of using starch, such as non-toxicity, safety, biocompatibility, and biodegradability [62, 73].

1.4 Conclusions

Research on the production of starch nanomaterials (SNMts) is recent and could have promissing applications. In general, starch nanocrystals (SNCs), nanofibers (SNFs), nanoparticles (SNPs), nanomicelles (SNMs), and nanovesicles (SNVs) can be obtained by using starch chains and the "top-down" and "bottom-up" approaches such as acid/enzymatic hydrolysis, antisolvent precipitation, electrospinning, self-assembly, among others. Particularly, SNCs and SNFs have been used as fillers in packaging materials, whereas, SNPs have been used to stabilize bioactive compounds to develop antioxidant and antimicrobial nanomaterials. SNMs and SNVs have been explored in medical applications to encapsulate hydrophobic materials and increase the solubility of compounds lower-polarity compounds. Further studies are required to optimize the production of SNMts and standardize their particle size and shape, and to explore food and food packaging applications.

Acknowledgments W.D.C. Chacon gratefully acknowledges the Fundação de Amparo à Pesquisa e Inovação do Estado de Santa Catarina (FAPESC) for the PhD fellowship (3003/2021). R.R. Dalsasso thanks the Coordenaçãoo de Aperfeiçoamento de Pessoal de Nível Superior (CAPES) for the PhD fellowship (finance code 001). G. A. Valencia would like to thank the National Council for Scientific and Technological Development (CNPq) (grant 302434/2022-4) for financial support. The authors gratefully acknowledge the Federal University of Santa Catarina for the support.

Conflicts of Interest The authors declare no conflicts of interest.

References

1. Bie P, Pu H, Zhang B, Su J, Chen L, Li X. Structural characteristics and rheological properties of plasma-treated starch. Innov Food Sci Emerg Technol [Internet]. 2016;34:196–204. Available from: https://doi.org/10.1016/j.ifset.2015.11.019.
2. dos Alves MJS, Chacon WDC, Gagliardi TR, Henao ACA, Monteiro AR, Valencia GA. Food applications of starch nanomaterials: a review. Starch/Stärke. 2021;73:2100046.
3. Zabala CCV. Starch-based nanomaterials [Internet]. Springer; 2020. 29 p. Available from: http://www.springer.com/series/10203
4. Schmiele M, Sampaio UM, Pedrosa Silva Clerici MT. Basic principles: composition and properties of starch. In: Silva Clerici MTP, Schmiele M, editors. Starches for food application [Internet]. Academic; 2019. p. 1–22. Available from: https://www.sciencedirect.com/science/article/pii/B9780128094402000010
5. Coelho G, Denise L, Laroque A, Rodrigues A, Bruno M, Carciofi M, et al. Current status and perspectives of starch powders modified by cold plasma: a review. J Polym Environ. 2024;32:510–23.
6. Rostamabadi H, Karaca AC, Deng L, Colussi R, Narita IMP, Kaur K, et al. Oat starch—how physical and chemical modifications affect the physicochemical attributes and digestibility? Carbohydr Polym [Internet]. 2022;296:119931. Available from: https://www.sciencedirect.com/science/article/pii/S0144861722008360

7. Le Corre D, Bras J, Dufresne A. Starch nanoparticles: a review. Biomacromolecules [Internet]. 2010;11(5):1139–53. Available from: https://doi.org/10.1021/bm901428y.
8. Campelo PH, Sant'Ana AS, Pedrosa Silva Clerici MT. Starch nanoparticles: production methods, structure, and properties for food applications. Curr Opin Food Sci [Internet]. 2020;33(i):136–40. Available from: https://doi.org/10.1016/j.cofs.2020.04.007.
9. Leandro GC, Cesca K, Monteiro AR, Valencia GA. Morphological and physicochemical properties of pine seed starch nanoparticles. Starch - Stärke [Internet]. 2023;2200225:1–7. Available from: https://onlinelibrary.wiley.com/doi/10.1002/star.202200225
10. Abid N, Khan AM, Shujait S, Chaudhary K, Ikram M, Imran M, et al. Synthesis of nanomaterials using various top-down and bottom-up approaches, influencing factors, advantages, and disadvantages: a review. Adv Colloid Interface Sci [Internet]. 2022;300:102597. Available from: https://www.sciencedirect.com/science/article/pii/S0001868621002384
11. Velásquez-Castillo LE, Leite MA, Ditchfield C, do Amaral Sobral PJ, Moraes ICF. Quinoa starch nanocrystals production by acid hydrolysis: kinetics and properties. Int J Biol Macromol. 2020;143:93–101.
12. Pinto VZ, Moomand K, Deon VG, Biduski B, da Rosa Zavareze E, Lenhani GC, et al. Effect of physical pretreatments on the hydrolysis kinetic, structural, and thermal properties of pinhão starch nanocrystals. Starch/Staerke. 2021;73(7–8):1–9.
13. Vasilyev G, Putaux J, Zussman E. Rheological properties and electrospinnability of high-amylose starch in formic acid. Biomacromolecules. 2015;16:2529–36.
14. Kong L, Ziegler GR. Fabrication of pure starch fibers by electrospinning. Food Hydrocoll. 2014;36:20–5.
15. Pires JB, Fonseca LM, Siebeneichler TJ, Crizel RL, dos Santos FN, dos Santos Hackbart HC, et al. Curcumin encapsulation in capsules and fibers of potato starch by electrospraying and electrospinning: thermal resistance and antioxidant activity. Food Res Int. 2022;162 (August):112111.
16. Kou Z, Dou D, Lan L, Zhang J, Lan P, Yu Q, et al. Preparation, characterization, and performance analysis of starch-based nanomicelles. Int J Biol Macromol [Internet]. 2020;145: 655–62. Available from: https://doi.org/10.1016/j.ijbiomac.2019.12.220.
17. Hao Y, Chen Y, Li Q, Gao Q. Preparation of starch nanocrystals through enzymatic pretreatment from waxy potato starch. Carbohydr Polym [Internet]. 2018;184(December 2017):171–7. Available from: https://doi.org/10.1016/j.carbpol.2017.12.042.
18. Qiu C, Wang C, Gong C, McClements DJ, Jin Z, Wang J. Advances in research on preparation, characterization, interaction with proteins, digestion and delivery systems of starch-based nanoparticles. Int J Biol Macromol. 2020;152:117–25.
19. Le Corre D, Angellier-Coussy H. Reactive and functional polymers preparation and application of starch nanoparticles for nanocomposites: a review. React Funct Polym. 2014;85:97–120.
20. Kaur J, Kaur G, Sharma S, Jeet K. Cereal starch nanoparticles—a prospective food additive: a review. Crit Rev Food Sci Nutr. 2018;58(7):1097–107.
21. Liu C, Li K, Li X, Zhang M, Li J. Formation and structural evolution of starch nanocrystals from waxy maize starch and waxy potato starch. Int J Biol Macromol. 2021;180:625–32.
22. Kim J-Y, Park D-J, Lim S-T. Fragmentation of waxy rice starch granules by enzymatic hydrolysis. Cereal Chem. 2008;85(2):182–7.
23. Kim HY, Park DJ, Kim JY, Lim ST. Preparation of crystalline starch nanoparticles using cold acid hydrolysis and ultrasonication. Carbohydr Polym. 2013;98(1):295–301.
24. Bel Haaj S, Thielemans W, Magnin A, Boufi S. Starch nanocrystals and starch nanoparticles from waxy maize as nanoreinforcement: a comparative study. Carbohydr Polym. 2016;143: 310–7.
25. Angellier H, Choisnard L, Molina-boisseau S, Ozil P, Dufresne A. Optimization of the preparation of aqueous suspensions of waxy maize starch nanocrystals using a response surface. Biomacromolecules. 2004;5:1545–51.

26. Costa KDC, de Souza CO, da Silva JB, Druzian JI. Hydrolysis of part of cassava starch into nanocrystals leads to increased reinforcement of nanocomposite films. J Appl Polym Sci. 2017;134(41):1–9.
27. Kim HY, Park SS, Lim ST. Preparation, characterization and utilization of starch nanoparticles. Colloids Surf B Biointerfaces. 2015;126:607–20.
28. Mohammad Amini A, Razavi SMA. A fast and efficient approach to prepare starch nanocrystals from normal corn starch. Food Hydrocoll [Internet]. 2016;57:132–8. Available from: https://doi.org/10.1016/j.foodhyd.2016.01.022.
29. Li S, Zhou W, Huang C, Hu Y, Gao Q, Chen Y. Rapid preparation of starch nanocrystals by the mixed acid of sulfuric acid and hydrochloric acid. Int J Biol Macromol. 2023;232(9):123402.
30. Dai L, Li C, Zhang J, Cheng F. Preparation and characterization of starch nanocrystals combining ball milling with acid hydrolysis. Carbohydr Polym. 2018;180(October 2017): 122–7.
31. Gao F, Li D, Bi C-h, Mao Z h, Adhikari B. Application of various drying methods to produce enzymatically hydrolyzed porous starch granules. Dry Technol. 2013;31(13–14):1627–34.
32. Caicedo Chacon WD, Verruck S, Monteiro AR, Valencia GA. The mechanism, biopolymers and active compounds for the production of nanoparticles by anti-solvent precipitation: a review. Food Res Int. 2023;168(September 2022):112728.
33. Fazeli M, Lipponen J. Developing Self-Assembled Starch Nanoparticles in Starch Nanocomposite Films. ACS Omega [Internet]. 2022;7(49):44962–71. Available from: https://doi.org/10.1021/acsomega.2c05251.
34. Chacon WDC, Valencia GA, Rojas GMA, Henao ACA. Mathematical models for prediction of water evaporation and thermal degradation kinetics of potato starch nanoparticles obtained by nanoprecipitation. Starch/Stärke. 2019;1800081:1–7.
35. dos Santos Alves M, Rodrigues Monteiro A, Ayala Valencia G. Antioxidant nanoparticles based on starch and the phenolic compounds from propolis extract: Production and physico-chemical properties. Starch - Stärke. n/a(n/a):2100289.
36. Bhatia M, Rohilla S. Formulation and optimization of quinoa starch nanoparticles: quality by design approach for solubility enhancement of piroxicam. Saudi Pharm J. 2020;28(8):927–35.
37. Qiu C, Chang R, Yang J, Ge S, Xiong L, Zhao M, et al. Preparation and characterization of essential oil-loaded starch nanoparticles formed by short glucan chains. Food Chem [Internet]. 2017;221:1426–33. Available from: https://doi.org/10.1016/j.foodchem.2016.11.009.
38. Raj V, Prabha G. Synthesis, characterization and in vitro drug release of cisplatin loaded Cassava starch acetate–PEG/gelatin nanocomposites. J Assoc Arab Univ Basic Appl Sci [Internet]. 2016;21:10–6. Available from: https://www.sciencedirect.com/science/article/pii/S1815385215000279
39. Marku D, Wahlgren M, Rayner M, Sjöö M, Timgren A. Characterization of starch Pickering emulsions for potential applications in topical formulations. Int J Pharm [Internet]. 2012;428(1): 1–7. Available from: https://www.sciencedirect.com/science/article/pii/S0378517312000658
40. Dai L, Qiu C, Xiong L, Sun Q. Characterisation of corn starch-based films reinforced with taro starch nanoparticles. Food Chem [Internet]. 2015;174:82–8. Available from: https://www.sciencedirect.com/science/article/pii/S0308814614017300
41. Valodkar M, Thakore S. Isocyanate crosslinked reactive starch nanoparticles for thermo-responsive conducting applications. Carbohydr Res [Internet]. 2010;345(16):2354–60. Available from: https://www.sciencedirect.com/science/article/pii/S0008621510003484
42. Chin SF, Yazid SNAM, Pang SC, Ng SM. Facile synthesis of fluorescent carbon nanodots from starch nanoparticles. Mater Lett [Internet]. 2012;85:50–2. Available from: https://www.sciencedirect.com/science/article/pii/S0167577X12009299
43. Li J, Shin GH, Lee IW, Chen X, Park HJ. Soluble starch formulated nanocomposite increases water solubility and stability of curcumin. Food Hydrocoll [Internet]. 2016;56:41–9. Available from: https://www.sciencedirect.com/science/article/pii/S0268005X15301569
44. Chen Y-Y, Liu K, Zha X-Q, Li Q-M, Pan L-H, Luo J-P. Encapsulation of luteolin using oxidized lotus root starch nanoparticles prepared by anti-solvent precipitation. Carbohydr

Polym [Internet]. 2021;273:118552. Available from: https://www.sciencedirect.com/science/article/pii/S0144861721009395
45. Ismail NS, Gopinath SCB. Enhanced antibacterial effect by antibiotic loaded starch nanoparticle. J Assoc Arab Univ Basic Appl Sci [Internet]. 2017;24:136–40. Available from: https://www.sciencedirect.com/science/article/pii/S1815385216300426
46. Dong H, Chen L, Zhang Q, Gao J, Vasanthan T. Optimization of processing parameters to produce nanoparticles prepared by rapid nanoprecipitation of pea starch. Food Hydrocoll [Internet]. 2021;121:106929. Available from: https://www.sciencedirect.com/science/article/pii/S0268005X21003453
47. Yan X, Diao M, Yu Y, Gao F, Wang E, Wang Z, et al. Characterization of resistant starch nanoparticles prepared via debranching and nanoprecipitation. Food Chem [Internet]. 2022;369:130824. Available from: https://www.sciencedirect.com/science/article/pii/S0308814621018306
48. Wu J, Huang Y, Yao R, Deng S, Li F, Bian X. Preparation and characterization of starch nanoparticles from potato starch by combined solid-state acid-catalyzed hydrolysis and nanoprecipitation. Starch - Stärke. 2019;71(9–10):1900095.
49. Ahmadian A, Shafiee A, Aliahmad N, Agarwal M. Overview of nano-fiber mats fabrication via electrospinning and morphology analysis. Textiles. 2021;1(2):206–26.
50. Gagaoua M, Pinto VZ, Göksen G, Alessandroni L, Lamri M, Dib AL, et al. Electrospinning as a promising process to preserve the quality and safety of meat and meat products. Coatings. 2022;12(5):644.
51. Hemamalini T, Giri Dev VR. Comprehensive review on electrospinning of starch polymer for biomedical applications. Int J Biol Macromol. 2018;106:712–8.
52. Dierings de Souza EJ, Kringel DH, Guerra Dias AR, da Rosa Zavareze E. Polysaccharides as wall material for the encapsulation of essential oils by electrospun technique. Carbohydr Polym. 2021;265(December 2020):118068.
53. Kong L, Ziegler GR. Patents on fiber spinning from starches. Recent Pat Food Nutr Agric. 2012;4(3):210–9.
54. Kong L, Ziegler GR. Role of molecular entanglements in starch fiber formation by electrospinning. Biomacromolecules. 2012;13:2247–53.
55. Kong L, Ziegler GR. Quantitative relationship between electrospinning parameters and starch fiber diameter. Carbohydr Polym. 2013;92(2):1416–22.
56. de Souza EJD, dos Santos FN, Pires JB, Kringel DH, da Silva WMF, Meinhart AD, et al. Production and optimization of ultrafine fiber from yam starch by electrospinning method using multivariate analysis. Starch/Staerke. 2021;73(3–4):1–8.
57. Wang H, Ziegler GR. Electrospun nanofiber mats from aqueous starch-pullulan dispersions: optimizing dispersion properties for electrospinning. Int J Biol Macromol. 2019;133:1168–74.
58. Liu Q, Yang X, Xu H, Pan K, Yang Y. Novel nanomicelles originating from hydroxyethyl starch-g-polylactide and their release behavior of docetaxel modulated by the PLA chain length. Eur Polym J. 2013;49(11):3522–9.
59. Wang T, Liu K, Zhang L, Zhang Y, Chen L, Li X. Development of M cell-targeting starch-based nanomicelles for oral delivery of immunoactive peptides. Ind Crops Prod. 2022;184(January):115074.
60. Wu C, Yang J, Xu X, Gao C, Lü S, Liu M. Redox-responsive core-cross-linked mPEGylated starch micelles as nanocarriers for intracellular anticancer drug release. Eur Polym J. 2016;83:230–43.
61. Wang R, Qin X, Du Y, Shan Z, Shi C, Huang K, et al. Dual-modified starch nanoparticles containing aromatic systems with highly efficient encapsulation of curcumin and their antibacterial applications. Food Res Int. 2022;162(PA):111926.
62. Liang S, Du J, Hong Y, Cheng L, Gu Z, Li Z, et al. Octenyl succinate anhydride debranched starch-based nanocarriers for curcumin with improved stability and antioxidant activity. Food Hydrocoll. 2023;135(January 2022):108118.

63. Lin Q, Liang R, Zhong F, Ye A, Hemar Y, Yang Z, et al. Self-assembled micelles based on OSA-modified starches for enhancing solubility of β-carotene: effect of starch macromolecular architecture. J Agric Food Chem. 2019;67(23):6614–24.
64. Liu K, Zhang HL, Pan LH, Li QM, Luo JP, Zha XQ. The nanomicelles consisting of lotus root amylopectin and quinoa protein: construction and encapsulation for quercetin. Food Chem. 2022;387(April):132924.
65. Sintov AC. AmyloLipid nanovesicles: a self-assembled lipid-modified starch hybrid system constructed for direct nose-to-brain delivery of curcumin. Int J Pharm [Internet]. 2020;588 (July):119725. Available from: https://doi.org/10.1016/j.ijpharm.2020.119725.
66. Hou H, Chi C, Wang T, He Y, Chen L, Li X. Multi-responsive starch-based nanocapsules for colon-targeting delivery of peptides: in vitro and in vivo evaluation. Int J Biol Macromol. 2023;242(P3):124953.
67. Xu T, Pan M h, Chiou Y s, Li Z, Wei S, Yin X, et al. Improvement of emulsifying properties of potato starch via complexation with nanoliposomes for stabilizing Pickering emulsion. Food Hydrocoll. 2023;136(PA):108271.
68. Li D, Liu B, Yang F, Wang X, Shen H, Wu D. Preparation of uniform starch microcapsules by premix membrane emulsion for controlled release of avermectin. Carbohydr Polym. 2016;136: 341–9.
69. Channab BE, El Idrissi A, Zahouily M, Essamlali Y, White JC. Starch-based controlled release fertilizers: a review. Int J Biol Macromol. 2023;238(January):124075.
70. Saberi Riseh R, Hassanisaadi M, Vatankhah M, Kennedy JF. Encapsulating biocontrol bacteria with starch as a safe and edible biopolymer to alleviate plant diseases: a review. Carbohydr Polym. 2023;302(November 2022):120384.
71. Qiu A, Wang Y, Zhang G, Wang H. Natural polysaccharide-based nanodrug delivery systems for treatment of diabetes. Polymers (Basel). 2022;14(15):1–21.
72. Carrion CC, Nasrollahzadeh M, Sajjadi M, Jaleh B, Soufi GJ, Iravani S. Lignin, lipid, protein, hyaluronic acid, starch, cellulose, gum, pectin, alginate and chitosan-based nanomaterials for cancer nanotherapy: challenges and opportunities. Int J Biol Macromol. 2021;178:193–228.
73. Li K, Liu H, Gao W, Chen M, Zeng Y, Liu J, et al. Mulberry-like dual-drug complicated nanocarriers assembled with apogossypolone amphiphilic starch micelles and doxorubicin hyaluronic acid nanoparticles for tumor combination and targeted therapy. Biomaterials. 2015;39:131–44.

Chapter 2
Starch Nanocrystals

Vania Zanella Pinto

2.1 Introduction

Nanomaterials and nanostructures are considered to have "at least one external dimension, or an internal or surface structure, in the nanoscale range (approximately 1 nm to 100 nm)" [1]. This definition is also supported in different areas and in regulatory affairs. However, some reports have also considered sub-micro or nanomaterials when the structures are larger than 100 nm [2–5]. The native starch granule structure is composed of different micro- and nanometric-scale architectures [6]. These granules can be converted to nanoscale materials using two main "top-down" or "bottom-up" approaches [2]. The starch nanocrystals (SNCs) are mainly developed by top-drown hydrolysis, which results in insoluble crystalline platelets derived from the thin crystalline lamella of starch granules (Fig. 2.1). They are isolated by removing the amorphous regions from the starch granules and preserving the crystallinity, under conditions below the gelatinization temperature [7, 8]. This crystalline structure and proprieties allow the development of different nanoscale materials for food and non-food uses.

The starch granule characteristics and properties are strongly dependent on of the starch botanic origin, especially granule size and shape, amylose, and amylopectin ratios, and amylopectin crystallinity structure. The amylose chains are related to amorphous and the amylopectin to crystalline domains [9]. Waxy starches are very often used to produce nanoscale structures due to the abundance of amylopectin long chains; however, several different starch sources have been studied to obtain promising SNCs sources with good yields and different properties. However, some other starches and hydrolysis processes have also been studied to produce starch

V. Z. Pinto (✉)
Engenharia de Alimentos, Programa de Pós-Graduação em Ciência e Tecnologia de Alimentos, Universidade Federal da Fronteira Sul, Paraná, Brazil
e-mail: vania.pinto@uffs.edu.br

G. Ayala Valencia (ed.), *Starch Nanomaterials and Food Applications*,
https://doi.org/10.1007/978-3-031-60086-9_2

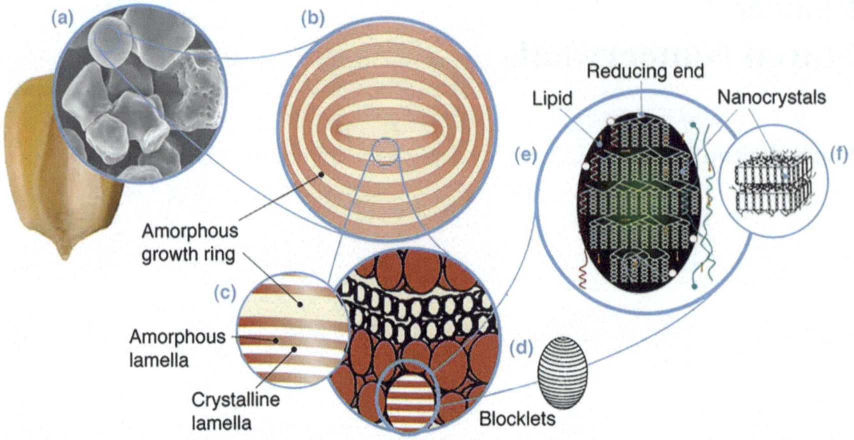

Fig. 2.1 Starch multiscale structure: (**a**) starch granules from normal maize (30 μm); (**b**) amorphous and semicrystalline growth rings (120–500 nm); (**c**) amorphous and crystalline lamellae (9 nm), magnified details of the semicrystalline growth ring; (**d**) blocklets (20–50 nm) constituting a unit of the growth rings; (**e**) amylopectin double helixes forming the crystalline lamellae of the blocklets; (**f**) nanocrystals. (Reused with permission from Lecorre et al. [6])

nanocrystals. Corn or maize starch is the main source used to produce SNCs, with several reports about rice, potato, wheat, and bean starches, as well as cassava, quinoa, amaranth, sago, and some less common sources such as starchy seeds from fruits and nuts [2, 3, 5, 10–14].

In addition, pre- and post-treatments have been tested for isolation, crystallinity, and SNCs propriety improvement [4, 5, 7, 8, 12, 15]. The pretreatments can be physical, enzymatic, and chemical processes that are described to speed up the hydrolysis, increase the yield, crystallinity, and thermal stability. Enzymatic hydrolysis, sonication-assisted isolation and ball milling processes are efficient pretreatments for producing this crystalline nanoscale materials [16–19]. Heat-moisture-treatment (HMT), annealing (ANN), also are considered promising in different ways for SNC producing [12, 15, 20]. Post-treatments are usually applied to reduce the isolated SNCs size, disrupt some agglomerates, and modify their structure and properties to drive them for different uses. Some new approaches, such as dielectric barrier discharge plasma and chemical modifications, have been used as pos-treatments [21–23].

SNCs applications are still under investigation, with few reports of commercial uses. The most popular commercial product is Mater-Bi™ from Novamont, which is used as carbon black replacement for tires, packaging solutions, and agriculture [24]. However, there are several publications about SNCs' use as fillers in packaging materials [25–28], and Pickering emulsion stabilizer [14, 29]. This chapter elucidates the SNCs developing approaches to isolation, pre- and post-treatment effects, their main properties, and advanced uses.

2.2 Acid SNCs Preparation

The main approach to isolate SNCs involves mild acid hydrolysis using H_2SO_4 or HCl, which substantially degrades the amorphous lamellae of native granular starch [30–32]. These lamellae are the region most susceptible to acid hydrolysis on starch granules [6]. Therefore, acid hydrolysis is used to isolate the highly crystalline insoluble starch fractions. The first report of SNCs was published in 1996 by Dufresne, Cavaille, Helbert [33]. This report describes a microcrystaline starch isolated from potato starch using 2.2 M HCl after 15 days of hydrolysis. Subsequently, H_2SO_4 was often described for SNCs isolation using waxy starch [31, 32], and later, other starches sources and amylose content were also studied.

The hierarchical organization of starch granules (Fig. 2.1) and their semicrystalline structure are critical factors in the preparing of SNCs via controlled acid hydrolysis [34]. The insoluble residue from acid hydrolysis is segregated by high-speed centrifugation and several steps of water-washing to remove the acid solution, adjusting the pH, and starch-soluble fractions. The most described processing yield ranges from 5% to 15%, and the particle size (thickness or diameter) ranges from 20 to 400 nm [6, 12, 25, 32, 35, 36].

The acid hydrolysis of starch granules is based on a two-stage hydrolysis pattern, which is related to amylose and amylopectin erosion. A relative rapid hydrolysis rate that occurs at 0–3 days or up to the first 8 days, followed by a slow rate of 4–7 days or 8–15 days. The first and fastest step results in a preferential erosion of the amorphous growth rings from starch granules, which is related to the amylose chains. In the second stage, the granule starch crystalline structure from amylopectin chain organization is slowly degraded by the acid attack [6, 37]. The speed rate and each stage duration are related to the acid type and concentration, starch type and source, as well as, hydrolysis conditions (temperature, pressure and stirring) [32, 35, 37]. This evidence was supported by scanning or transmission electron microscopy (SEM and TEM) images (Fig. 2.2).

The main mechanism of acid hydrolysis of starch granules is described by the H_3O^+ ion promoting an electrophilic attack in the α $(1 \rightarrow 4)$ glycosidic bonds. As follows, the electrons of the carbon–oxygen bond generate an intermediate and unstable Lewis' acid by moving the oxygen atom. This Lewis' acid subsequently reacts with water (Lewis' base), generating a hydroxyl group [37]. This acid process leads to a water-insoluble and acid-resistant fraction, which is the remaining crystalline parts of the starch granules [35, 37, 38]. The dense packing of this fraction is suggested to be responsible for the slow acid hydrolysis. Other hypotheses of this resistance to acid hydrolysis are related to the chair → half chair changing conformation of the D-glucopyranosyl unit from the well-packaged starch crystallites [6, 32, 37].

The acid hydrolysis time is a crucial parameter in the preparation of SNCs, owing to the great difference in the reaction time used at the optimum conditions and the first described SNCs preparation. The H_2SO_4 hydrolysis is described up to 5–15 days [32, 33], whereas HCl hydrolysis requires a longer hydrolysis time,

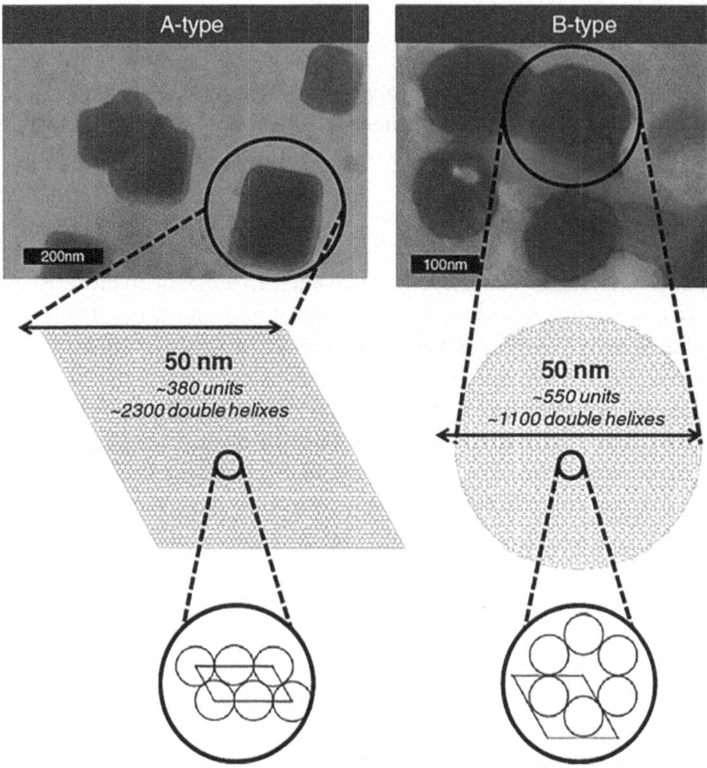

Fig. 2.2 Waxy maize starch (A-type) and high amylose starch (B-type). Starch nanocrystals and double helixes packing configuration illustration. (Reused with permission from LeCorre et al. [39])

ranging from 15 to 45 days [36]. The H_2SO_4 is the main approach described for SNCs production, while HCl is less used to produce SNCs, alone [31, 36] or associated with H_2SO_4 [19]. HCl treatment has similar behavior as H_2SO_4 [19], during the appropriate time, and very low yield (0.5%). The SNCs isolated using H_2SO_4 and HCl result in different surface chemical compositions. H_2SO_4 hydrolysis results in carboxyl groups and sulfate esters on the SNCs surface, whereas no sulfate esters were detected in SNCs after HCl hydrolysis [40]. The H_2SO_4 hydrolysis for SNCs isolation using corn waxy starch was optimized, yielding 15.7% (w/w) in 5 days of 3.16 mol/L of H_2SO_4 hydrolysis at 40 °C, and 100 rpm stirring [32]. These parameters are very highly replicated in the following studies.

The amylose content and starch botanical origin interfere with the production of SNCs by acid hydrolysis [32, 39]. Waxy maize starch is the most commonly used for the production of SNCs because of its amylose-free characteristic [41]. Higher crystallinity was expected for hydrolyzed amylopectin-rich (waxy) starch because there is less amorphous starch to remove. Indeed, amylopectin-rich starches are more crystalline than amylose-rich starches [39]. The initial starch crystallinity results in

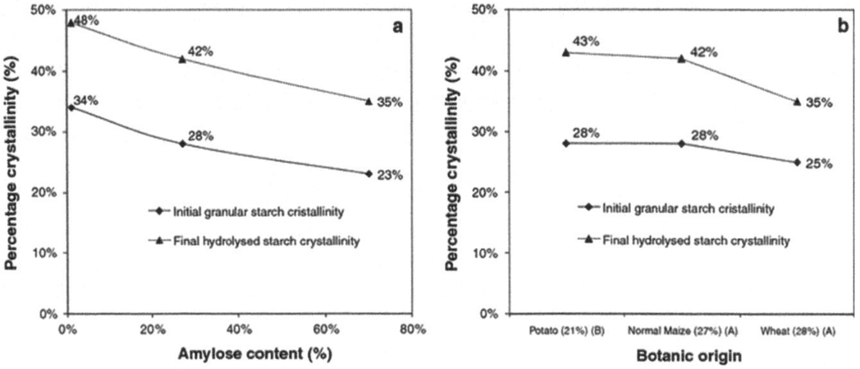

Fig. 2.3 Original crystallinity (diamond) and final crystallinity (triangle) for maize starches with different amylose contents (**a**) and from different starch sources (**b**). (Reused with permission from LeCorre et al. [39])

SNCs with high crystallinity, which is highly desirable for this nanomaterial (Fig. 2.3). The insoluble hydrolyzed residue obtained from waxy maize comprised crystalline nanoplatelets approximately 5–7 nm thick with a length of 20–40 nm and a width of 15–30 nm [30].

Non-waxy starches can also result in highly crystalline SNCs. The first results and optimization for SNCs preparation report only information about SNCs size and yield, with no SNCs crystallinity [32, 33, 39]. Since then, new research has aimed to understand how SNCs crystallinity is affected by different amylose content and starch botanical sources. SNCs with different relative crystallinity (RC) 43%, 42%, and 35%, were obtained from potato, normal maize, and wheat starches, with 21%, 27%, and 28% of amylose content, respectively. This discrete change indicates that as the original content increases, the crystallinity of SNCs decreases if the hydrolysis does not go longer than the optimum time.

SNCs crystallinity of amadumbe root starch increased by 21–29% depending on the varieties, upon 3.16 M H_2SO_4 hydrolysis [42]. Banana and tapioca starches SNCs reached 47.13% and 60.06% RC after 3.5 M H_2SO_4, during 7 and 10 h of hydrolysis, respectively [43]. Pinhão starch with 28.48% RC results in SNCs with 52.23% and 43.21% RC after 5 and 7 days of 3 M H_2SO_4 hydrolysis, respectively [12]. Cassava starch, regular corn starch, and carioca bean starch with 36.4%, 25.4%, 30.9% RC results in SNCs with 57.7%, 54% and 51.4% RC, respectively, after 5 days of 3.16 M H_2SO_4 hydrolysis [44]. These results support the hypothesis that non-waxy and normal amylose starches can also be used for SNCs production.

In addition, the crystallinity polymorphism can be changed or not after acid hydrolysis. The starch semicrystalline structure is described as A-, B-, or C-type based on its amylopectin double helices spatial arrangement. The A-type pattern is described as a packed and dense structure with only four water molecules in the monoclinic unit cell, whereas the B-type pattern is packed in a hexagonal unit cell with 36 water molecules per unit cell (Fig. 2.2) [45]. The C-type starches are a mixture of A- and B-type allomorphs that occur simultaneously. The B-type allomorph is arranged in the distal region of the eccentric hilum, whereas the A-type allomorph is organized in the around the hilum, and a mixed distribution of A- and B-type allomorph is observed in the granule center [46]. The starch V-type polymorphism is also described for starch with high amylose content and for amylose-lipid,—emulsifiers, and -butanol complex inclusions, which comprise a small proportion of non-ordered regions [47].

The waxy maize [29] and rice [48] starches show an A-type crystalline structure, and their SNCs show the same pattern. The same behavior is described for cassava or tapioca [44], and quinoa [5] starches. The native pinhão and carioca bean starches showed a C-type crystalline pattern and the resulting SNCs have an A-type pattern [12, 15, 44]. Waxy potato starch (B-type) resulted in SNCs A-type after 3.16 M H_2SO_4 hydrolysis because of a great reduction in the peak intensity around 5.5° 2θ (B-type feature) [7]. Regular rice starch (A-type) results in a A+V-type SNCs after 7 days of 3.16 M H_2SO_4 hydrolysis [48]. The crystallinity pattern change can be a result of the acid resistance of the starch crystalline organization. In general, native starches with A- or C-type polymorphs display enhanced resistance to acid hydrolysis than the B-type starches [12, 49], resulting in different hydrolysis patterns and SNCs yield. The A-type dense structure in the starch granules tends to hinder the acid attack, whereas the B-type structure is more susceptible to acid hydrolysis due to its less compact structure [50, 51], by affecting H_3O+ ion penetration and the difficult conformational changes in the D-glucopyranosyl units (chair-boat) [37].

The SNCs size scaling is influenced by the starch hydrolysis condition, when using various acid concentrations and hydrolysis times [52]. Waxy corn and amaranth starch with an A-type polymorphism pattern and different size and shape granules resulted in similar SNCs shape with different sizes (Fig. 2.4). Amaranth starch has small granules (~1.5 μm), whereas waxy corn starch has high ones (~16 μm). The smaller granules from amaranth were rapidly hydrolyzed and resulted in SCNs with mean sizes of 484, 374, and 376 nm from 3, 5, and 10 days of 3.16 M H_2SO_4 at 40 °C hydrolysis. On the other hand, waxy corn SNCs result in 577, 462, and 322 nm after the same hydrolysis conditions [52].

A daily time-scale SNCs optimization using batch preparation of waxy corn starch was performed over a range of 3–4.5 M H_2SO_4 hydrolysis from 1 to 15 h, at 25 to 40 °C and starch concentration of 15–40 (g/100 mL). The SNCs were approximately 172–550 nm after redispersion using Ultra-Turrax for 5 min at 13,000 rpm and coarse filtration (40 μm). Daily SNCs production was reached only after fine filtration (1 μm) at 40 °C, 4.5 M H_2SO_4, 15 h, and 40% of starch concentration [53].

The effects on SNCs crystallinity from different botanic origin starches are usually most related to the amylose content and polymorphism type. Waxy (A-type), regular (A-type), and high amylose (B-type) corn, potato (B-type), and mung bean (C-type) starches were used to produce SNCs after 3.16 M H_2SO_4 and 5 and 7 days [54]. The results indicate a slight peak intensity increase on waxy corn and mung bean SNCs, no changes for regular corn SNCs, and a decrease in the high amylose corn and potato SNCs upon the acid hydrolysis for 5 days [54].

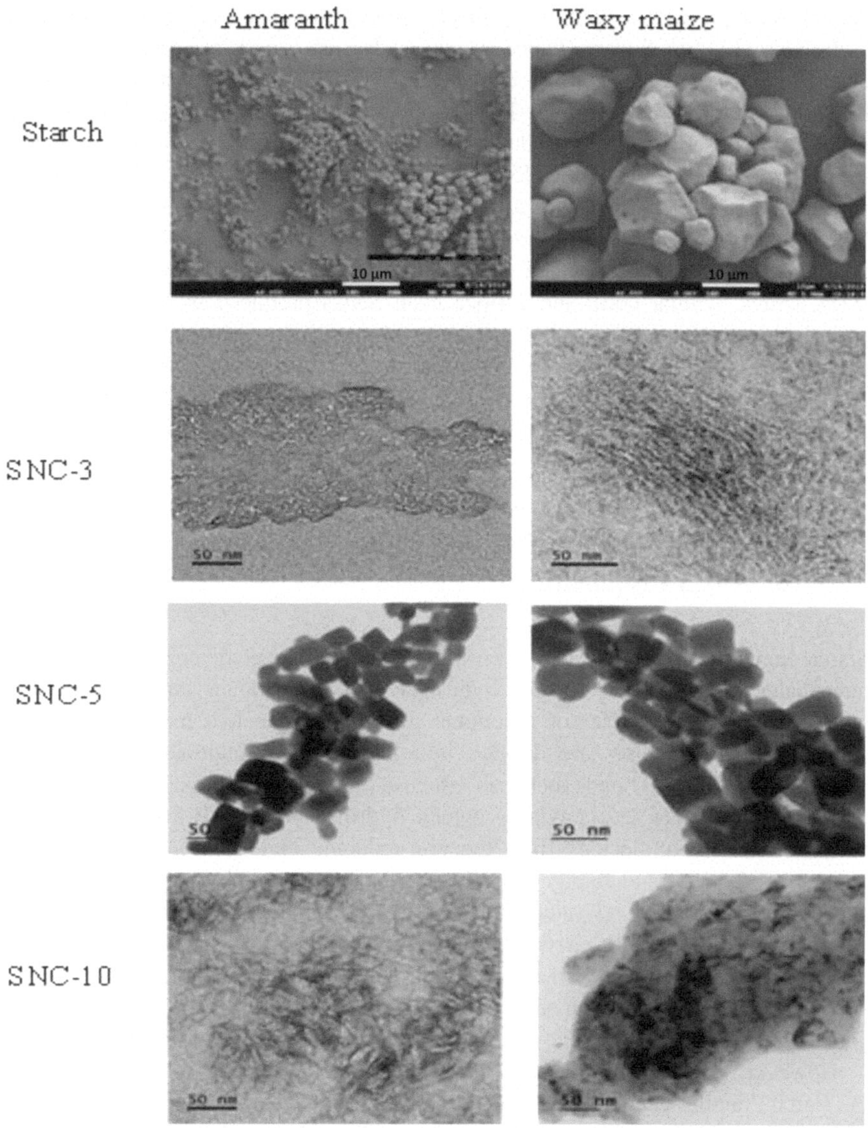

Fig. 2.4 Morphological structures of amaranth and waxy maize/corn starches by scanning electron microscopy and the resultant starch nanocrystals after 3, 5, and 10 days of H_2SO_4 hydrolysis by electron transmission microscopy. (Reused with permission from Concha et al. [52])

In addition, the shape of SNCs had no correlation with botanic origin or amylose content (Fig. 2.2). The main relation seems to be with the granule crystallinity type [39]. The main shape expected for SNCs is thin crystalline platelets [31, 32]. However, some parallelepipedic round or square-like or rounded-corner-square nanocrystals, as well as non-recognized shapes, have been described after acid hydrolysis [39, 54]. Square-shaped nanocrystals with a parallel arrangement were observed in waxy maize (Type-A) and wheat (Type-A) SNCs, whereas those derived from high amylose maize (Type-B) exhibited a round morphology [42]. Several studies have described SNCs production with 20–80 nm of size using waxy starches from maize [30–32, 55–58], rice [59], and potato [7]. Other regular or high amylose starches also result in SNCs, such as regular maize [39, 49, 58], and potato [7, 39, 49, 58], wheat [39, 58], and high amylose maize [39, 49, 54, 58], and other non-conventional starches, such as amadumbe [42], mung bean [49], quinoa [5]. Cassava, regular corn, and carioca bean SNCs produced by H_2SO_4 3.16 M hydrolysis were round or oval shaped with no correlation with the starch source, tending to agglomerate similarly to grape clusters [44]. There are other reports describing that individual SNCs are linked in larger aggregates after the recovering processes [7, 12].

2.3 Enzymatic, Ultrasound, and Other Approaches for SNCs Preparation

Starch enzymatic hydrolysis is the result of the action of specific enzymes from two main categories: amylases, which hydrolyze glycosidic bonds, and transferases, which catalyze the transfer of functional groups between two molecules (donor and acceptor). Amylases are divided into three groups: endo-amylases, which hydrolyze α– [1, 4] bonds between glucose molecules, internally resulting in α-anomeric products; exo-amylases, which hydrolyze α– [1, 4] and/or α– [1, 6] bonds of glucose molecules from the non-reducing end, resulting in α– or β– anomeric products; and finally, debranching enzymes, which include isoamylase and pululanase, exclusively attacking α– [1, 6] bonds, resulting in linear polysaccharide chains. The main enzymes used for SNCs processing are α–amylase, β–amylase and glucoamylase. Enzymatic hydrolysis of starch granules can result in porous starch, SNCs, or even starch nanoparticles [60]. The amylase hydrolysis of native starches is very efficient when the starch granules are gelatinized [18, 60].

However, for SNCs production, the crystalline lamellae must be preserved, and the starch granules cannot be gelatinized. Due it, porous corn starch were produced with a mixture of α-amylase and glucoamylase (ratio of 1:4) after 14 h at 40 °C. The original starch size was not reduced after the enzymatic hydrolysis [18]. In addition, a fast non-gelatinized starch fragmentation, using α-amylase and glucoamylase, mainly results in amorphous eroding simultaneously with low crystalline regions degradation. Granule fragmentation is induced by the amorphous erosion and the resulting fragments are considered to be crystalline blocklets [61, 62]. The waxy rice starch fragmentation by α-amylase hydrolysis during 3 h resulted in small-sized materials (0.5 μm) after ultrasonication in ethanol, which was assumed to consist of crystalline blockers [61].

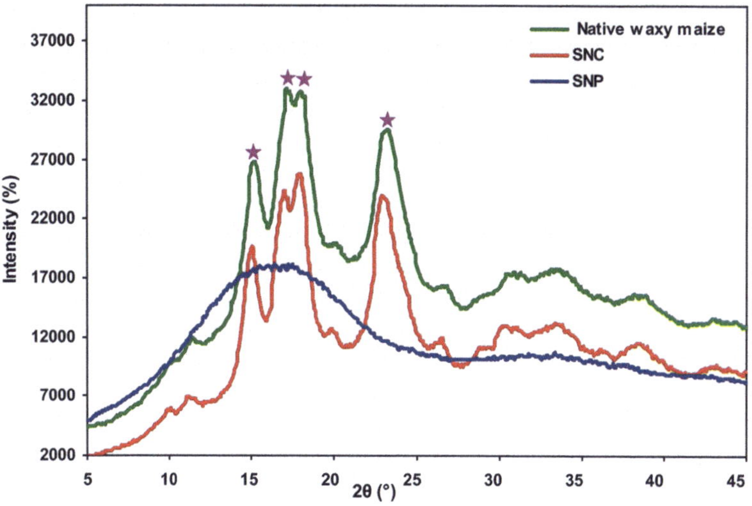

Fig. 2.5 Wide-angle X-ray scattering diffraction patterns of freeze-dried native starch, starch nanocrystals (SNCs), and starch nanoparticles (SNPs) from waxy maize (* starch A-type). (Reused with permission from Bel Haaj et al. [34])

Ultrasonication is a physical method that involves vibrations which come into direct contact with the fluids or suspensions through sonotrodes or probes. These are the formation of gas- or vapor-filled cavities, which reduce pressure. By the cavity collapse, there is an increase in pressure generating cavitation [63, 64]. During the cavitation collapse, there is intense heating in the bubbles (~5000 °C), high pressures (~500 atm) with a few microseconds of lifetime [65]. This leads to the breaking of C-C molecular bonds and the formation of radicals with long chains [66].

High- power low-frequency ultrasound (0–160 W, 20 kHz) processing on aqueous dispersions of potato starch damaged the granules after 30 min at 5 °C. These damage mechanisms are associated with shear forces and microjets induced by the cavitation of the sonication on solid surfaces [67]. Amorphous starch nanoparticles instead of SNCs (Fig. 2.5) were produced by high-power ultrasonication. This physical method results in nanoparticles between 30 and 100 nm in size after 75 min of processing in water and at low temperature (8 °C) [34, 68].

A new approach for waxy corn SNCs preparation was performed by mixing H_2SO_4 and HCl acids to provide H^+ at 1:0 (3, 4, 5, 6, 7, 8, 9, and 10 days of hydrolysis), 3:1 (6 h, 12 h, 24 h, 2, and 3 days of hydrolysis), 5:3 (2 h, 6 h, 12 h, 24 h, and 2 days of hydrolysis), and 1:1 (1 h, 1.25 h, 1.5 h, 1.75 h, 2 h, 2.5 h and 3 h of hydrolysis) ratios and H^+ concentration of 6.32 mol/L. It resulted in similar morphology, group structure, and surface elements to conventional H_2SO_4 (3;16 M, 5 days, 200 rpm for 40 °C). However, the hydrolysis time was slightly reduced by mixed acids (~0.83%), and the yield was reduced. The high concentration of HCl hydrolyzes starch very quickly into monosaccharides in a short time, with very small

SNCs output. The SNCs size was smaller after mixed acid hydrolysis, when 50% of H^+ was provided by the HCl, than conventional H_2SO_4, as do the crystallinity (38–56%) was higher, and the polymorphism was changed from A-type to V-type [19].

High-pressure homogenization or processing (HPP) is a non-thermal technology in which pressure acts in all directions, equally, instantaneously, and homogeneously. It promotes a decrease in the volume with a pressure enhance/lessen, and when the pressure is quickly dropped down to atmospheric pressure compression, there are alterations on starch non-covalent bonds. In addition, this processing results in damage of starch granules because the forces responsible for granular restriction are broken [69]. HPP followed by ultrasound (HPH-US) treatment on chestnut, corn, and potato starches to produce SNCs results in a substantial reduction in particle size, while partially preserving the RC of SNCs derived from these three distinct crystalline polymorph starches (A, B, and C). During SNCs processing, starch granules undergo into distinct stages of stress, aggregation, and agglomeration due to the intense mechanical forces generated by HPH-treatment, resulting in structural and physicochemical changes. The resulting SNCs exhibited non-regular morphologies with dimensions ranging from 135.36 to 203.47 nm, and crystallinity inferior to that of starch counterparts [70].

Another approach for corn starch SNCs production was a simultaneous treatment with ultrasound and H_2SO_4 hydrolysis. This results in round edge SNCs smaller than 100 nm within 45 min of processing with up to 21.6% yield and crystallinity below 40% [16]. The synergistic effects result in efficient extraction of crystallites and increase the yield of SNCs. This behavior may be due to the surface damage of the starch granules by ultrasound mechanical energy, which presumably facilitates the fast penetration of acid into the granule structure [16]. Regular corn SNCs were also prepared from maize starch using ultrasound-assisted H_2SO_4 (0.2 M) hydrolysis under magnetic stirring for 48 h. The oval platelet-shaped SNCs sizing around 253 nm were produced after processing for up to 2 h, with an improved yield of 36% and 16% of crystalline rising, compared to conventional H_2SO_4 hydrolysis [71].

2.4 Pretreatments on SNCs Isolation

Starch pretreatments can improve the yield of SNCs [7] and their properties [62], as well as speed up the process and change the morphology [5]. Few reports describe enzymatic and physic pretreatments to produce these crystalline nanoscale materials. Table 2.1 summarizes the main pretreatments used to produce SNCs.

Enzymatic pretreatment using glucoamylase on waxy potato starch was used to produce SNCs via H_2SO_4 hydrolysis. The enzymatic hydrolysis first fast phase was approximately 0–6 h, resulting in a pseudo-first-order process rate k significantly higher than that of the second stage (6–24 h). In the second stage, the starch granule crystalline lamellae were attacked by the enzyme, due to the prolonged hydrolysis

Table 2.1 Starch source, pretreatments, and starch nanocrystal processing parameters

Starch source	Pretreatment	SNCs processing	Reference
Waxy potato starch	Glucoamylase (1000 U/g of starch), 6 h, 60 °C, 150 rpm of stirring, pH 4.0	3.16 M H_2SO_4 hydrolysis, 5 days, 200 rpm of stirring at 40 °C, 15% (w/w) suspension	[7]
Waxy and regular corn starches	α-amylase form porcine pancreas (804 U/g of starch), fungal β-amylase (500 U/g of starch) and glucoamylase (125 U/g of starch), 2 h hydrolysis	3.16 M H_2SO_4 hydrolysis, 5 days, 200 rpm of stirring at 40 °C, 15% (w/w) suspension	[62]
Waxy corn starch	Glucoamylase (1000 U/g of starch), pH 4, 60 °C, 6 h	3.16 M H_2SO_4 hydrolysis, 5 days, 200 rpm of stirring at 40 °C, 15% (w/w) suspension	[72]
Waxy corn starch	Wett ball milling (anhydrous ethanol 1:3 w/v), rolling speed of 300 rpm for 15, 30, 45, 60, 75, and 90 min	3.16 M H_2SO_4 hydrolysis, 5 days, 200 rpm of stirring at 40 °C, 15% (w/w) suspension	[17]
Pinhão starch	HMT (22% of moisture, 1 h, 100 °C), ANN (16 h, 50 °C) and Ultrasound (50% of amplitude at 20 kHz, with on/off cycles of 30 s/5 s, 90 min)	3 M H_2SO_4 hydrolysis, 5 and 7 days, constant stirring at 40 °C, 15% (w/w) suspension	[12]
Common bean starch	HMT (22% of moisture, 1 h, 100 °C), ANN (16 h, 50 °C) and Ultrasound (50% of amplitude at 20 kHz, with on/off cycles of 30 s/5 s, 90 min)	3 M H_2SO_4 hydrolysis, 5 and 7 days, constant stirring at 40 °C, 15% (w/w) suspension	[15]
Waxy maize starch	HMT (99.2 °C, 1.8 h, and 18.2% of moisture)	3.16 M H_2SO_4 hydrolysis, 5 days, 200 rpm of stirring at 40 °C, 15% (w/w) suspension	[20]

ANN annealing, *HMT* heat-moisture-treatment, *SNC* starch nanocrystal

periods, resulting in undesired crystallinity lost. Therefore, the 6 h glucoamylase pretreatment was use to decrease the acid hydrolysis duration for production with SNCs crystallinity improvement [7].

The α-amylase, β-amylase and glucoamylase 2 h hydrolysis were used as pretreatments in waxy and regular corn starch for SNCs. Glucoamylase enzyme was the most efficient for creating a porous starch. It can hydrolyze (1 → 4) and (1 → 6) glycosidic bonds from amylopectin chains and erode the crystalline lamellae when the pretreatment is too long. After enzymatic pretreatment, only 6 h of 3.16 M H_2SO_4 hydrolysis was required to produce SNCs. In addition, after 45 h of acid hydrolysis, a regular yield of 15% of small-sized SNCs was reached. This behavior was due to the pretreatment, which creates a microporous diffusion pathway on starch granules, thereby enhancing the acid attack [62]. The 6 h of glucoamylase pretreatment on waxy corn starch was pretreated for SNCs producing. SNCs were obtained after 2 days of 1000 U/(g starch) enzyme pretreatment combined with H_2SO_4 hydrolysis with approximately 34% of crystallinity and mean sizing of 112.1 ± 8.9 nm [72].

Wet ball milling pretreatment was used in waxy corn starch for SNCs production using H_2SO_4 hydrolysis. The RC in the starches decreased after 30 min of ball milling. The SNCs crystallinity of obtained from starch ball milled for 30 and 45 min was similar, despite the initial lower crystallinity of 45 min ball milled starch. The 45 min of ball milling pretreatment accelerated the first stage of acid hydrolysis from 3 to 2 days because of partially damaged starch granules, which probably allowed fast acid hydrolysis. After 3 days of hydrolysis, the ball-milled pretreated samples were nanosized with a preserved crystalline structure [17].

Physical pretreatments (HMT, ANN, and sonication) were used to produce SNCs from unconventional C-type pinhão with 26% of amylose [12] and common bean starches with 38% of amylose content by H_2SO_4 hydrolysis for 5 and 7 days [15]. Each pretreatment affects the starch crystallinity in different ways. HMT might be responsible for some lateral and adjacent movements of amylopectin double helices in the crystalline lamella, reflecting the change in the B-type to A-type crystallinity pattern [73, 74] and consequent RC decrease in SNCs [15]. The ANN can induce corrections in the defects of the double helices within amylopectin chains, facilitating improvements in the perfection or augmentation of crystalline structures [75]. On the other hand, sonication generates a cavitation process that promotes a non-selective disruption in both the crystalline and amorphous regions of starch granules, especially with intense and prolonged treatment [12].

Despite some pretreated starch having lower crystallinity than the non-treated one, after 5 days of hydrolysis, HMT, ANN, and sonication physical pretreatments resulted in SNCs with higher crystallinity than the non-treated one. Also, for pinhão starch sonication speeding up the first stage of acid hydrolysis HMT can increase the hydrolysis yield and ANN resulted in SNCs with high thermal stability (ΔH) [12]. On the other hand, for common bean HMT and sonication proved to be effective in improving hydrolysis yield and thermal stability, while ANN slightly accelerated the production of SNCs [15].

The HMT also was studied as a pretreatment in waxy corn starch to produce SNCs by H_2SO_4 hydrolysis for 5 days. On the 4th day of hydrolysis, the maximum crystallinity was reached for pretreated and non-treated SNCs, resulting in nanosized structures. The HMT as pretreatment improved SNCs yield compared to the non-treated one resulting in one-third higher crystallinity than the original pretreated starch [20].

2.5 Post-treatment on SNCs

After SNCs isolation, their size and properties can be shaped according to different uses. Physical SNCs processing can be applied to reduce their sizes, while chemical modifications are especially described for reinforcement materials, drugs delivery systems, purifying and treatment systems, stabilizers and emulsifiers [76, 77].

Ultrasound and high shear Ultra-Turrax are very often described for SNCs individualization particles. A dispersion method for submicron SNCs (prepared by H_2SO_4 hydrolysis) was used by cross-linking using sodium hexametaphosphate and ultrasound cleaner for 4 h in water medium. It resulted in crystalline SNCs with improved water dispersion and gave rise to a stable and uniform SNCs aqueous suspension [78]. Also, some SNCs were processed to improve the dispersibility and reduce the size. Waxy corn SNCs were treated using Ultra-Turrax for 5 min at 13000 rpm to break the aggregates, followed by a filtering on tissue (40 μm) [53]. Pinhão and Carioca beans SNCs produced by H_2SO_4 hydrolysis were dispersed using Ultra-Turrax (13,000 rpm) for 5 min and sonicated for 2 min (50% amplitude at 20 kHz; on/off cycles of 30 s/5 s).

On the other hand, high pressure processing (HPP) is very efficient to break particles aggregates; however, usually promotes alterations on the crystalline structure [79], being severe at intermedium (~450 MPa) and high pressures (~600 MPa), which results in the destruction of the crystalline region [69].

Ultrasound assisted H_2SO_4 hydrolysis promotes crystallinity structure modifications and results in potato starch amorphous nanoparticles instead of SNCs due to the cavitation intensity [80]. Potato SNCs were produced using H_2SO_4 (3.16–5 M) hydrolysis, during 24–72 h and ultrasonic as post-treatment. The ultrasound pulse mode facilitates better formation of starch nanoparticles with average particle size of 13–370 nm after 3 days of hydrolysis and sonication. The crystallinity was enhanced with increasing H_2SO_4 concentration, and leads to a change in this structure, due to a possible rearrangement during retrogradation. After 3 days of 5 M H_2SO_4 hydrolysis followed by sonication the crystallinity was compromised [22].

The potato SNCs were prepared by H_2SO_4 (3.16 M, 200 rpm, at 40 °C for 5 days) hydrolysis followed by dielectric barrier discharge plasma endowed with two parallel electrodes of 1.3 cm, 40 V of input voltage, and 1 A of discharge for 2, 4, 6 and 8 min. The RC (26–30%) of SNCs plasma-treated decreased compared to non-treated SNCs (31.47%) with no polymorphism pattern changing (A-type). The size of SNCs increased after the short- plasma treatment (2–6 min), followed a decreasing after 8 min of treatment. The surface charge of SNCs influences their dispersion and aggregation, which can be increased by the plasma-treatment [21].

Surface chemical modification on waxy maize starch SNCs was done using alkenyl succinic anhydride and phenyl isocyanate to produce more hydrophobic particles. The SNCs were not altered by the chemical modifications, while particles morphologies were changed from platelets SNCs to uneven shape. The contact angle measurement confirmed the hydrophobic surface from nonpolar isocyanate or anhydride functions in the SNCs. This characteristic enables the SNCs utilization in nonpolar polymers matrices for composite materials development [23].

SNCs were successfully modified using stearic acid chloride and poly(ethylene glycol) methyl ether (PEGME), later graft the PEGME, to aiming to produce binding sites in drug delivery systems or for purifying and treatment systems. These modifications did not affect the SNCs crystallinity and resulted in crystalline hydrophobic shell around the hydrophilic SNCs. These modified SNCs can be used as toxic agent trapping and in co-continuous composites developing [81].

Octenyl succinic anhydride (OSA) SNCs was prepared by waxy maize starch H_2SO_4 hydrolysis and modified by ∼3 mg/mL of OSA. It results in an esterification reaction between glucose monomers hydroxyl groups and OSA molecules, with a

DS about ~0.013 of OSAs per glucose unit, which increased the SNCs hydrophobicity. This modification changes de particle morphology and does not alter the crystallinity polymorphism in the X-ray diffraction. The well-formed and stable Pickering emulsion were produced using from 0.01 to 0.50 wt% of entire emulsion system. With only the amount of 0.15 wt% of OSA-SNCs was effectively to stabilized and saturated Pickering emulsion [82]. Also, tapioca SNCs had high crystallinity (45.1%) and the succinylation (succinic anhydride, pH 9.0, 2 h, DS 0.1009) reduce the crystallinity (42.0%) [83].

SNCs from cassava, carioca beans and corn starch were isolated using H_2SO_4 hydrolysis and further oxidized using sodium hypochlorite under alkaline conditions (pH = 9.5) [44]. The oxidization reduced the SNCs crystallinity from cassava and corn starches and increased from carioca bean one. This reduction was associated to the amylopectin crystalline agglomerates depolymerization [44], while the carioca bean increasing can be consequence of the sodium hypochlorite acted in the reminded amorphous fraction and carboxyl groups arisen with a partial reorganization of those regions [40]. The SNCs oxidation was efficient to improve the colloidal stability of the Pickering emulsions up to 21 days of storage, being the cassava SNCs the best one.

The SNCs- graft -Polycaprolactone (PCL) are synthesized via assisted ring-opening polymerization and used as filler to improve the tensile strength and elongation properties of waterborne polyurethane matrix. Only 5% of SNCs-g-PCL increased the waterborne properties [84]. However, SNCs-g-PCL self-aggregate as crystalline domains impeded better improvement with higher concentrations, but enhanced elongation modulus.

2.6 Advances in SNCs Uses

Several studies have shown that SNCs could be used as fillers to improve mechanical and barrier properties of biocomposites [25–28] and as black carbon replacement for tires [25]. Nanobiocomposites are combined biomaterials based on two phases: a continuous matrix and a nanosized filler or reinforcement discontinuous ones. SNCs is generally used as the discontinuous phase due to their small and homogeneous size [25, 27]. The main interest in these nano sized materials is to use lower filler concentration (0.1–10%) rather than traditional material (5–50%) and improve properties for each specific application [25–27, 85, 86]. The materials combining may result in synergic properties from each individual materials, especially its processability and mechanical properties [85, 86]. The nanocomposites reinforced by SNCs are also promising materials for petroleum-based polymers and plastics, at least in part, replacing food and non-food applications [6, 25, 87, 88].

Mango kernel SCNs were isolated and used to produce mango kernel and corn starch bionanocomposite films in different concentrations (0–10%). These starch/ SNCs nanocomposite films tensile strength increased up to 90% by 5% of SNCs adding. The elongation modulus also increased about 120% at 7.5 wt% of SNCs.

The tensile strength and elongation modulus also increased by addition of amaranth SNCs (3–12%) [13] and cassava SNCs (1%) [89]. Furthermore, the water vapor permeability decreased about 15% with 7.5–10 wt% of SNCs [28]. The water vapor permeability of cassava starch and SNCs composites decreasing (~40%) was described with the addition of 2.5 wt% cassava SNCs [36], while 0.1% of rice SNCs reduced ~64% [90].

A 380% of increasement in the rubbery storage modulus (at 50 °C) and a 40% in the water vapor permeability decreasing were described in nanocomposites of cassava starch and waxy SNCs [91]. The increase in tensile properties of the films may be a consequence of the more compact structure and stiffness of the nanocrystals [6]. The SNCs are related to forming tortuous paths through the film due to the platelet structure of nanocrystals for water molecules to permeates, which can reduce the water vapor permeability [92]. Regular corn SNCs suspensions were used to reinforce commercial polyurethane (PU) films prepared by casting. SNCs-PU nano composite had better mechanical properties and lower water solubility as compared to a pure PU films [71].

There are some indications that high amylose starches can result in SNCs with weakly bonded residual amylose fragments which had negative influence in a nano filler networking formation [93]. This behavior can influence in the water absorption and reinforcement on natural rubber. The natural rubber reinforced (20 wt%) with corn waxy SNCs also results in relaxed modulus at room temperature of 75 times higher than the one of the unfilled matrix, being the particle-particle interactions the main role in these properties [94].

The effects of SNCs on the stability and rheological properties of oil-in-water (o/w) emulsions are very often highlighted. These nano materials can be used as an alternative solid emulsifier. Emulsions o/w with oil ratios of 5–35% (v/v) and sago SNCs (1–4% w/v) results in mean droplet diameter ranged from19.12 to 35.96 μm, and 4% of SNCs results in emulsions stable against coalescence [14]. Waxy maize SCNs exhibit a great potential to act as the sole stabilizer for o/w high internal phase emulsions [29]. The SNCs present higher crystallinity and hydrophobicity than native starches, which are relevance to stabilize these emulsions and are appropriate to food formulations.

Cassava, carioca bean and regular corn SNCs at 1, 3, and 5% (w/w) were used to stabilize o/w Pickering emulsions at water:oil ratios 1:1, 3:1 and 1:3 (v/v). The oxidized SNCs also were tested and resulted in stable emulsion after 1 week of storage. The emulsion index was improved by SNCs oxidation, and was higher with larger amounts of oil, even though the average droplets increasing [44]. Stable paraffin liquid: water o/w Pickering emulsions were produce using waxy maize SNCs (0.02, 0.05, 0.1, 1.0, 3.0, and 6.0 wt%) as emulsifier and storage up to 2 months. The SNCs increasing in o/w emulsion resulted in droplet size average decreasing. It probably occurs due to more available SNCs to stabilize smaller oil droplets [95]. Gastrointestinal digestion using simulated fluids *in vitro* study of Pickering emulsions stabilized by OSA-SNCs indicates that above of 0.15 (wt%) stabilize the oil droplets and did not impact the gastrointestinal digestion [82].

Also, SNCs from normal chestnut, corn, and potato starches were produced by high-pressure homogenizer followed by ultrasound treatment for fat replacers in ice cream. The increased fat substitution by SNCs (20%, 40%, 60%, 80%, and 100%), notable improves the ice cream hardness and viscosity, accompanied by a reduction in the melting rate. The overall sensory evaluation indicates corn SNCs as a viable alternative fat replacer for enhancing the quality of ice cream. Therefore, there is a crucial need for further research to explore the intricate interplay between SNCs proteins and small molecules and sugars present in ice cream [70].

There are some promising results on SNCs heavy metal removing from water. The adsorptions of metal ions Zn(II), Cu(II), and Fe(II) were effective when the initial concentration of metal ions were 200 mg/L. The adsorption capacity of SNCs reached equilibrium at 60 min in an optimal pH = 5 [72]. The Cu(II) and methylene blue adsorption were done with SNCs and succinylated-SNCs in aqueous solutions. The Cu(II) and methylene blue probably occurred on SNCs and succinylated-SNCs surface. The adsorption equilibrium was reached in short time. The adsorption capacity increased by more than two times after the SNCs modification [83].

2.7 Conclusions

The starch nanocrystals (SNCs) are successfully produced by different approaches; however, the H_2SO_4 hydrolysis is still the main processing method used to isolate insoluble crystalline platelets with sizes ranging from 20 to 400 nm. The research efforts improved the SNCs processing by increasing the SCNs recovering yield and decreasing the processing time. There was an important SNCs production optimization by Angellier et al. [32] that drives most of the research. Although the SNCs preparation efficiency has gradually improved, some lacunes seeking solutions. Is there any other environmentally friendly solution that can substitute concentrate acid hydrolysis for SNCs isolation, which results in highly crystalline nano-sized structures? How to produce SNCs with a competitive cost for wide commercial uses? How new technology as pulse electrical field and plasma-water-activated starch pretreatments or SCNs post-treatments can improve SCNs isolation or applications? Among these questions, it is clear that the SNCs' applications are vast of possibilities and in need of investigation.

Acknowledgments The author gratefully acknowledges the Universidade Federal da Fronteira Sul for the support.

Conflicts of Interest The author declares no conflicts of interest.

References

1. Food and Drug Administration. Considering whether an FDA-regulated product involves the application of nanotechnology [Internet]. 2014 [cited 2023 Aug 27]. Available from: https://www.fda.gov/RegulatoryInformation/Guidances/ucm257698.htm
2. Alves MJ dos S, Chacon WDC, Gagliardi TR, Agudelo Henao AC, Monteiro AR, Ayala Valencia G, et al. Food applications of starch nanomaterials: a review. Starch/Staerke. 2021;73(11–12):2100046.
3. Campelo PH, Sant'Ana A de S, Clerici MTP da S. Starch nanoparticles: production methods, structure, and properties for food applications. Curr Opin Food Sci [Internet]. 2020;33 (i):136–40. Available from: https://doi.org/10.1016/j.cofs.2020.04.007.
4. Xiong X, Huang M, Zhou X, Zhou H, Zeng C, Zhao Z, et al. Physicochemical studies of nanocrystals of starches from two rice (Oryza sativa L.) types and their characteristics using various modern instrument techniques. J Sci Food Agric. 2021;101(3):1038–46.
5. Velásquez-Castillo LE, Leite MA, Ditchfield C, Sobral PJ do A, Moraes ICF. Quinoa starch nanocrystals production by acid hydrolysis: kinetics and properties. Int J Biol Macromol. 2020;143:93–101.
6. Lecorre DS, Bras J, Dufresne A. Starch nanoparticles: a review. Biomacromolecules. 2010;11: 1139–53.
7. Hao Y, Chen Y, Li Q, Gao Q. Preparation of starch nanocrystals through enzymatic pretreatment from waxy potato starch. Carbohydr Polym [Internet]. 2018;184(September 2017):171–7. Available from: https://doi.org/10.1016/j.carbpol.2017.12.042.
8. Qiu C, Wang C, Gong C, McClements DJ, Jin Z, Wang J. Advances in research on preparation, characterization, interaction with proteins, digestion and delivery systems of starch-based nanoparticles. Int J Biol Macromol [Internet]. 2020;152:117–25. Available from: https://doi.org/10.1016/j.ijbiomac.2020.02.156.
9. Kulkarni PR. Starch—a review. Compr Rev Food Sci Food Saf. 2006;5(1):1–17. (Figure 2).
10. Costa ÉK de C, de Souza CO, da Silva JBA, Druzian JI. Hydrolysis of part of cassava starch into nanocrystals leads to increased reinforcement of nanocomposite films. J Appl Polym Sci. 2017;134(41):1–9.
11. Kim HY, Park SS, Lim ST. Preparation, characterization and utilization of starch nanoparticles. Colloids Surf B Biointerfaces [Internet]. 2015;126:607–20. Available from: https://doi.org/10.1016/j.colsurfb.2014.11.011.
12. Pinto VZ, Moomand K, Deon VG, Biduski B, Zavareze E da R, Lenhani GC, et al. Effect of physical pretreatments on the hydrolysis kinetic, structural, and thermal properties of pinhão starch nanocrystals. Starch/Staerke. 2021;73(7–8):1–9.
13. Añón MC, Dufresne A, Noemi A, Cecilia M. Composite and nanocomposite films based on amaranth biopolymers. Food Hydrocoll. 2018;74:159–67.
14. Azfaralariff A, Farahfaiqah F, Joe LS, Fazry S, Mohamed M, Nazar MF, et al. Sago starch nanocrystal-stabilized Pickering emulsions: stability and rheological behavior. Int J Biol Macromol [Internet]. 2021;182:197–206. Available from: https://doi.org/10.1016/j.ijbiomac.2021.03.132.
15. Pinto VZ, Pinto CC, Souza SM De, Moomand K, Biduski B, Santos GHF dos, et al. Physical pretreatment on common bean starch at acid hydrolysed nanocrystals structure and properties. Starch – Stärke. 2024;76:5–6 (May 2024) 2300204. Avalilable from https://doi.org/10.1002/star.202300204
16. Mohammad Amini A, Razavi SMA. A fast and efficient approach to prepare starch nanocrystals from normal corn starch. Food Hydrocoll [Internet]. 2016;57:132–8. Available from: https://doi.org/10.1016/j.foodhyd.2016.01.022.
17. Dai L, Li C, Zhang J, Cheng F. Preparation and characterization of starch nanocrystals combining ball milling with acid hydrolysis. Carbohydr Polym [Internet]. 2018;180(October 2017):122–7. Available from: https://doi.org/10.1016/j.carbpol.2017.10.015.

18. Gao F, Li D, Bi C h, Mao Z h, Adhikari B. Application of various drying methods to produce enzymatically hydrolyzed porous starch granules. Dry Technol. 2013;31(13–14):1627–34.
19. Li S, Zhou W, Huang C, Hu Y, Gao Q, Chen Y. Rapid preparation of starch nanocrystals by the mixed acid of sulfuric acid and hydrochloric acid. Int J Biol Macromol [Internet]. 2023;232(9): 123402. Available from: https://doi.org/10.1016/j.ijbiomac.2023.123402.
20. Dai L, Zhang J, Cheng F. Succeeded starch nanocrystals preparation combining heat-moisture treatment with acid hydrolysis. Food Chem [Internet]. 2019;278(November 2018):350–6. Available from: https://doi.org/10.1016/j.foodchem.2018.11.018.
21. Shen H, Ge X, Zhang B, Su C, Zhang Q, Jiang H, et al. Preparing potato starch nanocrystals assisted by dielectric barrier discharge plasma and its multiscale structure, physicochemical and rheological properties. Food Chem [Internet]. 2022;372(September 2021):131240. Available from: https://doi.org/10.1016/j.foodchem.2021.131240.
22. Baruah KN, Singha S, Uppaluri RVS. Preparation of potato starch nanoparticles using acid hydrolysis and ultrasonic post-treatment. ACS Food Sci Technol. 2023;3(4):626–34.
23. Molina-boisseau S, Belgacem MN, Dufresne A. Surface chemical modification of waxy maize starch. Langmuir. 2005;22:2425–33.
24. Bastioli C, Floridi G, Tredici GD. U.S. Patent 2009/03114455 A1, 2009.
25. Le D, Angellier-Coussy H, Le Corre D, Angellier-Coussy H. Preparation and application of starch nanoparticles for nanocomposites: a review. React Funct Polym. 2014;85:97–120.
26. Lagaron JM. Multifunctional and nanoreinforced polymers for food packaging, vol. 1. Cambridge: Woodhead Publishing; 2019. p. 1–476.
27. Palanisamy CP, Cui B, Zhang H, Jayaraman S, Muthukaliannan GK. A comprehensive review on corn starch-based nanomaterials: properties, simulations, and applications. Polymers (Basel). 2020;12(9):1–27.
28. Oliveira AV, da Silva APM, Barros MO, de sá M. Souza Filho M, Rosa MF, Azeredo HMC. Nanocomposite films from mango kernel or corn starch with starch nanocrystals. Starch/Staerke. 2018;70(11–12):1–7.
29. Yang T, Zheng J, Zheng BS, Liu F, Wang S, Tang CH. High internal phase emulsions stabilized by starch nanocrystals. Food Hydrocoll [Internet]. 2018;82:230–8. Available from: https://doi.org/10.1016/j.foodhyd.2018.04.006.
30. Molina-boisseau S, Dufresne A. Waxy maize starch nanocrystals as filler in natural rubber. Macromol Symp. 2006;233:132–6.
31. Putaux JL, Molina-Boisseau S, Momaur T, Dufresne A. Platelet nanocrystals resulting from the disruption of waxy maize starch granules by acid hydrolysis. Biomacromolecules. 2003;4(5): 1198–202.
32. Angellier H, Choisnard L, Molina-Boisseau S, Ozil P, Dufresne A. Optimization of the preparation of aqueous suspensions of waxy maize starch nanocrystals using a response surface methodology. Biomacromolecules. 2004;5(4):1545–51.
33. Dufresne A, Cavaille JY, Helbert W. New nanocomposite materials: microcrystalline starch reinforced thermoplastic. Macromolecules. 1996;29(1):7624–6. Communications
34. Bel Haaj S, Thielemans W, Magnin A, Boufi S. Starch nanocrystals and starch nanoparticles from waxy maize as nanoreinforcement: a comparative study. Carbohydr Polym [Internet]. 2016;143:310–7. Available from: https://doi.org/10.1016/j.carbpol.2016.01.061.
35. Lin N, Huang J, Chang PR, Anderson DP, Yu J. Preparation, modification, and application of starch nanocrystals in nanomaterials: a review. J Nanomater. 2011;2011:1–13.
36. Kristo E, Biliaderis CG. Physical properties of starch nanocrystal-reinforced pullulan films. Carbohydr Polym. 2007;68:146–58.
37. Hoover R. Acid-treated starches. Food Rev Int. 2000;16(3):369–92.
38. Wang L, Wang Y j. Structures and physicochemical properties of acid-thinned corn, potato and rice starches. Starch - Stärke. 2001;53:570–6.
39. LeCorre D, Bras J, Dufresne A. Influence of botanic origin and amylose content on the morphology of starch nanocrystals. J Nanopart Res. 2011;13(12):7193–208.

40. Wei B, Xu X, Jin Z, Tian Y. Surface chemical compositions and dispersity of starch nanocrystals formed by sulfuric and hydrochloric acid hydrolysis. PLoS ONE [Internet]. 2014;9(2):1–7.
41. Wei B, Hu X, Li H, Wu C, Xu X, Jin Z, et al. Effect of pHs on dispersity of maize starch nanocrystals in aqueous medium. Food Hydrocoll [Internet]. 2014;36:369–73. Available from: https://doi.org/10.1016/j.foodhyd.2013.08.015.
42. Mukurumbira A, Mariano M, Dufresne A, Mellem JJ, Amonsou EO. Microstructure, thermal properties and crystallinity of amadumbe starch nanocrystals. Int J Biol Macromol [Internet]. 2017;102:241–7. Available from: https://doi.org/10.1016/j.ijbiomac.2017.04.030.
43. Saeng-on J, Aht-Ong D. Production of starch nanocrystals from agricultural materials using mild acid hydrolysis method: optimization and characterization. Polym Renew Resour. 2017;8 (3):91–116.
44. Daniel THG, Bedin AC, Souza ECF, Lacerda LG, Demiate IM. Pickering emulsions produced with starch nanocrystals from cassava (Manihot esculenta Crantz), beans (Phaseolus vulgaris L.), and corn (Zea mays L.). Starch - Stärke. 2020;72:1900326.
45. Buléon A, Colonna P, Planchot V, Ball S. Starch granules: structure and biosynthesis. Int J Biol Macromol. 1998;23:85–112.
46. Cai J, Cai C, Man J, Yang Y, Zhang F, Wei C. Crystalline and structural properties of acid-modified lotus rhizome C-type starch. Carbohydr Polym [Internet]. 2014;102(1):799–807. Available from: https://doi.org/10.1016/j.carbpol.2013.10.088.
47. Morrison WR, Tester RF, Gidley MJ, Karkalas J. Resistance to acid hydrolysis of lipid-complexed amylose and lipid-free amylose in lintnerised waxy and non-waxy barley starches. Carbohydr Res. 1993;245(2):289–302.
48. Xiong X, Huang M, Zhou X, Zhou H, Zeng C, Zhao Z, et al. Physicochemical studies of nanocrystals of starches from two rice (Oryza sativa L.) types and their characteristics using various modern instrument techniques. J Sci Food Agric. 2020;101(3):1036–46.
49. Kim HY, Lee JH, Kim JY, Lim WJ, Lim ST. Characterization of nanoparticles prepared by acid hydrolysis of various starches. Starch/Staerke. 2012;64(5):367–73.
50. Wang S, Copeland L. Effect of acid hydrolysis on starch structure and functionality: a review. Crit Rev Food Sci Nutr. 2015;55(8):1081–97.
51. Nakazawa Y, Wang YJ. Acid hydrolysis of native and annealed starches and branch-structure of their Naegeli dextrins. Carbohydr Res. 2003;338(24):2871–82.
52. Concha BBS de la, Agama-Acevedo E, Nuñez-Santiago MC, Bello-Pérez LA, Alvarez-Ramirez J. Acid hydrolysis of waxy starches with different granule size for nanocrystal production. J Cereal Sci. 2018;79:193–200.
53. Le Corre D, Bras J, Choisnard L, Dufresne A. Optimization of the batch preparation of starch nanocrystals to reach daily time-scale. Starch/Staerke. 2012;64(6):489–96.
54. Kim HY, Lee JH, Kim JY, Lim WJ, Lim ST. Characterization of nanoparticles prepared by acid hydrolysis of various starches. Starch - Stärke. 2012;64(5):367–73.
55. Molina-boisseau S, Dole P, Dufresne A. Thermoplastic starch - waxy maize starch nanocrystals nanocomposites. Biomacromolecules. 2006;7:531–9.
56. Molina-boisseau S, Lebrun L, Dufresne A. Processing and structural properties of waxy maize starch nanocrystals reinforced natural rubber. Macromolecules. 2005;38(9):3783–92.
57. Molina-boisseau S, Dufresne A. Processing and characterization of waxy maize starch films plasticized by sorbitol and reinforced with starch nanocrystals. Macromol Biosci. 2007;7(11): 1206–16.
58. Lecorre D, Bras J, Dufresne A. Influence of native starch's properties on starch nanocrystals thermal properties. Carbohydr Polym. 2012;87(1):658–66.
59. Ji N, Ge S, Li M, Wang Y, Xiong L, Qiu L, et al. Effect of annealing on the structural and physicochemical properties of waxy rice starch nanoparticles: effect of annealing on the properties of starch nanoparticles. Food Chem. 2019;286(January):17–21.
60. Wu Y, Du X, Ge H, Lv Z. Preparation of microporous starch by glucoamylase and ultrasound. Starch/Staerke. 2011;63(4):217–25.

61. Kim JY, Park DJ, Lim ST. Fragmentation of waxy rice starch granules by enzymatic hydrolysis. Cereal Chem. 2008;85(2):182–7.
62. Lecorre D, Vahanian E, Dufresne A, Bras J. Enzymatic pretreatment for preparing starch nanocrystals. Biomacromolecules. 2012;13(1):132–7.
63. Feng H, Barbosa-Cánovas GV, Weiss J, editors. Ultrasound technologies for food and bioprocessing, Food engineering series. Urbana: Springer London; 2011.
64. Herceg Z, Šubaric D, Babic J, Brnc M, Rez A, Bosiljkov T, et al. Ultrasound effect on physical properties of corn starch. 2010;79:91–100.
65. Suslick KS. Sonochemistry. Science (80-). 1990;247(4949):1439–45.
66. Minakawa AFK, Faria-Tischer PCS, Mali S. Simple ultrasound method to obtain starch micro- and nanoparticles from cassava, corn and yam starches. Food Chem [Internet]. 2019;283 (October 2018):11–8. Available from: https://doi.org/10.1016/j.foodchem.2019.01.015.
67. Yue Y, Zuo J, Hébraud P, Hemar Y, Ashokkumar M. Quantification of high-power ultrasound induced damage on potato starch granules using light microscopy. Ultrason - Sonochemistry [Internet]. 2012;19(3):421–6. Available from: https://doi.org/10.1016/j.ultsonch.2011.08.006.
68. Haaj SB, Magnin A, Pétrier C, Boufi S. Starch nanoparticles formation via high power ultrasonication. Carbohydr Polym. 2013;92:1625–32.
69. Castro LMG, Alexandre EMC, Saraiva JA, Pintado M. Impact of high pressure on starch properties: a review. Food Hydrocoll [Internet]. 2020;106(October 2019):105877. Available from: https://doi.org/10.1016/j.foodhyd.2020.105877.
70. Lin D, Zhao J, Fan H, Qin W, Wu Z. Enhancing starch nanocrystal production and evaluating their efficacy as fat replacers in ice cream: investigating the influence of high pressure and ultrasonication. Int J Biol Macromol [Internet]. 2023;251(June):126385. Available from: https://doi.org/10.1016/j.ijbiomac.2023.126385.
71. Hakke VS, Bagale UD, Boufi S, Uday Bhaskar Babu G, Sonawane SH. Ultrasound assisted synthesis of starch nanocrystals and it's applications with polyurethane for packaging film. J Renew Mater. 2020;8(3):239–50.
72. Chen Q, Liang C, Gao X, Zhao Y, You N, Wang F, et al. Removal of heavy metals from aqueous solution using starch nanocrystals. J Polym Environ [Internet]. 2022;30(10):4304–19. Available from: https://doi.org/10.1007/s10924-022-02513-2.
73. Hoover R. The impact of heat-moisture treatment on molecular structures and properties of starches isolated from different botanical sources. Crit Rev Food Sci Nutr. 2010;50(9):835–47.
74. Vermeylen R, Goderis B, Delcour JA. An X-ray study of hydrothermally treated potato starch. Carbohydr Polym. 2006;64(2):364–75.
75. Vamadevan V, Bertoft E, Soldatov D V, Seetharaman K. Impact on molecular organization of amylopectin in starch granules upon annealing. Carbohydr Polym [Internet]. 2013;98(1): 1045–55. Available from: https://doi.org/10.1016/j.carbpol.2013.07.006.
76. Li C, Guo Y, Chen M, Wang S, Gong H, Zuo J, et al. Recent preparation, modification and application progress of starch nanocrystals: a review. Int J Biol Macromol [Internet]. 2023;250 (July):126122. Available from: https://doi.org/10.1016/j.ijbiomac.2023.126122.
77. Kumari S, Yadav BS, Yadav RB. Synthesis and modification approaches for starch nanoparticles for their emerging food industrial applications: a review. Food Res Int [Internet]. 2020;128(May 2019):108765. Available from: https://doi.org/10.1016/j.foodres.2019.108765.
78. Ren L, Jiang M, Wang L, Zhou J, Tong J. A method for improving dispersion of starch nanocrystals in water through crosslinking modification with sodium hexametaphosphate. Carbohydr Polym [Internet]. 2012;87(2):1874–6. Available from: https://doi.org/10.1016/j.carbpol.2011.08.070.
79. Liu Y, Selomulyo VO, Zhou W. Effect of high pressure on some physicochemical properties of several native starches. J Food Eng. 2008;88(1):126–36.
80. Shabana S, Prasansha R, Kalinina I, Potoroko I, Bagale U, Shirish SH. Ultrasound assisted acid hydrolyzed structure modification and loading of antioxidants on potato starch nanoparticles. Ultrason - Sonochemistry [Internet]. 2019;51(March 2018):444–50. Available from: https://doi.org/10.1016/j.ultsonch.2018.07.023.

81. Thielemans W, Belgacem MN, Dufresne A. Starch nanocrystals with large chain surface modifications. Langmuir. 2006;6:4804–10.
82. Jo M, Ban C, Goh KKT, Choi YJ. Gastrointestinal digestion and stability of submicron-sized emulsions stabilized using waxy maize starch crystals. Food Hydrocoll [Internet]. 2018;84 (May):343–52. Available from: https://doi.org/10.1016/j.foodhyd.2018.06.026.
83. Chen Q jie, Zheng X ming, Zhou L ling, Zhang Y fei. Adsorption of Cu (II) and Methylene Blue by Succinylated Starch Nanocrystals. 2019;1800266(Ii):1–7.
84. Chang PR, Ai F, Chen Y, Dufresne A, Huang J. Effects of Starch Nanocrystal- graft -Polycaprolactone on Mechanical Properties of Waterborne polyurethane-based nanocomposites. J Appl Polym Sci. 2008;111:619–27.
85. Dufresne A. Processing of polymer nanocomposites reinforced with polysaccharide nanocrystals. Molecules. 2010;15(6):4111–28.
86. Dufresne A. Comparing the mechanical properties of high performances polymer nanocomposites from biological sources. J Nanosci Nanotechnol. 2006;6:322–30.
87. Vieira MGA, Da Silva MA, Dos Santos LO, Beppu MM. Natural-based plasticizers and biopolymer films: a review. Eur Polym J [Internet]. 2011;47(3):254–63. Available from: https://doi.org/10.1016/j.eurpolymj.2010.12.011.
88. Yu L, Dean K, Li L. Polymer blends and composites from renewable resources. Prog Polym Sci. 2006;31(6):576–602.
89. Santana JS, Costa ÉKDC, Rodrigues PR, Correia PRC, Cruz RS, Druzian JI. Morphological, barrier, and mechanical properties of cassava starch films reinforced with cellulose and starch nanoparticles. J Appl Polym Sci 2018;47001:14–17.
90. Martins PC, Latorres JM, Martins VG. Impact of starch nanocrystals on the physicochemical, thermal and structural characteristics of starch-based films. Lwt. 2022:156.
91. Garcia NL, Ribba L, Dufresne A, Aranguren MI, Goyanes S, Garcı NL, et al. Physico-mechanical properties of biodegradable starch nanocomposites. Macromol Mater Eng. 2009;294(3):169–77.
92. Li X, Qiu C, Ji N, Sun C, Xiong L, Sun Q. Mechanical, barrier and morphological properties of starch nanocrystals-reinforced pea starch films. Carbohydr Polym. 2015;121:155–62.
93. Lecorre DS, Bras J, Dufresne A. Influence of the botanic origin of starch nanocrystals on the morphological and mechanical properties of natural rubber nanocomposites. Macromol Mater Eng. 2012;297(10):969–78.
94. Angellier H, Molina-Boisseau S, Dufresne A. Mechanical properties of waxy maize starch nanocrystal reinforced natural rubber. Macromolecules. 2005;38(22):9161–70.
95. Li C, Sun P, Yang C. Emulsion stabilized by starch nanocrystals. Starch/Staerke. 2012;64(6): 497–502.

Chapter 3
Starch Nanoparticles

Wilson Daniel Caicedo Chacon and Germán Ayala Valencia

3.1 Introduction

Nanotechnology has emerged as a transformative discipline in various scientific areas and industrial applications, offering revolutionary potential for designing and developing nanoscale materials with enhanced properties and innovative applications. Within this field, the production of starch nanoparticles (SNPs) has stood out as a promising research line, driving significant advances in addressing specific challenges [1]. It has been observed that reducing starch granules to the nanoscale has improved various properties, such as solubility and dispersion in aqueous media, addressing the intrinsic tendency of conventional starch to form gels and aggregates [2, 3]. In addition, the increased specific surface area of SNPs has enhanced their ability to interact with other components in complex systems, improving their functionality in applications such as food, pharmacology, and biocompatible materials. The application of nanotechnology to starch has paved the way for increased colloidal stability, controlled bioactivity, and sustained release of stabilized compounds, positioning SNPs as a versatile and effective material [4–6].

In this context, the purpose has been to study various methods to fragment starch to produce SNPs, with anti-solvent precipitation (ASP) or nanoprecipitation, ball milling (BM), ultrasonication (UT), and high-pressure homogenization (HPH) being some of the most used techniques. ASP is a "bottom-up" nanoparticle production method that has proven to be highly effective in overcoming limitations inherent in conventional approaches, allowing precise control over the size, morphology, and distribution of particles [7]. ASP involves the controlled addition of an anti-solvent, in this case, the addition of ethanol (as a non-miscible component with starch) to the

W. D. C. Chacon · G. Ayala Valencia (✉)
Department of Chemical and Food Engineering, Federal University of Santa Catarina, Florianópolis, SC, Brazil
e-mail: g.ayala.valencia@ufsc.br

G. Ayala Valencia (ed.), *Starch Nanomaterials and Food Applications*,
https://doi.org/10.1007/978-3-031-60086-9_3

reaction medium (gelatinized starch), triggering nanoparticle formation through insolubilization [7]. "Top-down" techniques such as BM, UT, and HPH have positioned themselves as valuable tools in this field, each with its advantages and challenges. UT, based on ultrasonic waves, stands out for its ability to disperse particles, and improve dispersion, enhancing properties such as solubility and colloidal stability [8, 9]. On the other hand, BM harnesses mechanical energy to reduce particle size, influencing the surface properties of starch [10, 11]. HPH, using intense hydrodynamic forces, produces stable and uniform dispersions [12, 13].

However, despite significant advances, each technique presents specific challenges and inherent limitations. Optimizing process parameters and a deep understanding of starch properties are essential to maximize the benefits of these methodologies and overcome potential barriers.

This chapter aimed to review the production of SNPs by anti-solvent, ultrasonication, ball milling, and high-pressure homogenization methods, as well as to analyze the effect of each production method on the physicochemical properties of the resulting nanomaterials. This chapter mainly considered the literature published during the last 6 years (2019–2023).

3.2 Methods to Produce Starch Nanoparticles

3.2.1 Anti-solvent Precipitation

The ASP method (nanoprecipitation or solvent displacement) is a direct process for SNP production that differs from emulsion-based methods (emulsification diffusion, emulsion evaporation, and salting-out techniques) in that it does not require an emulsifying agent. The ASP method is characterized by its speed, ease, and simplicity [14–17]. Due to its advantages, several research studies have produced robust SNPs, stabilized oils through nanoemulsions, and films from SNPs [18, 19]. However, it is crucial to understand the mechanism of nanoparticle production using the anti-solvent precipitation method to ensure stable, monodisperse, and nanoscale SNP production [7].

In practice, the hydrophobic solute (biopolymer molecules or lipids) is first dissolved in a polar organic solvent, such as ethanol, methanol, or acetone, typically producing binary solutions. Then, this solution is added to a specified amount of an anti-solvent (this insolubility should be relative to the solute), usually water, which must be miscible with the polar solvent in all proportions. At this stage, the binary solution begins to become insoluble for the solute as the anti-solvent concentration increases, and the system evolves toward phase separation [7]. When the solute is hydrophilic, as is the case with starch at temperatures above 70 °C (requiring a gelatinization process), it is generally solubilized in water, making the polar organic compound the anti-solvent, usually using ethanol (Table 3.1). Although it seems simple to execute, several studies show that process conditions affect SNP production and stabilization, even under standardized production conditions [7, 32, 33],

Table 3.1 Starch nanoparticles produced by nanoprecipitation using anti-solvent

Starch source	Pre-nanoprecipitation process	SLR	Anti-solvent	Solvent	ASSR	Post-nanoprecipitation process	Major findings	Reference
Potato and cassava	None	0.05	Acidified hydroethanolic solution (50%, 60%, 70%, 80%)	Water	1:1	Constant stirring for 12 h. The SNPs were centrifuged three times with ethanol in water (80% v/v) and finally washed with absolute ethanol (99.6%). The excess ethanol from the samples was evaporated with a forced air convection oven at 60 °C for 10 min	The SNPs have particle size lower than 1000 nm and V-type crystalline structure. Furthermore, these nanomaterials showed high solubility values in water at 30 and 90 ° C	[20]
High amylose maize starch	None	0.05	Absolute ethanol	Water	1:1	Constant stirring for 2 h and ultrasound treatment (15, 30, 45, 60, 75 min) amplitude of 40%	A successful three-step process was employed to reduce the size of starch particles, transforming non-uniform agglomerated nanoparticles into uniform particles of approximately 170 nm	[21]
Lotus root	Oxidized lotus root starch (OLRS) with NaClO as oxidant agent	0.005	Absolute ethanol	Water	10:1	Was added into luteolin stock solution with the injection rate of 5 mL/min and stirred for 1 h	The particle size, polydispersity index and zeta-potential of luteolin-OLRS SNPs were 305 nm, 0.173	[22]

(continued)

Table 3.1 (continued)

Starch source	Pre-nanoprecipitation process	SLR	Anti-solvent	Solvent	ASSR	Post-nanoprecipitation process	Major findings	Reference
							and 20.8 mV, respectively. An encapsulation efficiency of 87.2% of lutenin were given	
Field pea	The gelatinized starch paste was then sonicated for 5–30 min using a 20 kHz ultrasonic processer, carried out at specified amplitudes (10%, 20%, 30%, 40%, 50%, 60%, and 70%)	0.001, 0.005, 0.01	Absolute ethanol	Water	1:1	Constant stirring (500 rpm) for 1 h	The dispersed SNPs showed good thermal stability in water and retained their particle characteristics at a temperature range of 25–60 °C	[15]
Corn starch	Starch gelatinization with 6.2 g of glycerol in 200 mL of water	0.054	Absolute ethanol	Water	1:1	Was stirred constantly for 1 min and poured into Petri dishes at 50 °C for 24 h	As a result of adding starch nanoparticles to the starch film, its wettability is reduced, and the existence of the plasticizer resulted in more homogeneous and more hydrophobic film	[23]
Normal maize (NMS) and waxy maize	None	0.06	Absolute ethanol	Water	2:1	The sediments of starch were washed twice with absolute ethanol, centrifuged,	NMS and WMS presented oil loading capacity of 0.72 g/g, lower than the SNPs	[24]

starch (WMS)						and then oven-dried at 40 °C overnight. The obtained samples were groundthrough a 100-mesh sieve. The SNPs were used to encapsulate oil	produced by nanoprecipitation of 1.65 g/g	
Maize and amilose	None	0.01, 0.03, 0.05	Luteonin ethanol solution (0, 0.3, 0.6, 0.92 mg/mL)	Water	1:1	The precipitate was washed per 50 mL of ethanol to move free lutein and water. The starch-lutein nanoparticles were dried with hot air (35 °C) for 6 h, and ground to pass through a 125-mesh sieve	The lutein content in the nanoparticles reaches 13.2 mg/g when 90 mg of lutein is added to 50 ml of 1% (w/v) starch paste. SNPs protected lutein from chemical oxidation	[25]
Quinoa starch	None	0.05	Absolute ethanol	Water	1:1	Homogenized at a high-speed for 5 min at 15,000 rpm. Pickering emulsions with QSNPs concentrations of 0.5–2.5% (based on emulsion volume) were prepared at a fixed oil volume fraction of 0.50	The oxidative stability of the emulsions increased with increasing starch concentration and ionic strength	[26]
Waxy maize starch	Short amylose was prepared using by pullulanase at 58 C for 6 h	0.01	Absolute ethanol	Water	3:1, 4:1, 5:1	The precipite was stirred continually using a magnetic stirrer at a constant rate of	The results showed that the optimal ratio of ethanol to short amylose solution and	[27]

(continued)

Table 3.1 (continued)

Starch source	Pre-nanoprecipitation process	SLR	Anti-solvent	Solvent	ASSR	Post-nanoprecipitation process	Major findings	Reference
						600 rpm and 50 °C at the same time. Suspensions of SNPs were subjected to ultrasound (amplitude 40%). The curcumin solution (0.3 g/30 mL ethanol 80%) added dropwise into the different SNPs samples suspensions	ultrasonic time was 4:1 and 4 min, respectively. The nanoparticles showed a small size (82.43 nm), relatively high loading capacity (11.57%), and a peak gelatinization temperature of 97.74 °C	
Corn starch	None	0.02	Absolute ethanol	Water (0.5 M sodium hydroxide solution and 20% w/w Tween 80)	1:1, 5:1, 9:1	None	The results observed in nanoprepitation in spinning disc reactor (SDR) indicate that an increase in flow rate and rotation speed leads to smaller nanoparticles and narrower size distributions, which is attributed to greater shear and instabilities within the liquid film	[28]
Normal maize	None	0.03	Ethanol stearic acid solution (0.1 mg/mL)	Water	1:1	Immediately after precipitation immersed in liquid nitrogen to stop	The stearic acid content and complexation index of the sample increased to 2.8 mg/g	[29]

						further physicochemical changes in the sample	of compound and 83.4%, respectively	
Normal maize	None	0.03	Ethanolic fatty solution (1 mg/mL)	Water	1:1	The precipitate was washed with ethanol (100 mL) to remove free fatty acids and were lyophilized using a freeze-dryer	As the chain length of fatty acids increased from 12 to 22, fatty acids content in the composites increased from 1.69 to 14.85 mg/g composite	[30]
Sweet potato starch	The gelatinized starch paste was then sonicated for 5–25 min using a 20 kHz probe sonicator	0.02, 0.05, 0.08	Absolute ethanol	Water	1:1, 2:1, 5:1	The plasma treatment conditions were as follows: the gas source was air, the distance of the two parallel electrodes was 1.3 cm, the input voltage was 90 ± 1 V, and the discharge was 1 A. The SNPs were treated for 90 s	SNPs with an average particle size of 64.51 ± 0.15 nm and polydispersity index of 0.23 ± 0.01 were obtained	[31]
Cassava starch	The gelatinized starch paste was then sonicated for 15 min and acetylation	0.035	Absolute ethanol	Water	10:1	Starch paste ultrasonicated and hydrophobized through acetylation. Pickering emulsion was prepared with soybean oil	The hydrophobicity of the SNPs was improved by acetylation. Oil-in-water emulsions (3:7, v/v) containing 1.5 wt% SNPs were stabilized for more than 35 days without creaming	[19]

SLR solid-liquid ratio, *ASSR* anti-solvent/solvent ratio

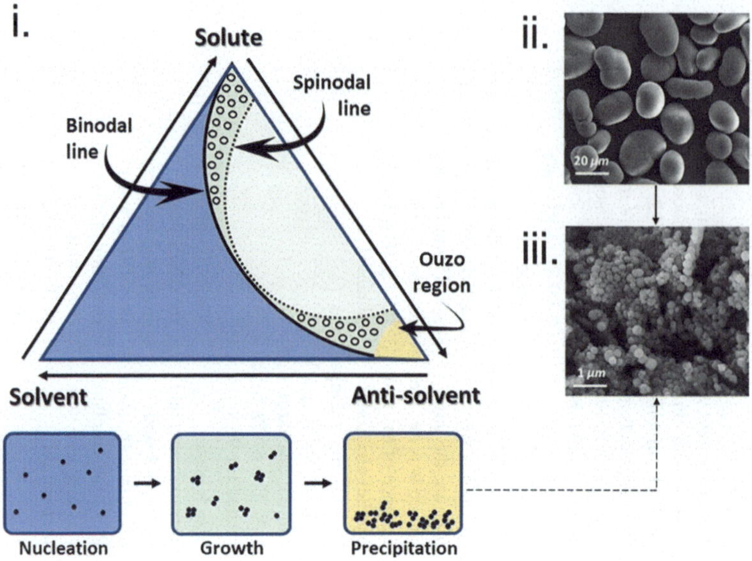

Fig. 3.1 Representation of the ternary system of starch nanoprecipitation (i) [7]. Native pea starch granules (ii) and SNPs after the nanoprecipitation process (iii). (Adapted from Dong et al. [34])

sometimes resulting in particles outside the nanoscale range (between 1 and 1000 nm) and/or the possible formation of large starch aggregates or very high polydispersity indices.

Many studies have focused on identifying the most relevant experimental parameters to control the size and polydispersity of SNPs produced by anti-solvent precipitation (Table 3.1). Some of these investigations have focused on using phase diagrams of ternary systems, in which the solute/solvent/anti-solvent interaction is studied, being, in the case of SNP production, the starch/water/ethanol interaction (Fig. 3.1). This knowledge is of utmost importance to prepare monodisperse SNPs efficiently and reproducibly, using a rational approach rather than a trial-and-error process. Research such as that of Alves et al. [20] focused on studying the effect of the percentage of acidified hydroethanolic solutions as an anti-solvent, noting that at high ethanol concentrations, smaller SNPs were observed, resulting in significantly increased solubility of starch particles between temperatures of 30 to 90 °C compared with unmodified starch, which requires higher temperatures (70 °C) for complete solubilization. Likewise, studies conducted by Apostolidis et al. [21] and Sana et al. [28], through optimized processes, managed to produce nanoparticles with average sizes of 170 nm and 248 nm, respectively, allowing for nanoscale SNPs with reduced agglomeration, studying different anti-solvent/solvent ratios, and observing that SNPs can be obtained at lower ethanol amounts.

It is important to control particle polydispersity to obtain SNPs in the desired size ranges, for which it is vital to control their nucleation and growth processes over time. For the formation of SNPs with low polydispersity, a single nucleation burst is required, which can be achieved through processes with very short mixing times to ensure uniform oversaturation [7]. Most studies use relatively fast dripping rates and stirring speeds that are fast enough to avoid aggregates and decrease polydispersity. Good results have been observed by adding ethanol as an anti-solvent at rates of 2 and 5 mL/min [22, 29, 30]. Similarly, ASP with a spinning disk reactor has been successful, operating at 6, 12, and 18 mL/s [28]. According to these studies, the lower the rate at which the anti-solvent is added, the more likely the production of aggregates, as it promotes fewer initial nucleus productions. On the other hand, another critical parameter is the mixing speed between the moment of adding the anti-solvent to the solvent, as higher speeds reduce the probability of producing aggregates and likewise decrease the polydispersity index. It has been observed that agitation rates between 450 and 600 rpm are sufficient [24, 27, 29, 30].

Once SNPs are produced, it is essential to remove the anti-solvent solution because the system is oversaturated, which could favor SNP agglomeration. In this regard, immediately after the precipitation process, drying with an oven or freeze-drying has been used [20, 21, 34], as well as immersion in liquid nitrogen to halt physicochemical changes [30].

To stabilize SNPs, nanoprecipitation can be performed in the presence of a small amount of stabilizer. The stabilizer is generally not necessary if SNPs exhibit hydrophilic fragments or have a non-zero zeta potential [35], as in the case of the research conducted by Wang et al. [31], which produced nanoscale SNPs with an average particle size of 64.51 ± 0.15 nm, rendering the use of a surfactant unnecessary. However, some studies have resorted to surfactants such as Tween 80 to prevent flocculation or aggregation of SNPs [28].

The production of SNPs has been attractive because this polymer has proven to be a potential carrier of bioactive compounds or drugs. Shin et al. [30] achieved, through nanoprecipitation, the loading of fatty acids into normal corn starch up to a concentration of 14.85%. In this case, the lipid phase was dissolved in ethanol (anti-solvent) to be adsorbed onto starch during precipitation. Similarly, Fu et al. [25] stabilized luteolin by nanoprecipitation, initially dissolving it in absolute ethanol, successfully stabilizing luteolin in SNPs under the best conditions at a concentration of 13.2 mg/g.

Processes before and after the nanoprecipitation process have been employed to favor the production of starch nanoparticles. For example, physical methods such as ultrasound before and after the nanoprecipitation process have been used to reduce agglomerations, decrease particle sizes, and reduce polydispersity indices. Dong et al. [15] used ultrasound in gelatinized starch solutions at 20 kHz with amplitudes ranging from 10% to 70%, promoting a decrease in solution viscosity, which enhances the homogeneous dispersion of the anti-solvent. Similarly, Dong et al. [34] and Wang et al. [31] used ultrasound at 20 kHz to promote a decrease in SNP sizes. Additionally, Wang et al. [31] used plasma treatment after nanoprecipitation with voltages of 90 ± 1 V and discharges of 1 A to promote particle size reduction. Pre-nanoprecipitation chemical treatments have also been carried out mainly to increase the loading efficiency of SNPs produced by ASP, as in the study by Chen et al. [22], which oxidized lotus root starch to improve its hydrophilic affinity, thus increasing luteolin stabilization. On the other hand, Yao et al. [19] acetylated cassava starch to produce hydrophobic SNPs to achieve Pickering emulsions after the nanoprecipitation process.

3.2.2 Ultrasonication, Ball Milling, and High-Pressure Homogenization

SNPs can be obtained using physical methods such as UT, BM, and HPH. These methods present the advantage of producing clean label SNPs, thereby guarantying the absence of no chemical residues in the final products [36]. In general, high surface damage can be obtained in starches treated with UT, BM, and HPH (Fig. 3.2), affecting the crystalline structure and physicochemical properties of the resulting nanomaterials (Table 3.2).

Very few studies have evaluated the use of ultrasound to produce SNPs (Table 3.2). Ultrasound is a green technique that induces physical and chemical alterations in biopolymers due to the cavitation effects. Acoustic cavitation causes

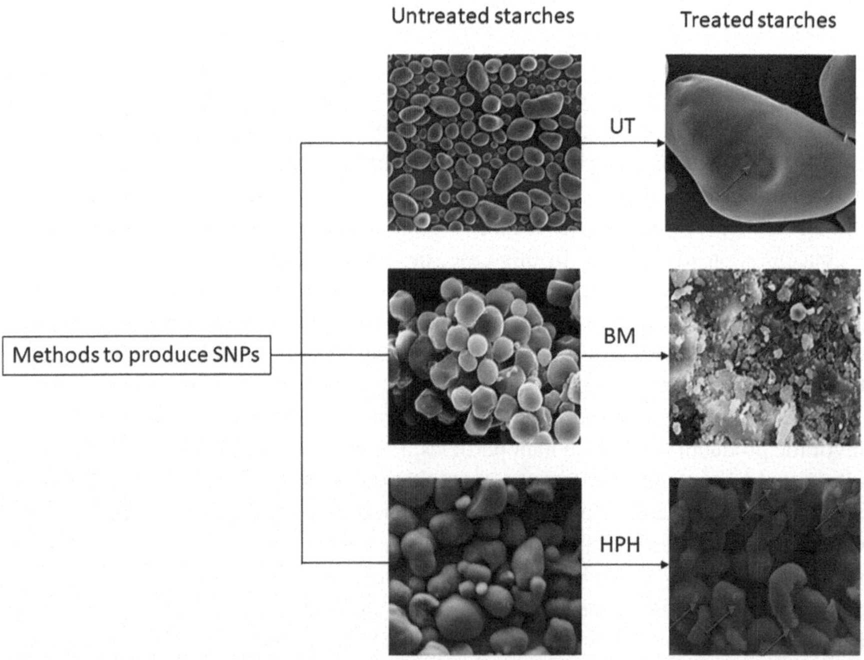

Fig. 3.2 Morphology of starch granules treated with ultrasonication (UT), ball milling (BM), and high-pressure homogenization (HPH). (Adapted from Chorfa et al. [37], Jhan et al. [10], and Apostolidis and Mandala [36])

Table 3.2 Starch nanoparticles produced by UT, BM, and HPH methods

Starch source	Production method	Method of production	Major findings	Reference
Lotus stem (*Nelumbo nucifera*)	UT	Starch (resistant starch) suspensions were sonicated using an ultrasonic probe ultrasonicator for 20 min (400 W power, pulse mode set at 1 s on/ 1 s off). SNPs were dried by lyophilization	SNPs with particle size of 345 nm and PI of 0.25 were obtained by UT	[38]
Potato	UT	Starch suspension in water (50 g/L) or not was sonicated with a high-intensity ultrasonic probe processor for 60 min (450 W power and 10% amplitude). SNPs obtained under dry conditions were directly sieved. In contrast, SNPs obtained in aqueous suspensions were dried in a vacuum oven at 40 °C for 12 h or dried via lyophilization, then ground and sieved	Ultrasonication reduced starch crystallinity in dry conditions, and a complete loss of starch crystallinity when starch was treated in aqueous suspensions	[37]
Corn	UT	Starch suspensions (concentration not informed) were sonicated for 90 min with an ultrasonic probe processor (500 W power and 90% amplitude). SNPs were dried by lyophilization	Ultrasonication reduced starch crystallinity	[39]
Water chestnut, lotus stem, and horse chestnut	BM	Starches were milled using a planetary ball mill with rotational speed of 600 rpm for 5 h	SNPs had particle size oscillating between 343 and 855 nm. The relative crystallinity was decreased from 26.2 to 8.0, 22.3 to 8.0, and 13.1 to 6.2 in water chestnut, lotus stem, and horse chestnut starches, respectively	[11]
Pearl and proso millets	BM	Starches were milled using a mill pot made of ceramic, using 10 mill balls of diameter at a rolling speed of 500 rpm for 35 min	SNPs with particle size between 417 and 636 were produced by milling. The resulting SNPs had a reduction in	[10]

(continued)

Table 3.2 (continued)

Starch source	Production method	Method of production	Major findings	Reference
			relative crystallinity and thermal properties	
Lotus stem	BM	Starch was milled using a using a planetary monomill at 750 rpm for 30, 60, and 90 min	SNPs with particle size between 123 and 543 nm. The reduction in particle size elevated the glycemic index values in SNPs	[40]
Rice	BM	Starch suspensions (3%) were milled using a MiniSeries media mill (Netzsch LLC, PA, U.S. A.) with milling speed of 3000 rpm for 4 h	The particle size of SNPs was reduced from approximately 1200–700 nm by increasing the UT from 1 to 4 h	[41]
Corn	HPH	Starch suspensions (2%) were submitted at 140, 200, and 250 MPa for 1, 2, 3, and 4 cycles	Native starch granules had particle sizes of approximately 7 μm. The particle size in SNPs decreased with the increasing in pressure and treatment cycles	[36]
Rice	HPH	Starch suspensions (3%) were submitted at 60, 80, 100, 120, 140, and 160 MPa for 5 cycles	All treatment produced SNPs and the increased of pressure reduced the particle size. The mean particle diameter of SNPs ranged between 154 and 260 nm	[42]

BM ball milling, *HPH* high-pressure homogenization, *PI* polydispersity index, *UT* ultrasonication

the rapid generation of bubbles in the liquid, generating high pressure and temperature. The collapse of these bubbles may break the covalent bonds in the polymer material [43]. Recently, Noor et al. [38] produced SNPs with a particle size of approximately 345 nm by UT; however, this research aimed to develop hydrogels designed with this nanomaterial and not to understand the effect of UT on the SNP production (Table 3.2). The dispersion of starch in water is fundamental for achieving the modification of this macromolecule at the nanoscale. In this way, Chorfa et al. [37] compared the UT treatment in potato starch dispersed or not in water and concluded that a complete amorphization was obtained only in starch dispersions (Table 3.2). This result was due to the better propagation of bubbles in starch dispersions when compared with starch granules having only water molecules associated with their moisture content. In contrast, Chutia and Mahanta [39] observed a reduction in the crystallinity of potato starch granules dispersed in water and treated by UT (Table 3.2). However, the starch dispersion concentration was not determined in this research. Other factors such as starch granule size and crystallinity could also affect the production of SNPs by UT.

Most studies have used UT as an alternative solution for reducing the processing time required to generate SNPs and increasing the yield production [9, 44–47]. In this way, Lin et al. [48] applied UT after the use of enzymatic/acid hydrolysis used to produce SNPs from lotus seeds. The authors investigated the effect of ultrasonic power (200, 600, and 1000 W), time (5, 15, and 25 min), and liquid ratio (starch: buffer solution: 1%, 3%, and 5%). The same authors concluded that ultrasonic waves first acted on the amorphous region of starch granules, and the average molecular weight gradually decreased as ultrasonic power and time increased. Only, SNPs treated with 600, and 1000 W power had particle sizes lower than 700 nm. Similarly, Apostolidis et al. [49] concluded that UT was fundamental for dismantling the nonuniform agglomerate nanoparticles produced by the antisolvent nanoprecipitation, furnishing uniform nanoparticles with particle sizes of approximately 170 nm. Finally, UT have also been used as pretreatment in the production of SNPs. Wang et al. [50] used UT with 600 W and on/off pulse of 3/5 s to treat gelatinized starch (3%) isolated from tiger nut (*Cyperus esculentus* L.), which later was precipitated with ethanol to produce SNPs. The authors investigated the UT time (10–30 min) and power (60–100%) used as pretreatment. When the ultrasonic power was lower than 80%, incomplete UT was produced and particle sizes over 1 μm were observed in the final material, suggesting the presence of fragmented starch granules at the microscale. In contrast, when the ultrasonic power was higher than 80%, some amylose and amylopectin chains may aggregate again, leading to an increase in the particle size of SNPs. The same authors concluded that the best UT power–time combination was 80% for 20 min when SNPs were obtained using particle sizes lower than 200 nm.

BM is an ecofriendly physical technique in which the combinatorial effects of friction, collision, impingement, and shear alter the structure of starch granules [51]. This approach is very interesting because the use of solvent is not necessary, in this way, starches are processed in the absence of water and no drying is necessary after processing [10, 11, 40]. Furthermore, SNPs obtained by BM have particle sizes lower than 1000 nm (Table 3.2), which is similar to those informed in SNPs obtained by the antisolvent method (Sect. 3.2.1). SNPs produced by BM have higher adsorption capacity and viscosity when compared with native starches [11]. In addition, these nanomaterials have better antioxidant properties than native starches [10], but they also have higher glycemic index values, limiting their food applications [40]. As an alternative, BM can also be applied in starch dispersions; however, 4 h is required to obtain particle sizes of approximately 700 nm in starch dispersions [41] (Table 3.2). The resulting SNPs in water were applied to stabilize soybean oil using Pickering emulsions. The authors concluded that SNP concentration of 0.8–3.5% in the emulsion could form stable systems where the oxidation rate decreased with the presence of the SNPs.

HPH is another physical method used to produce SNPs in recent years. HPH treatment induces physicochemical alteration in starch granules because pressure causes phenomena of cavitation, shear, and turbulence, and temperature increases during the treatment [52]. In this approach, starch granules are dispersed in water and subjected to high pressure. Apostolidis and Mandala [36] investigated the effect of pressure (140, 200, and 250 MPa) and treatment cycles (HPH time: 1, 2, 3, and 4 cycles) on the particle size and microstructure of SNPs from corn starch. The authors concluded that pressure and treatment cycles are important to reduce the size in SNPs. In this way, the smallest particles with a size of approximately 540 nm were obtained after 4 cycles of homogenization at 250 MPa. The pressure increase provoked changes such as pores, deep slots and aggregates in the nanoparticles and destroyed the amylopectin structure. However, SNPs had similar crystalline structure when compared with untreated corn starch (B-type crystalline structure) (Table 3.2). The authors concluded that HPH is a promising method for obtaining SNPs with high amylose content [36]. Similarly, Sun et al. [42] observed a reduction in the particle size and crystallinity of SNPs the pressure increased from 60 to 160 MPa in HPH (Table 3.2). Amorphous regions in were observed in SNPs between 60 and 80 MPa, and destruction of the crystalline regions was evidenced with pressures higher than 100 MPa. Crystalline modification affected the thermal and rheological properties of SNPs. Hence, the melting temperature associated with starch gelatinization decreased as the pressure increased during HPH. The viscosity of SNP samples increased between 60 and 120 MPa because of damage to the amorphous region and the leaching of amylose, and decreased when the pressure was higher than 140 MPa because of the damage to intermolecular hydrogen bonds and crystalline regions in SNPs [42].

Finally, HPP has been used in dual treatments used to obtain SNPs. The combination of US followed by HPH have allowed the production of SNPs with particle size lower than 60 nm, which were used to stabilize carotenoids in Pickering emulsions [12]. As a pretreatment, HPH has been used to damage starch granules, which were later enzymatically hydrolyzed to produce SNPs with high crystallinity [53].

3.3 Future Trends

Most studies have used the antisolvent or nanoprecipitation method as an alternative to produce SNPs [54, 55]. This method is preferred because of its ease of implementation at the laboratory scale; however, nanoparticles are precipitated by the addition of ethanol dropwise without standardization. Future research studies must investigate the main parameters in the antisolvent method (starch concentration, solvent/antisolvent ratio, and mixing rate) and their effect on the particle size and physicochemical properties of SNPs, as well as to optimize the best conditions to produce these nanomaterials.

Similarly, it is necessary to optimize the production of SNPs by UT and BM. Future studies must focus on applying UT to produce SNPs and not only as an alternative to complement nanoprecipitation or enzymatic/acid hydrolysis. SNPs with particle sizes lower than 1000 nm can be obtained by UT, BM, and HPH; however, it is important to understand the effect of starch granule size and its amylose/amylose ratio on the obtention of SNPs in each method. Furthermore, the use of dual treatment (e.g. UT plus HPH) is an alternative to reduce the particle size and polydispersity index.

Finally, exhaustive physicochemical characterization of SNPS produced by UT, BM, and HPH is necessary to understand their real food applications. The glycemic index and digestibility of SNPs must be investigated. It is expected that the production of SNPs using these methods will increase in the next few years due to the absence of chemical residues (clean label SNPs).

3.4 Conclusions

Starch nanoparticles (SNPs) are obtained by the disruption of starch granules with "top-down" methods such as antisolvent or nanoprecipitation, as ultrasonication (UT), ball milling (MB), and high-pressure homogenization (HPH). The SNPs obtained by these methods have particle sizes lower than 1000 nm and different physicochemical properties when compared with native starches. In general, the production of SNPs reduces their crystallinity and gelatinization temperature and increases water adsorption and viscosity.

ASP, UT, BM, and HPH have shown their potential to destroy starch granules, affecting the starch crystallinity. These nanomaterials could be used to encapsulate and stabilize bioactive compounds and oils in food applications, or as additives to modify food texture.

Acknowledgments W.D.C. Chacon gratefully acknowledges the Fundação de Amparo à Pesquisa e Inovação do Estado de Santa Catarina (FAPESC) for the PhD fellowship (3003/2021), as well as the Hermes 57983 project of the Universidad Nacional de Colombia sede Palmira, the opportunity to learn skills for academic writing. G. A. Valencia would like to thank the National Council for Scientific and Technological Development (CNPq) (grant 302434/2022-4) for financial support. The authors gratefully acknowledge the Federal University of Santa Catarina for the support.

Conflicts of Interest The authors declare no conflicts of interest.

References

1. Caldonazo A, Almeida SL, Bonetti AF, Lazo REL, Mengarda M, Murakami FS. Pharmaceutical applications of starch nanoparticles: a scoping review. Int J Biol Macromol. 2021;181:697–704. Elsevier B.V.
2. Gujral H, Sinhmar A, Nehra M, Nain V, Thory R, Pathera AK, et al. Synthesis, characterization, and utilization of potato starch nanoparticles as a filler in nanocomposite films. Int J Biol Macromol. 2021;186:155–62.
3. Das A, Sit N. Modification of taro starch and starch nanoparticles by various physical methods and their characterization. Starch/Staerke. 2021;73(5–6)
4. Bhatia M, Rohilla S. Formulation and optimization of quinoa starch nanoparticles: quality by design approach for solubility enhancement of piroxicam. Saudi Pharm J. 2020;28(8):927–35.

5. Ahmad M, Mudgil P, Gani A, Hamed F, Masoodi FA, Maqsood S. Nano-encapsulation of catechin in starch nanoparticles: characterization, release behavior and bioactivity retention during simulated in-vitro digestion. Food Chem. 2019;270:95–104.
6. Morán D, Gutiérrez G, Blanco-López MC, Marefati A, Rayner M, Matos M. Synthesis of starch nanoparticles and their applications for bioactive compound encapsulation. Appl Sci (Switzerland). 2021;11(10):4547.
7. Caicedo Chacon WD, Verruck S, Monteiro AR, Valencia GA. The mechanism, biopolymers and active compounds for the production of nanoparticles by anti-solvent precipitation: a review. Food Res Int. 2023;168:112728. Elsevier Ltd.
8. Hassan NA, Darwesh OM, Smuda SS, Altemimi AB, Hu A, Cacciola F, et al. Recent trends in the preparation of nano-starch particles. Molecules. 2022;27:5497. MDPI.
9. Bajer D. Nano-starch for food applications obtained by hydrolysis and ultrasonication methods. Food Chem. 2023;402:134489.
10. Jhan F, Shah A, Gani A, Ahmad M, Noor N. Nano-reduction of starch from underutilised millets: effect on structural, thermal, morphological and nutraceutical properties. Int J Biol Macromol. 2020;159:1113–21.
11. Ahmad M, Gani A, Masoodi FA, Rizvi SH. Influence of ball milling on the production of starch nanoparticles and its effect on structural, thermal and functional properties. Int J Biol Macromol. 2020;151:85–91.
12. Chutia H, Mahanta CL. Properties of starch nanoparticle obtained by ultrasonication and high pressure homogenization for developing carotenoids-enriched powder and Pickering nanoemulsion. Innov Food Sci Emerg Technol. 2021;74:102822.
13. Tagliapietra BL, de Melo BG, Sanches EA, Plata-Oviedo M, Campelo PH, Clerici MTPS. From micro to nanoscale: a critical review on the concept, production, characterization, and application of starch nanostructure. Starch/Staerke. 2021;73. John Wiley and Sons Inc.
14. Barthold S, Hittinger M, Primavessy D, Zapp A, Groß H, Schneider M. Preparation of maltodextrin nanoparticles and encapsulation of bovine serum albumin—influence of formulation parameters. Eur J Pharm Biopharm. 2019;142:405–10.
15. Dong H, Chen L, Zhang Q, Gao J, Vasanthan T. Optimization of processing parameters to produce nanoparticles prepared by rapid nanoprecipitation of pea starch. Food Hydrocoll. 2021;121:106929.
16. Jaison D, Chandrasekaran G, Mothilal M. pH-sensitive natural almond gum hydrocolloid based magnetic nanocomposites for theragnostic applications. Int J Biol Macromol. 2020;154:256–66.
17. Liu C, Xu B, McClements DJ, Xu X, Cui S, Gao L, et al. Properties of curcumin-loaded zein-tea saponin nanoparticles prepared by antisolvent co-precipitation and precipitation. Food Chem. 2022;391:133224.
18. Lammari N, Louaer O, Meniai AH, Elaissari A. Encapsulation of essential oils via nanoprecipitation process: overview, progress, challenges and prospects. Pharmaceutics. 2020;12:431. MDPI AG.
19. Yao X, Lin R, Liang Y, Jiao S, Zhong L. Characterization of acetylated starch nanoparticles for potential use as an emulsion stabilizer. Food Chem. 2023;400:133873.
20. dos Santos Alves MJ, Calvo Torres de Freitas PM, Monteiro AR, Ayala Valencia G. Impact of the acidified hydroethanolic solution on the physicochemical properties of starch nanoparticles produced by anti-solvent precipitation. Starch/Staerke. 2021;73(11–12).
21. Apostolidis E, Stergiou A, Kioupis D, Amin S, Paximada P, Kakali G, et al. Production of nanoparticles from resistant starch via a simple three-step physical treatment. Food Hydrocoll. 2023:137, 108412.
22. Chen YY, Liu K, Zha XQ, Li QM, Pan LH, Luo JP. Encapsulation of luteolin using oxidized lotus root starch nanoparticles prepared by anti-solvent precipitation. Carbohydr Polym. 2021;273:118552.
23. Fazeli M, Lipponen J. Developing self-assembled starch nanoparticles in starch nanocomposite films. ACS Omega. 2022;7(49):44962–71.

24. Feng Y, Wu C, Junejo SA, Zhang B, Fu X, Tan CP, et al. Effect of V-type crystallinity and starch particle structure on the oil loading capacity and anti-oxidation. Carbohydr Polym. 2022;297:120015.
25. Fu Y, Yang J, Jiang L, Ren L, Zhou J. Encapsulation of lutein into starch nanoparticles to improve its dispersity in water and enhance stability of chemical oxidation. Starch/Staerke. 2019;71(5–6):1800248.
26. Jiang F, Zhu Y, Hu WX, Li M, Liu Y, Feng J, et al. Characterization of quinoa starch nanoparticles as a stabilizer for oil in water Pickering emulsion. Food Chem. 2023;427:136697.
27. Li X, Wang S, Zhong J, Li T, Fan G, Zhou D, et al. Preparation and characterization of fine and stable short amylose nanocarriers for curcumin using a highly efficient and convenient method. Int J Biol Macromol [Internet]. 2024;257:128738. Available from: https://linkinghub.elsevier.com/retrieve/pii/S0141813023056374
28. Sana S, Boodhoo K, Zivkovic V. Production of starch nanoparticles through solvent-antisolvent precipitation in a spinning disc reactor. Green Process Synth. 2019;8(1):507–15.
29. Shin HY, Lee JH, Kim JY. Formation mechanism of nanocomposites between starch and stearic acid via nanoprecipitation. Food Hydrocoll. 2022;131:107780.
30. Shin HY, Kim JH, Kim JY. Effect of fatty acid chain length on physicochemical properties of starch nanocomposites obtained via nanoprecipitation. Int J Biol Macromol. 2023;230:123415.
31. Wang J, Yu YD, Zhang ZG, Wu WC, Sun PL, Cai M, et al. Formation of sweet potato starch nanoparticles by ultrasonic—assisted nanoprecipitation: effect of cold plasma treatment. Front Bioeng Biotechnol. 2022 Sep;16:10.
32. Santos S, Puna J, Gomes J. A brief review of the supercritical antisolvent (SAS) technique for the preparation of nanocatalysts to be used in biodiesel production. Energies. 2022;15: 9355. MDPI.
33. Zhang J, Liu M, Zeng Z. The antisolvent coprecipitation method for enhanced bioavailability of poorly water-soluble drugs. Int J Pharm. 2022;626;122043. Elsevier B.V.
34. Dong H, Zhang Q, Gao J, Chen L, Vasanthan T. Comparison of morphology and rheology of starch nanoparticles prepared from pulse and cereal starches by rapid antisolvent nanoprecipitation. Food Hydrocoll. 2021;119:106828.
35. Johnson BK, Prud'homme RK. Mechanism for rapid self-assembly of block copolymer nanoparticles. Phys Rev Lett. 2003;91(11):118302.
36. Apostolidis E, Mandala I. Modification of resistant starch nanoparticles using high-pressure homogenization treatment. Food Hydrocoll. 2020;103(September 2019):105677.
37. Chorfa N, Nlandu H, Belkacemi K, Hamoudi S. Physical and enzymatic hydrolysis modifications of potato starch granules. Polymers (Basel). 2022;14(10)
38. Noor N, Jhan F, Gani A, Raina IA, Shah MA. Nutraceutical and toxicological evaluation of hydrogels architected using resistant starch nanoparticles and gum acacia for controlled release of kaempferol. Food Struct. 2023;35(September 2022):100307.
39. Chutia H, Mahanta CL. Kinetics and thermodynamics studies on thermal degradation of differently treated starch nanoparticles and estimation of the OH groups and H2O content on particle surface. J Appl Polym Sci. 2023;140(6):1–11.
40. Noor N, Gani A, Jhan F, Fazli AA, Shah A, Huang Q. Effect of nanoreduction on functional and structural properties of resistant-starch from lotus stem. ACS Food Sci Technol. 2021;1(8): 1444–55.
41. Lu X, Huang Q. Nano/submicrometer milled red rice particles-stabilized Pickering emulsions and their antioxidative properties. J Agric Food Chem. 2020;68(1):292–300.
42. Sun C, Hu Y, Yu X, Zhu Z, Hao S, Du X. Morphological, structural and physicochemical properties of rice starch nanoparticles prepared via ultra-high pressure homogenization. Int J Food Eng. 2021;17(12):981–8.
43. Mallakpour S, Khodadadzadeh L. Ultrasonic-assisted fabrication of starch/MWCNT-glucose nanocomposites for drug delivery. Ultrason Sonochem. 2018;40:402–9.
44. Baruah KN, Singha S, Uppaluri RVS. Preparation of potato starch nanoparticles using acid hydrolysis and ultrasonic post-treatment. ACS Food Sci Technol. 2023;3(4):626–34.

45. Du C, Jiang F, Hu W, Ge W, Yu X, Du S k. Comparison of properties and application of starch nanoparticles optimized prepared from different crystalline starches. Int J Biol Macromol. 2023;235(February):123735.
46. Hashemilar H, Jafarizadeh-Malmiri H, Ahmadi O, Jodeiri N. Enzymatically preparation of starch nanoparticles using freeze drying technique—gelatinization, optimization and characterization. Int J Biol Macromol. 2023;237(March):124137.
47. Zhou L, He X, Ji N, Dai L, Li Y, Yang J, et al. Preparation and characterization of waxy maize starch nanoparticles via hydrochloric acid vapor hydrolysis combined with ultrasonication treatment. Ultrason Sonochem. 2021;80:105836.
48. Lin X, Sun S, Wang B, Zheng B, Guo Z. Structural and physicochemical properties of lotus seed starch nanoparticles prepared using ultrasonic-assisted enzymatic hydrolysis. Ultrason Sonochem. 2020;68(May):105199.
49. Apostolidis E, Stergiou A, Kioupis D, Amin S, Paximada P, Kakali G, et al. Production of nanoparticles from resistant starch via a simple three-step physical treatment. Food Hydrocoll. 2023;137(September 2022):108412.
50. Wang J, Zhang R, Huang Z, Cai M, Lou W, Wang Y, et al. Preparation and characterization of octenyl succinic anhydride nano starch from tiger nut meals. Food Quality Saf. 2023;7(August):1–13.
51. Li N, Niu M, Zhang B, Zhao S, Xiong S, Xie F. Effects of concurrent ball milling and octenyl succinylation on structure and physicochemical properties of starch. Carbohydr Polym. 2017;155:109–16.
52. Shahbazi M, Majzoobi M, Farahnaky A. Physical modification of starch by high-pressure homogenization for improving functional properties of κ-carrageenan/starch blend film. Food Hydrocoll. 2018;85:204–14.
53. Wang B, Lin X, Zheng Y, Zeng M, Huang M, Guo Z. Effect of homogenization-pressure-assisted enzymatic hydrolysis on the structural and physicochemical properties of lotus-seed starch nanoparticles. Int J Biol Macromol. 2021;167:1579–86.
54. dos Santos Alves MJ, de Sousa MHO, de Moura NF, Cesca K, Verruck S, Monteiro AR, et al. Starch nanoparticles containing phenolic compounds from green propolis: characterization and evaluation of antioxidant, antimicrobial and digestibility properties. Int J Biol Macromol. 2024;255:128079.
55. Yan X, Liu R, Bai J, Wang Y, Fu J. Preparation of starch-palmitic acid complex nanoparticles and their effect on properties of the starch composite film. Int J Biol Macromol. 2023;251:126154.

Chapter 4
Starch Nanofibers

Luis A. Bello-Perez and Marco A. Maldonado-Campos

4.1 Introduction

Nanofibers are one-dimensional nanomaterials with a diameter of less than 1000 nm [1]. This morphology is preferred over others because of its versatile properties: large surface area, better porosity, effective mechanical characteristics, simplicity, and cost-effective fabrication process [2].

These nanofibers can be produced from various natural, synthetic, inorganic, organic, and polymeric materials. The synthetic polymers used in nanofiber production are polyvinyl alcohol (PVA), polycaprolactone, polylactic acid, and polyethylene oxide, among others, whereas natural polymers include starch, chitosan, and cellulose [2].

There are several methods to produce nanofibers, and they are generally divided into two technique groups: electrospinning and non-electrospinning. The first one is divided into "electrospinning techniques based on design modifications in the electrospinning setup" and "electrospinning techniques based on the modification of the shape of the collector" [1, 3]. Electrospinning and its variations (coaxial electrospinning, temperature-assisted electrospinning, one-step temperature-assisted electrospinning, drum rotating electrospinning, water electrospinning, dry-electrospinning) are widely used. Electrospinning was invented in the early 1900s on the basis of previous observations of water behavior under the influence of electrostatics. It has developed rapidly in recent years to create nanofibers from polymeric materials [4].

Electrospinning is a straightforward, versatile, and cost-effective process that produces solid, continuous nanofibers with efficient morphology. During this process, polymeric liquids are elongated under electrostatic forces, and the size of the

L. A. Bello-Perez (✉) · M. A. Maldonado-Campos
Instituto Politécnico Nacional, CEPROBI, Yautepec, Mexico
e-mail: labellop@ipn.mx

G. Ayala Valencia (ed.), *Starch Nanomaterials and Food Applications*,
https://doi.org/10.1007/978-3-031-60086-9_4

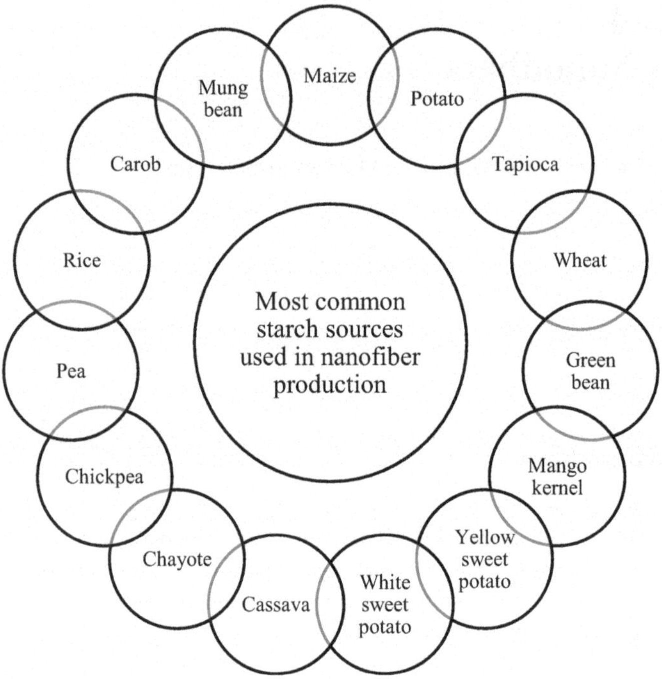

Fig. 4.1 Typical sources of starches used in nanofiber production

resulting fibers ranges between nanometers and micrometers. The process involves an electrically charged jet of a polymer solution where the electrical forces stretch and thin the jet due to the electrostatic repulsion between the surface charges and evaporate the solvent [1, 2, 5, 6].

The group of natural polymers includes starch, the second largest source of biomass on earth after cellulose, and one of the most important renewable resources. Starch and its derivatives have been widely used in various industries because of their advantages: They are natural, renewable, highly abundant, affordable, biodegradable, and biocompatible [7]. Several botanical sources have been used for starch isolation (Fig. 4.1); as conventional and non-conventional starches, they show varied structural, physicochemical, and functional characteristics and have been studied for nanofiber production.

Based on current knowledge, starch has a remarkable potential for nanofiber production through electrospinning [7, 8]. The first pure starch nanofibers (SNFs) were reported by Kong and Ziegler [9], using electrospinning. The use of pure or native starch to produce nanofibers has drawbacks that must be overcome to produce materials with good properties. Starch requires modifications (chemical, physical, enzymatic or a combination of them) or blending with other natural or synthetic polymers to produce nanofibers that meet the requirements for applications.

Realistic applications of natural and synthetic polymers have been examined, proving that synthetic polymers are more suitable because of their biodegradability, biocompatibility, non-toxicity, minimal immunogenicity, and efficient bioactivity responses. Furthermore, these natural polymers are used in tissue engineering, biomedicine, and filtration. However, they show poor solubility, low tensile strength, and large surface tension [2]. SNFs made with native or modified starch, or starch blended with other polymers can be used to protect and carry bioactive compounds and maintain the bioaccessibility and bioavailability. They can also reinforce biomaterials as films that can be used as smart or antimicrobial packaging to increase the shelf-life of products.

The present chapter aims to review the methodologies used to produce SNFs and their food applications.

4.2 Starch Sources of Nanofibers

Starch is present in cereals and pseudocereals, tubers, legumes, and unripe fruits. It is isolated from different sources and is used as an ingredient in diverse industries, such as food, pharmaceutical, cosmetics, paper, adhesive, and paint, among others. Usually, isolated or native starch does not meet the physicochemical and functional characteristics required by products because it is modified. Commercial starch can be found as native and modified, and some sources such as maize, potato, wheat, rice, and cassava are available. Both commercially available native and modified starches have been used to prepare nanofibers. Table 4.1 shows the starches used in the preparation of nanofibers using different methods, and the main insights of those studies are described. Starches with large (potato) and medium-sized (maize, wheat, and tapioca) granules have been reported. Starches with small granules (rice, amaranth, quinoa, and taro) are not suitable for nanofiber preparation. Commercially available modified starches such as Purity Gum®, Hi-Cap®, Hi-Maize 260®, and Gelose 80® are available. The use of biopolymers such as starch, cellulose, chitosan, and alginate is challenging because of their low conductivity. Starch modification or its blend with synthetic polymers is an area of opportunity for the production of nanofibers with improved functionality.

As stated before, electrospinning has been preferred for starch nanofiber preparation because of the processability of this polysaccharide when this technique is used.

4.3 Effect of Amylose (AM)/Amylopectin (AP) Ratio

Starch granules contain the biopolymers AM and AP. AM is essentially linear, with a DP between 1000–2000 (monomers in one chain), and it is responsible for starch gelation and retrogradation. AP is the branched component with a molar mass of

Table 4.1 Starches currently used in nanofiber production, preparation methods, and applications

Starch source	Amylose content (%)	Preparation method	Nanofiber application	Major findings	Reference
Ni	Ni	Electrospinning	Wound dressing	The authors obtained bead-free nanofibers	[10]
Ni	20	Coaxial electrospinning	Skin wound healing	Tensile strength increased 10-fold	[11]
Potato	32.5	Electrospinning	Encapsulation of carvacrol for antimicrobial activity against *S. aureus*	Nanofibers presented antimicrobial activity against *S. aureus*	[12]
HI-MAIZE™ 260	60	One-step temperature-assisted electrospinning	Nanofibrous films for food packaging	The authors constructed polyfunctional starch/tea polyphenols nanofibrous films (STNFs) with antioxidant activity by TP. STNFs showed optimum mechanical properties confirmed by tensile test	[13]
Maize	Ni	Drum rotating electrospinning	Supercapacitor	A hybrid supercapacitor with improved energy density was prepared using electrospinning, dip-coating, and post-heat treatment. The nanofiber structure was highly porous, contributing to superior electrochemical properties	[14]
Maize	27	Water electrospinning	Elimination of methylene blue from dye wastewater	The authors developed a polyvinyl alcohol (PVA)/starch hydrogel nanofiber membrane via wet electrospinning and cross-linking reaction by thermal treatment. Hydrogel nanofiber membrane	[15]

				showed large surface area and three-dimensional structure to remove methylene blue (MB) from wastewater	
Potato	Ni	Electrospinning	Wound dressing	Nanofibers presented enhanced wound healing vs. traditional dressings as cotton gauze	[16]
Potato	Ni	Electrospinning	Antimicrobial wound dressing	Nanofibers presented a high liquid capability absorption and suitable vapor permeability	[17, 18]
Potato	32.5	Electrospinning	Essential oil encapsulation	Nanofibers showed high encapsulation efficiency from 99.1–99.8%.	[19]
Maize	55–72	Electrospinning	Carriers of encapsulated active compounds in food and packaging	The authors tested different native and anionic corn starches modified with sodium tripolyphosphate. Fibers from Hylon VII and Hylon V starches showed homogeneous morphologies Fibers from regular corn starches exhibited droplets and heterogeneous morphologies; diameters were 70–264 nm. Native and anionic corn starches produced smooth continuous fibers	[20]
Potato	32.5	Electrospinning	Potential biomedical applications	Crosslinked neutral polysaccharide nanofibers exhibited high water resistance and excellent mechanical properties due to high	[21]

(continued)

Table 4.1 (continued)

Starch source	Amylose content (%)	Preparation method	Nanofiber application	Major findings	Reference
				reactivity of Schiff base crosslinking reaction. Crosslinked neutral polysaccharide nanofibers showed good biocompatibility	
Gelose 80	80	Electrospinning	Potential applications in food, textile, and biomedical industries	The authors created pure starch nanofibers	[22]
HI-MAIZE 260	53	Temperature-assisted electrospinning	Hydrophobic starch nanofibrous film for food packaging	The authors produced pure starch nanofibrous film with improved hydrophobicity using immersion. Fiber coating of stearic acid (STA) was inspired by biological organisms	[23]
Maize	25–28	Electrospinning	Biomimetic scaffolds for bone tissue engineering	Poly (3-hydroxybutyrate) (PHB) and starch (5–15 wt%) were used to produce a novel biomimetic scaffold. Nanofiber was bead-free, showed three-dimensional homogenous structure, and highly interconnected pores PHB-starch complex contained hydrogen bonds between them. Added starch improved tensile strength, degradation, and hydrophobicity. PHB-starch scaffolds increased viability and cell proliferation.	[24]

Maize	Ni	Electrospinning	Drug delivery system for cephalexin	The authors prepared cephalexin-loaded electrospun nanofibers with core-shell structures from polyvinyl alcohol (PVA) blended with carboxymethyl starch (CMS) in PVA:CMS at 90:10. Molecular interactions between polymeric chains and cephalexin were. Drug was gradually released in 8 h by diffusion. Nanofiber was suitable for wound treatment	[25]
Purity Gum@ Ultra (PGU), Purity Gum@ 2000 (PG2000), and Hi-Cap@ 100 (HC100)	Ni	Dry electrospinning	Green process to produce starch-based nanofibers for drug delivery, wound dressing, and tissue engineering	The authors developed a greener process for dry-electrospinning food-grade modified starch eliminating organic solvents Octenylsuccinylated (OS) starches with higher molecular weights (Mw's) produced nanofibers with beads or defects. Defects where overcome by adding pullulan (PUL). Fibers were smooth, continuous, and bead-free; diameter was 147–250 nm. PUL established molecular entanglement	[26]
Maize, potato, tapioca, wheat, and green bean	Maize: (0–18.8); Potato (30.5); Tapioca (20.7); Wheat (24.0); Green Bean: (34.6)	Electrospinning	None. New knowledge on starch electrospinnability	The authors described how AM and AP molecular structures affect starch electrospinning and starch nanofiber characteristics (average droplet number, average	[27]

(continued)

Table 4.1 (continued)

Starch source	Amylose content (%)	Preparation method	Nanofiber application	Major findings	Reference
				bead number, and average fiber diameter)	

AM amylose, *AP* amylopectin, *Ni* not informed

approximately 1×10^8 g/mol and is responsible for the starch solution viscosity [28]. Both components play important roles in the physicochemical and functional characteristics of the polysaccharide. The use of AM in nanofiber production was patented by dissolving AM in NaOH solution and extruding the mixture into concentrated aqueous ammonium sulfate to produce ultra-fine and tough fibers during coagulation [29]. This process is laborious because it involves AM separation from the starch granule.

AP does not produce fiber formation during electrospinning, which can be a challenge. The use of different methods to produce AP (waxy starch) nanofibers is a likely alternative.

There are starch sources with an AM content between 25% and 35%, these are named normal starch, while waxy starches show AP content around 98–99%, and high-AM starches present AM concentrations of 50–70%. The starch used for nanofiber production considered the AM content as an important variable. The starches with high AM content have more linear chains that produce more entanglement with the production of electrospun fibers with good properties [30]. Starches with different AM/AP ratios were used to evaluate the electrospinnability of dope solutions. It was found that the solutions were more conducive to electrospinning and the nanofibers showed high strength, stiffness, and ductility when AM concentrations were higher [31].

The effect of AM/AP ratio on nanofiber production by electrospinning remains unclear. AM produces good nanofibers by electrospinning while starch retrogradation negatively affects the characteristics of nanofibers. It is well known that A is responsible for starch retrogradation, a fact that must be considered to improve the production of SNFs and their functionality.

On the other hand, the role of AM in starch granules during electrospinning for nanofiber production is well understood; however, the role of AM and AP molecular structures deserves further research [27]. To evaluate the effect of the molecular structure of both starch components on electrospinnability indices such as average droplet number, average bead number, and average fiber diameter, as well as on dope properties such as shear viscosity, conductivity, and surface tension, 11 commercial starches with different AM/AP ratios (normal, high-AM and waxy starches) were analyzed in branched and debranched forms. Variables of the molecular starch structure (AM content, AM and AP hydrodynamic radius, AP chain-length distribution, and average degree of branching) were correlated with the electrospinnability indices and dope properties. The features of the AP molecular structure (chain length distribution, polymerization degree of debranched AP, and average degree of branching) influenced the formation of electrospun starch fibers. In addition, the AM molecular structure (hydrodynamic radius and polymerization degree of debranched AM) is important for the fiber diameter. This information is key to choosing the starch molecular structure that produces a specific nanofiber with improved functionality [27]. Some studies have reported the AM concentration in starch for nanofiber production (Table 4.1). The content ranges between 0% (waxy starch) and 80% (high-AM starch). The effect of the AM/AP ratio influences the physicochemical, functional, and mechanical properties of the nanofibers.

4.4 Modified Starch

Starch should be modified to improve its physicochemical and functional characteristics and its end uses. Diverse methods for starch modification have been reported [32, 33]. Physical methods are currently preferred because they do not introduce reagents and reduce the environmental pollution generated by chemical disposal. In addition, the organocatalytic modification of starch is suggested since the modified starch shows a high degree of substitution (DS) and the process generates less chemical residues [34]. Friendly methods for starch modification provide an opportunity to improve starch functionality and produce SNFs for new and improved applications.

Starches modified by chemical methods, such as esterification (acetylation) and oxidation, have been used to produce nanofibers by electrospinning. High-AM maize starch (70% AM) was acetylated with two DS (1.1 and 2.3), and formic acid/water as the solvent was used to produce the dope solution. SNFs were used as carriers for drug release. The effect of DS was evaluated on the tenacity of the nanofibers. The analysis of the results showed a slight but significant difference in tenacity both in both the dry and wet states in the two DS systems. Wet tenacity is important because nanofibers are usually in contact with water in most biomedical applications, and acetylated starch is hydrophobic. Water stability reached its peak at the highest DS (2.3). Starch with the highest DS presented higher drug loading than the lowest DS, but a quick drug release was obtained in SNFs with the lowest DS. This pattern can be explained by the low affinity (higher hydrophilic character) of SNFs with the lowest DS [35].

Potato starch was oxidized with HCl and ammonium persulfate under acidic conditions. The medium contained ammonium persulfate (oxidant agent) and HCl as a catalyzer. The oxidation produced carbonyl and carboxyl groups that could be improved through hydrolysis in the amorphous zones of starch granules by HCl. The oxidized starch was dissolved in DMSO (dimethyl sulfoxide) and electrospun. Beads, beaded fibers, and smooth fibers were produced depending on the modified starch concentration [36].

The use of carboxymethyl starch has been proposed to create composite nanofibers blending with a synthetic polymer such as poly(vinyl alcohol) (PVA) as carriers for drug delivery. The formation of a core-shell was confirmed by confocal laser scanning microscopy and transmission electron microscopy. The encapsulation of the drug into the composite and the molecular interactions between the two polymers was by inter- and intra molecular hydrogen bonds. The drug was released gradually within the first 8 h [21]. The nanofibers of carboxymethyl starch and PVA showed potential for the encapsulation, carrier, and release of bioactive compounds used in foods.

4.5 Techniques for Nanofiber Production

Obtaining starch in its fibrous form is difficult due to its low strength, water resistance, thermal instability, and poor processability [37]. Several techniques have been used to produce nanofibers, and they are classified into electrospinning, where electrostatic force is used to produce fibers, and non-electrospinning, where mechanical force is applied (Fig. 4.2). The desired alignment, cost, amount, and fibrous material necessary are factors to consider when choosing a method. Because of its versatility, electrospinning is preferred over other nanofiber production techniques, such as template synthesis, phase separation, drawing, and self-assembly [37].

Electrospinning is the most common method for producing nanofibers with submicron diameters through the combination of mechanical and electrical forces using a high-voltage power supply. This technique has been studied, and its basic configuration has been modified over the years to increase productivity and overcome some of its limitations. Electrospinning methods can be classified in two groups according to the configuration of the electrospinning setup or modifications to the collector shape [3]. Table 4.1 summarizes the different techniques for nanofiber production; a detailed discussion of the techniques can be found in Alghoraibi [3] and Leena et al. [1]. The variables involved in electrospinning, such as solution properties (concentration, molecular weight, viscosity), equipment properties (applied voltage, feed-rate, needle diameter), and environmental conditions (temperature, humidity) [2] can be determined to produce nanofibers with varying diameters from micrometers to nanometers, larger surface area to volume ratio, high porosity, small pore size, superior mechanical properties, and higher flexibility.

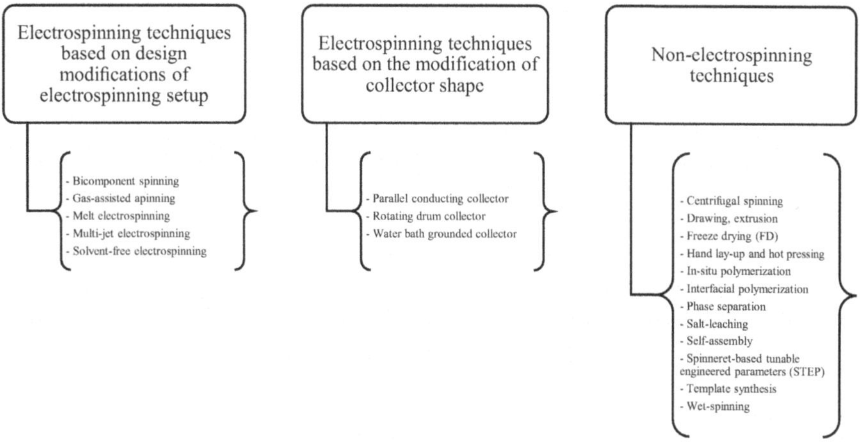

Fig. 4.2 Electrospinning and non-electrospinning techniques for nanofiber production. (Adapted from Leena et al. [1] and Alghoraibi [3])

4.5.1 Electrospinning for Starch Nanofiber Production

As stated previously, electrospinning is the most widely used method to prepare SNFs. Table 4.2 summarizes the studies using starch and its blends with polymers to produce nanofibers for diverse end uses. Kong and Zeigler [9] developed the first pure SNFs through a modified collector shape in the electrospinning setup by adding a coagulation bath (now known as electro-wet spinning). The idea behind this work was to electrospin pure starch fibers without including other biopolymers. The authors prepared a spinning dope solution by heating Gelose 80 (AM content ≈ 80%) in 95% DMSO until the starch was gelatinized. Ethanol was chosen as a solvent for the coagulation bath since it is miscible with DMSO but not with starch. The technique was modified to obtain solid fibers as evaporation of DMSO was difficult due to its non-volatility. They obtained electrospun starch fibers with an average size of 2.6 μm. This marked the beginning of studies on starch nanofiber production [22, 37]. Nowadays, several works have shown that AAM content plays a key role in starch electrospinning. For instance, higher AM leads to more molecular entanglement and results in the formation of electrospun fibers with good properties [9, 20, 30]. More recently, Cao [27] demonstrated the effects of AM and AP molecular structures on starch electrospinnability.

4.5.2 Further Techniques for Starch Nanofiber Production

As previously mentioned, electrospinning is the most common technique for nanofiber production. However, starch-based solutions are difficult to electrospin due to their high AP content. Therefore, several studies have used other techniques to create SNFs.

Recently, the thermomechanical process of extrusion has been used in starch gelatinization to break bonds and alter the properties of starch. Yu et al. [46] studied the formation of starch/poly(ethylene oxide) composite dispersions in a twin-screw extruder. Extrudates were air-cooled before to palletization, and nanofibers were characterized. Although this work did not explain how the synthesized composite scaffold could be applied, it provides an alternative to further explore the application of starch-based composites in various fields.

Mahdieh et al. [47] also explored the application of the functionalized starch-based nanocomposite synthesized by extrusion in tissue-engineering applications. Extrusion and injection molding were the main processes in the construction and production of the starch/ethylene vinyl alcohol/nanoforsterite composite scaffold. Wang et al. [48] studied the use of a twin-screw extruder to generate biodegradable composites of thermoplastic starch. Extrusion is an alternative to electrospinning when native starch-based nanofibers are required.

Salt-leaching is another technique that has been employed. Nasri-Nasrabadi et al. [49] synthesized starch/cellulose composite nanofibers by combining casting,

Table 4.2 Starches used to produce starch nanofibers by electrospinning

Botanical source	Total starch (%)	Amylose content (%)	Use	Reference
Mung bean	Ni	35	The authors fabricated pure mung bean starch nanofiber. The correlation between dispersion of mung bean starch rheological properties and electrospinnability was analyzed extrapolating critical entanglement concentration	[9]
Potato	93	30.5	The authors used potato starch to produce a nanofiber as encapsulation agent for carvacrol	[12]
			The authors used potato starch to produce starch-thermoplastic polyurethane bandages	[16]
			The authors used potato starch in nanofiber production to understand how AM and AP molecular structures affect starch electrospinnability	[27]
			The authors used potato starch to produce a nanofiber as delivery carrier for carvacrol	[38]
			The authors used potato starch to create thermoplastic starch multilayered mats containing carvacrol- loaded poly-(ε-caprolactone)	[39]
Maize	89.9	18.8	The authors used maize starch in nanofiber production to understand how AM and AP molecular structures affect starch electrospinnability	[27]
Wheat	95.2	24	The authors used wheat starch in nanofiber production to understand how AM and AP molecular structures affect starch electrospinnability	
Green bean	89.8	34.6	The authors used green bean starch in nanofiber production to understand how AM and AP molecular structures affect starch electrospinnability	
Tapioca	95.2	20.7	The authors developed a poly (vinylidene fluoride)/tapioca starch nanofiber with improved mechanical properties for tissue engineering	[40]
Mango kernel	60	Ni	The authors prepared nanofibers from natural mango kernel starch and evaluated their morphology	[41]
Cassava	Ni	18	The authors explored the pre-treatment of cassava starch by reactive extrusion before creating a nanofiber. Rheological properties of starches and AM availability improved and favored nanofiber creation	[42]
Pea	Ni	Ni	The authors evaluated electrospinnability of different pea starch concentrations in nanofiber production, morphology,	

(continued)

Table 4.2 (continued)

Botanical source	Total starch (%)	Amylose content (%)	Use	Reference
			topography, and pea starch/polyvinyl alcohol chemical interactions in nanofibers	
Chayote (*Sechium edule*)	Ni	89.1	The authors evaluated electrospinnability of different chayote starch concentrations in nanofiber production, morphology, topography, and chayote starch/polyvinyl alcohol chemical interactions in nanofibers	[43]
Chickpea	Ni	Ni	The authors used chickpea starch/polyethylene glycol nanofibers to encapsulate curcumin. Morphology, water permeability, degradation temperatures, and antimicrobial activities in *E. coli* and *S. aureus* were evaluated	[44]
Rice	Ni	Ni	The authors studied the effects of microwave pre-treatment on a rice-starch solution before electrospinning and its effects on nanofiber physicochemical characteristics. Microwave heating is a promising pre-treatment	[45]
Carob	Ni	Ni	The authors studied the effects of microwave pre-treatment on a rice-starch solution before electrospinning and its effects on nanofiber physicochemical characteristics. Microwave heating is a promising pre-treatment	

AM amylose, *AP* amylopectin, *Ni* not informed

salt-leaching, and freeze-drying. The resulting nanostructure revealed enhanced Young's modulus and tenacity, which was clearly due to the nanofibers blending with the cellulose. Moreover, salt-leaching after casting resulted in more porous fibers with diameters of 40–90 nm. After the morphological analysis and characterization, the scaffold was further investigated for application in tissue engineering.

The wet-spinning technique, an improvization on conventional electrospinning, has also been explored. Susano et al. [50] used this technique to prepare a starch-based composite scaffold by blending it with poly(ethylene-co-vinyl alcohol) and using DMSO as a solvent. After the production, scaffolds were characterized by scanning electron microscopy (SEM) for morphological analysis, micro CT for the overall scaffold topography, and cell culture studies. The results revealed that the scaffold had good structural properties, including a highly porous structure and efficient degradation properties. Thus, it constitutes a promising alternative scaffold for tissue engineering. The use of the wet-spinning method to encapsulate bioactive substances is not widely used and can be explored for those compounds.

In situ polymerization has been explored by Hosseinzadeh and Ramin [51] to produce starch-based composite hydrogels. The authors studied hydrogel functionalization using graphene oxide and hydroxyapatite, significantly contributing toward applications in bone tissue regeneration. The resulting starch-graft poly (acrylamide)/graphene oxide, along with hydroxyapatite as an adsorbent, was subjected to various characterizations such as SEM, infrared spectroscopy, X-ray diffraction, and thermogravimetric analysis. Moreover, the biodegradability and biocompatibility analyses produced the desired experimental results.

Hand lay-up and hot pressing are other techniques used. Patil et al. [52] applied them to defatted mango seed cake waste for the preparation of resins and modified sisal fibers. The as-spun sisal fibers were chemically treated at high temperatures, immersed in gelatinized starch, and subsequently squeezed several times. The resulting nanofibers were dried at 40 °C, followed by hydraulic pressing at 130 °C and 6 MPa pressure, and finally subjected to characterization. These composite nanofibers exhibited better tenacity and fiber characteristics and efficient resin-fiber interaction. These composites are potential green materials for various applications because of their low cost and simple two-step processing.

One of the most recently developed techniques is centrifugal spinning where the centrifugal force drives polymeric jets out of spinnerets. Li et al. [53] analyzed the morphologies of corn and potato starch fibers using sodium hydroxide as a solvent. The effect of the AP content on the fiber morphology was identified, and a hot blast temperature treatment during the process improved the appearance of the fibers at the nanoscale. Here, the authors only investigated the effect of AP and high temperature on the starch nanofiber fabrication process.

4.6 Electrospinning of Starch Composites

As previously mentioned, starch modification improves the physicochemical and functional properties to produce nanofibers that can be substance carriers, tissue regeneration scaffolds, and agent encapsulators, among others. However, the blend of synthetic polymers and biopolymers produces nanofibers with improved functionality, mechanical properties, tenacity, and electrospinning process. The nanocomposite of starch and synthetic polymers has been poorly explored. Commercially available glutinous rice starch (GRS) has been blended with polyvinyl alcohol (PVA) to produce nanofibers used to carry and release chlorpheniramine maleate (CM), an antihistamine. Electrospinning solutions include a blend of PVA and GRS, at different concentrations, and CM. The composites showed smooth nanofibers with PVA concentrations between 8% and 10% w/v, whereas beaded nanofibers were obtained at low PVA concentrations (5% w/v). The functionality of the PVA and GRS composite was evaluated according to swelling power and solubility. The swelling power of the nanofibers decreased at the highest temperature (60 °C), and no difference was observed at 37 and 45 °C where the highest swelling power value was determined. Solubility showed an inverse pattern where the highest water solubility index was obtained at 37 and 45 °C. The properties of the composite nanofibers suggest their potential use as drug delivery carriers and drug-release controls [54].

Polycaprolactone has a crystalline and hydrophobic nature, good solubility, and high biocompatibility. It can be blended with biopolymers (starch, chitosan, and cellulose) to expand its applications. A blend of polycaprolactone and modified porous starch has been used to prepare nanofiber scaffolds for bone tissue engineering. Other composites such as starch/polyethylene oxide, starch/polylactic acid, and starch/polylactide-co-glycolide have also been employed as scaffolds. Indeed, the utilization of starch-based nanofibers as carriers encapsulating bioactive substances and packaging materials constitutes a promising application area [2].

4.7 Starch Nanofibers as Carriers

Bioactive compounds are obtained from plant tissues, such as flowers, leaves, stems, fruits, among others. Bioactive substances are sensitive to environmental factors (temperature, light, humidity), conditions in the food matrix (pH, processing temperature, light), and other substances to which they can react, thus decreasing their activity. The conditions in the gastrointestinal tract, such as the pH of the stomach, produce compounds during digestion that can react with the bioactive compounds in the small intestine because they lose activity. The bioactive compounds should be protected to resist the aforementioned conditions. The use of nanofibers to encapsulate and carry bioactive compounds represents an interesting application field (Fig. 4.3). There are several reports on nanofibers as drug and chemical carriers within the cosmetic sector. However, this use is restricted and is commonly focused on the delivery of antimicrobial compounds as carvacrol and those used in traditional medicine, such as curcumin, which is used to treat arthritis, anxiety, and hyperlipidemia (Table 4.3). SNFs are usually employed to protect, carrier, and deliver active compounds. Therefore, they are likely to carry bioactive compounds obtained from plants and used in traditional medicine. Among these compounds are polyphenols, anthocyanins, and flavonoids, which prevent some food-related diseases.

4.8 Starch Nanofibers as Plastics

SNFs have nanometric dimensions and present functionality that can be applied to reinforce the plastic materials of biopolymers. It is well known that starch-only based plastics do not have good mechanical and barrier properties, therefore, is why they are reinforced with materials with nanometric dimensions such as nanocellulose, nanoclay, starch nanocrystals, as well as with other polymers such as chitosan, polyvinyl alcohol, polylactic acid, and polycaprolactone to produce plastic composites. SNFs prepared from starches with different AM/AP ratios, modified starches, and blends with other biopolymers were used to create plastic materials with improved properties. A blend of poly (3-hydroxybutyrate) and starch was used to produce a nanofiber film with improved hydrophobicity that was coated with stearic

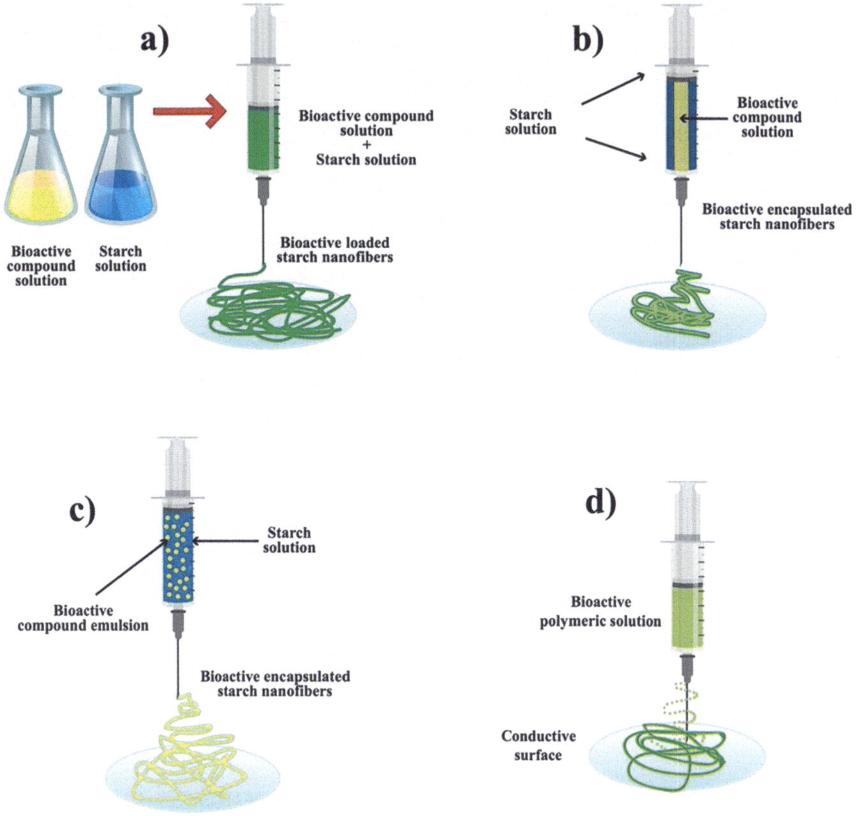

Fig. 4.3 Schematic illustration of different types of electrospinning used to elaborate starch nanofiber carriers. (**a**) Blend electrospinning, (**b**) Coaxial electrospinning, (**c**) Emulsion electrospinning, and (**d**) Surface modification electrospinning. (Adapted from Farokhi et al. [55])

acid. These films can be used as biodegradable food packaging materials [61]. Plastic materials from SNFs show additional antimicrobial activity due to the addition of thymol and have been used as packaging materials for foods [62].

4.9 Future Trends

The versatility of starch due to its physicochemical and functional properties extends the end uses of this biopolymer. The production of SNFs is a recent application in food and non-food products. Food applications of nanofibers include carrying and delivering bioactive compounds and being used to produce biodegradable materials for packaging. The increase in the consumption of functional foods is an area of

Table 4.3 Starch nanofibers as carriers

Starch source	Amylose content (%)	Use	Reference
Potato	30.5	The authors used potato SNFs as for carvacrol encapsulation	[12]
		The authors produced potato SNFs as delivery carrier for carvacrol	[38]
		The authors produced potato SNFs mats from thermoplastic starch multilayers containing carvacrol-loaded poly-(ε-caprolactone)	[39]
Potato	Ni	The authors incorporated silver nanoparticles (AgNPs) into potato starch nanofiber mats by *in situ* reduction, and investigated structure and properties of composite nanofiber mats. Cytotoxicity and antibacterial activity of starch/AgNP composite nanofiber mats were affected by release behavior of silver from mats. Antibacterial activity was related to silver release rate and AgNP content. Optimized AgNP size and content produced starch/AgNP composite nanofiber mats with good antibacterial activity and excellent cytocompatibility. Nanofibers mats showed good mechanical properties	[17]
Chickpea	Ni	The authors used chickpea starch/polyethylene glycol nanofibers for curcumin encapsulation. Morphology, water permeability, degradation temperatures, and antimicrobial activities of *E. coli* and *S. aureus* were evaluated	[44]
Succinic anhydride starch (OSA)	Ni	The authors investigated active food substance delivery. A curcumin loaded starch-based fast dissolving nanofiber was prepared by electrospinning. Hydroxypropyl-beta-cyclodextrin eliminated poor water-solubility and curcumin bioavailability to form inclusion complexes. Curcumin was well encapsulated in nanofibers with fine morphology. Fast-dissolving characters were successfully prepared	[56]
Cationic starch	Ni	The authors used cationic starch nanofibers loaded with carvacrol and casein nanoparticles to construct a novel enzyme-responsive nanomaterial to prevent contamination by *Bacillus cereus (B. cereus)* in soybean products. Nanofibers prevented texture and chroma changes under refrigeration and maintained sensory quality of soy products	[57]
Carboxymethyl starch	Ni	The authors developed a carboxymethyl starch nanofiber as a more efficient colon cancer treatment. The fiber was made from alginate and carboxymethyl starch via ionic crosslinking. The nanofiber matrix contained 41.68% 5-fluorouracil and 95.8% curcumin. The polysaccharide matrix showed good biocompatibility and reduced drug	[58]

(continued)

Table 4.3 (continued)

Starch source	Amylose content (%)	Use	Reference
		release in simulated stomach media, and improved in simulated intestinal and colonic fluids	
Glutinous rice starch	> 1	The authors produced GRS with PVA nanofiber via electrospinning. Highly water-soluble chlorphenir-amine maleate proved that nanofibers are effective delivery carriers and a platform for controlled drug release. Nanofibers presented smooth morphology and average diameter of 191 ± 25 nm	[54]
Hylon VII	70	The authors developed starch-based core-sheath compound fibers as bacterial cell carriers by coaxial electrospinning. Starch formate was the encapsulation agent and fiber core was made from glycerol as cell suspension medium. Mean diameter of starch formate/glycerol compound fibers was 4.13 ± 1.05 μm. Fibers encapsulated *Lactobacillus paracasei*, retaining bacterial viability. Starch nanofibers were effective as biotherapeutic encapsulation platform	[59]
Corn	Ni	The authors fabricated synthetic ampicillin (AM)/ maize starch (MST)/polymer composite nanofibers by electrospinning. Drug release of nanofibers was efficient and controlled. AM release properties of ST-based composite nanofibers were controlled by drug loading, ST/polymers ratio, polymer species, and tablet weight	[60]

AM amylose, *AP* amylopectin, *GRS* glutinous rice starch, *Ni* not informed, *PVA* polyvinyl alcohol, *SNFs* starch nanofibers

opportunity for SNFs because many bioactive compounds are sensitive to external factors, such as temperature, light, and oxygen. They also react to the internal conditions of the food matrix, including pH, processing temperature, storage, and reactions to components, among others. Smart SNFs can be an alternative to increase the bioaccessibility and bioavailability of bioactive compounds and, if used as a wall material, encapsulate compounds such as vitamins, flavors, and probiotics with improved functionality. The end uses of SNFs may be extended when they are included in the design and elaboration of biodegradable materials. The mechanical and barrier characteristics of these materials are enhanced because the nanofibers reinforce biopolymers such as chitosan and starch. Furthermore, biodegradable packaging with antimicrobial substances might also include SNFs. Smart packaging with sensors is another product that may also benefit from the use of SNFs (Table 4.4).

Table 4.4 Starch nanofibers as bioplastics

Starch source	Amylose content (%)	Use	Reference
HI-MAIZE™ 260	60	The authors developed polyfunctional starch/tea polyphenols nanofibrous films by one-step temperature-assisted electrospinning. Effects of tea polyphenols introduction in mechanical and antioxidant activity were evaluated. Nanofibers exhibited good antioxidant activity, optimal mechanical properties, and improved hydrophobicity. Nanofibers can be functional starch-based materials for food packaging	[13]
		The authors developed poly (3-hydroxybutyrate) and starch nanofibrous film with improved hydrophobicity using immersion approach to create stearic acid fiber coating. The final nanofiber was bead-free, had three-dimensional homogenous structure and highly interconnected pores. The film is a potential biodegradable food packaging material	[23]
		The authors developed crosslinked starch nanofibrous films by electrospinning and crosslinking starch solution in glutaraldehyde vapor. Electrospun starch nanofibrous films showed superior flexibility but poor mechanical strength and water sensitivity vs solution-cast starch films. Tensile strength and water contact angle of SNFs were enhanced post glutaraldehyde crosslinking. This is a affordable route to fabricate starch-based film materials for food packaging	[61]
High AM corn starch (HAS) Normal corn starch (NCS) Waxy corn starch (WCS)	HAS: 71 NCS: 30 WCS: 1	The authors prepared carboxymethyl starches/ pullulan nanofibers to encapsulate thymol by electrospinning. Thymol is non-toxic, biodegradable, and edible, promoting nanofibers as active food packing material	[62]
Corn	Ni	The authors developed Polyethylene/Thermoplastic Starch nanofibers reinforced with CNFs. Increased in CNFs content reduced tensile strength and elongation at break due to weak compatibility between CNFs and Polyethylene/Thermoplastic Starch matrix. Young's modulus improved by increasing CNFs content due to high stiffness of CNFs. Nanofibers can potentially replace current non-degradable synthetic products as plastics	[63]
Waxy corn starch (WS) Common corn starch (CS) High AM corn starch (HAS)	WS: 0–1 CS: 27 HAS: 71	The authors prepared hydroxypropyl starches and HPS/polyvinyl alcohol nanofibers with different AM contents to improve native starches. Some nanofibers presented continuous, homogeneous morphology and good mechanical properties. Citric acid treatment improved nanofiber hydrophobicity. Nanofibers were good candidates for food packaging	[64]

(continued)

Table 4.4 (continued)

Starch source	Amylose content (%)	Use	Reference
Purity Gum Ultra HiCap 100 Purity Gum 2000	Ni	The authors developed electrospun octenylsuccinylated starch-pullulan nanofibers as a potential method to absorb oyster peptide odor. Nanofiber mats showed excellent white appearance: smooth, continuous, bead-free, and randomly oriented nanofiber with diameters between 70 and 240 nm. They presented good odor adsorption rate, improving seafood flavor	[65]

AM amylose, *CNFs* cellulose nanofibers, *Ni* not informed

4.10 Conclusions

Starch nanofibers (SNFs) present specific properties due to the nanometric scale, and starch is considered a green material because it is biodegradable. The use of native and modified starches along with blends including other polymers has been reported in the preparation of nanofibers, mainly by electrospinning. Native starch does not meet the functional and physicochemical properties of the nanofibers; therefore, chemically modified starch was proposed. Chemical methods are no longer popular due to the pollution generated by the final disposal of the reagents used. Therefore, physical methods for starch modification are a suitable option that can provide new and improved functional and physicochemical characteristics to nanofibers. SNFs carry, deliver, and reinforce bioplastics, thus improving mechanical and barrier characteristics. Studies on the end uses of SNFs should focus on specific applications where the nanofibers meet their goals.

Acknowledgments The authors gratefully acknowledge the Instituto Politécnico Nacional, CEPROBI for the support.

Conflicts of Interest The authors declare no conflicts of interest.

References

1. Leena MM, Yoha KS, Moses JA, Anandharamakrishnan C. 3.37—Nanofibers in food applications. In: Knoerzer K, Muthukumarappan K, editors. Innovative food processing technologies. Oxford: Elsevier; 2021. p. 634–50.
2. Ashraf R, Sofi HS, Malik A, Beigh MA, Hamid R, Sheikh FA. Recent trends in the fabrication of starch nanofibers: electrospinning and non-electrospinning routes and their applications in biotechnology. Appl Biochem Biotechnol. 2019;187(1):47–74.
3. Alghoraibi I. Different methods for nanofibers design and fabrication. In: Barhoum A, Bechelany M, Makhlouf A, editors. Handbook of nanofibers. Cham: Springer; 2018.

4. Ma H, Hsiao BS, Chu B. Nanofiber. In: Drioli E, Giorno L, editors. Encyclopedia of membranes. Berlin/Heidelberg: Springer Berlin Heidelberg; 2016. p. 1358–9.
5. Angammana CJ, Jayaram SH. Fundamentals of electrospinning and processing technologies. Part Sci Technol. 2016;34(1):72–82.
6. Das A, Balakrishnan NTM, Joyner JD, Medhavi N, Manaf O, Jabeen Fatima MJ, et al. Electrospinning: the state of art technique for the production of nanofibers and nanofibrous membranes for advanced engineering applications. In: Balakrishnan NTM, Prasanth R, editors. Electrospinning for advanced energy storage applications. Singapore: Springer Singapore; 2021. p. 23–71.
7. Liu G, Gu Z, Hong Y, Cheng L, Li C. Electrospun starch nanofibers: recent advances, challenges, and strategies for potential pharmaceutical applications. J Control Release. 2017;252:95–107.
8. Lancuški A, Vasilyev G, Putaux J-L, Zussman E. Rheological properties and Electrospinnability of high-amylose starch in formic acid. Biomacromolecules. 2015;16(8): 2529–36.
9. Kong L, Ziegler GR. Role of molecular entanglements in starch fiber formation by electrospinning. Biomacromolecules. 2012;13(8):2247–53.
10. Adeli H, Khorasani MT, Parvazinia M. Wound dressing based on electrospun PVA/chitosan/ starch nanofibrous mats: fabrication, antibacterial and cytocompatibility evaluation and in vitro healing assay. Int J Biol Macromol. 2019;122:238–54.
11. Movahedi M, Asefnejad A, Rafienia M, Khorasani MT. Potential of novel electrospun core-shell structured polyurethane/starch (hyaluronic acid) nanofibers for skin tissue engineering: in vitro and in vivo evaluation. Int J Biol Macromol. 2020;146:627–37.
12. Fonseca LM, Cruxen CES, Bruni GP, Fiorentini ÂM, Zavareze ER, Lim L-T, et al. Development of antimicrobial and antioxidant electrospun soluble potato starch nanofibers loaded with carvacrol. Int J Biol Macromol. 2019;139:1182–90.
13. Zhang D, Chen L, Cai J, Dong Q, Din Z-u, Hu Z-Z, et al. Starch/tea polyphenols nanofibrous films for food packaging application: from facile construction to enhance mechanical, antioxidant and hydrophobic properties. Food Chem. 2021;360:129922.
14. Kebabsa L, Kim J, Lee D, Lee B. Highly porous cobalt oxide-decorated carbon nanofibers fabricated from starch as free-standing electrodes for supercapacitors. Appl Surf Sci. 2020;511: 145313.
15. Moradi E, Ebrahimzadeh H, Mehrani Z, Asgharinezhad AA. The efficient removal of methylene blue from water samples using three-dimensional poly (vinyl alcohol)/starch nanofiber membrane as a green nanosorbent. Environ Sci Pollut Res. 2019;26(34):35071–81.
16. Mistry P, Chhabra R, Muke S, Narvekar A, Sathaye S, Jain R, et al. Fabrication and characterization of starch-TPU based nanofibers for wound healing applications. Mater Sci Eng C. 2021;119:111316.
17. Lv H, Cui S, Yang Q, Song X, Wang D, Hu J, et al. AgNPs-incorporated nanofiber mats: relationship between AgNPs size/content, silver release, cytotoxicity, and antibacterial activity. Mater Sci Eng C. 2021;118:111331.
18. Lv H, Cui S, Zhang H, Pei X, Gao Z, Hu J, et al. Crosslinked starch nanofibers with high mechanical strength and excellent water resistance for biomedical applications. Biomed Mater. 2020;15(2):025007.
19. Fonseca LM, Radünz M, dos Santos Hackbart HC, da Silva FT, Camargo TM, Bruni GP, et al. Electrospun potato starch nanofibers for thyme essential oil encapsulation: antioxidant activity and thermal resistance. J Sci Food Agric. 2020;100(11):4263–71.
20. Fonseca LM, de Oliveira JP, de Oliveira PD, da Rosa Zavareze E, Dias ARG, Lim L-T. Electrospinning of native and anionic corn starch fibers with different amylose contents. Food Res Int. 2019;116:1318–26.
21. Zhang H, Cui S, Lv H, Pei X, Gao M, Chen S, et al. A crosslinking strategy to make neutral polysaccharide nanofibers robust and biocompatible: with konjac glucomannan as an example. Carbohydr Polym. 2019;215:130–6.

22. Kong L, Ziegler GR. Fabrication of pure starch fibers by electrospinning. Food Hydrocoll. 2014;36:20–5.
23. Cai J, Zhang D, Zhou R, Zhu R, Fei P, Zhu Z-Z, et al. Hydrophobic interface starch nanofibrous film for food packaging: from bioinspired design to self-cleaning action. J Agric Food Chem. 2021;69(17):5067–75.
24. Asl MA, Karbasi S, Beigi-Boroujeni S, Zamanlui Benisi S, Saeed M. Evaluation of the effects of starch on polyhydroxybutyrate electrospun scaffolds for bone tissue engineering applications. Int J Biol Macromol. 2021;191:500–13.
25. Abdul Hameed MM, Mohamed Khan SAP, Thamer BM, Al-Enizi A, Aldalbahi A, El-Hamshary H, et al. Core-shell nanofibers from poly(vinyl alcohol) based biopolymers using emulsion electrospinning as drug delivery system for cephalexin drug. J Macromol Sci A. 2021;58(2):130–44.
26. Li S, Kong L, Ziegler GR. Electrospinning of octenylsuccinylated starch-pullulan nanofibers from aqueous dispersions. Carbohydr Polym. 2021;258:116933.
27. Cao P, Wu G, Yao Z, Wang Z, Li E, Yu S, et al. Effects of amylose and amylopectin molecular structures on starch electrospinning. Carbohydr Polym. 2022;296:119959.
28. Bello-Perez LA, Agama-Acevedo E, Lopez-Silva M, Alvarez-Ramirez J. Molecular characterization of corn starches by HPSEC-MALS-RI: a comparison with AF4-MALS-RI system. Food Hydrocoll. 2019;96:373–6.
29. Pieter H, Johannes M. Process for the production of amylose articles by extrusion of aqueous sodium hydroxide solution thereof into concentrated aqueous ammonium sulphate solution. Google Patents; 1959.
30. Angel N, Li S, Yan F, Kong L. Recent advances in electrospinning of nanofibers from bio-based carbohydrate polymers and their applications. Trends Food Sci Technol. 2022;120:308–24.
31. Vasilyev G, Vilensky R, Zussman E. The ternary system amylose-amylopectin-formic acid as precursor for electrospun fibers with tunable mechanical properties. Carbohydr Polym. 2019;214:186–94.
32. BeMiller JN. Chapter 5—Physical modification of starch. In: Sjöö M, Nilsson L, editors. Starch in food. 2nd ed. Duxford: Woodhead Publishing; 2018. p. 223–53.
33. Chen Y-F, Kaur L, Singh J. Chapter 7—Chemical modification of starch. In: Sjöö M, Nilsson L, editors. Starch in food. 2nd ed. Duxford: Woodhead Publishing; 2018. p. 283–321.
34. Hernandez-Hernandez O, Julio-Gonzalez LC, Doyagüez EG, Gutiérrez TJ. Structure-digestibility relationship from noodles based on organocatalytically esterified regular and waxy corn starch obtained by reactive extrusion using sodium propionate. Food Hydrocoll. 2022;131:107825.
35. Xu W, Yang W, Yang Y. Electrospun starch acetate nanofibers: development, properties, and potential application in drug delivery. Biotechnol Prog. 2009;25(6):1788–95.
36. Wang HJ, Jin X, Wang WY, Xiao CF, Tong L. Preparation and electrospinning of acidified-oxidized potato starch. Adv Mater Res. 2012;535-537:2340–4.
37. Hemamalini T, Giri Dev VR. Comprehensive review on electrospinning of starch polymer for biomedical applications. Int J Biol Macromol. 2018;106:712–8.
38. Fonseca LM, Bona NP, Crizel RL, Pedra NS, Stefanello FM, Lim L-T, et al. Electrospun starch nanofibers as a delivery carrier for carvacrol as anti-glioma agent. Starch - Stärke. 2022;74 (1–2):2100115.
39. Tampau A, González-Martínez C, Chiralt A. Biodegradability and disintegration of multilayer starch films with electrospun PCL fibres encapsulating carvacrol. Polym Degrad Stab. 2020;173:109100.
40. Amini M, Arash Haddadi S, Ghaderi S, Ramazani SAA, Hassan Ansarizadeh M. Preparation and characterization of PVDF/starch nanocomposite nanofibers using electrospinning method. Materi Today Proc. 2018;5(7, Part 3):15613–9.
41. Gomez-Caturla J, Ivorra-Martinez J, Lascano D, Balart R, García-García D, Dominici F, et al. Development and evaluation of novel nanofibers based on mango kernel starch obtained by electrospinning. Polym Test. 2022;106:107462.

42. Tochihuitl-Vázquez D, Ramírez-Bon R, Yáñez-Limón JM, Martínez-Bustos F. Reactive extrusion as a pretreatment in cassava (Manihot esculenta Crantz) and pea (Pisum sativum L.) starches to improve spinnability properties for obtaining fibers. Molecules [Internet]. 2022; 27(18).
43. Porras-Saavedra J, Ricaurte L, Pérez-Pérez NC, Quintanilla-Carvajal MX. Development and characterization of Sechium edule starch and polyvinyl alcohol nanofibers obtained by electrospinning. Colloids Surf A Physicochem Eng Asp. 2022;649:129456.
44. Yildiz E, Sumnu G, Kahyaoglu LN. Assessment of curcumin incorporated chickpea flour/PEO (polyethylene oxide) based electrospun nanofiber as an antioxidant and antimicrobial food package. Food Bioprod Process. 2022;135:205–16.
45. Uygun E, Yildiz E, Sumnu G, Sahin S. Microwave Pretreatment for the Improvement of Physicochemical Properties of Carob Flour and Rice Starch–Based Electrospun Nanofilms. Food Bioprocess Technol 2020;13(5):838–850.
46. Yu F, Prashantha K, Soulestin J, Lacrampe M-F, Krawczak P. Plasticized-starch/poly(ethylene oxide) blends prepared by extrusion. Carbohydr Polym. 2013;91(1):253–61.
47. Mahdieh Z, Bagheri R, Eslami M, Amiri M, Shokrgozar MA, Mehrjoo M. Thermoplastic starch/ethylene vinyl alcohol/forsterite nanocomposite as a candidate material for bone tissue engineering. Mater Sci Eng C. 2016;69:301–10.
48. Wang P, Chen F, Zhang H, Meng W, Sun Y, Liu C. Large-scale preparation of jute-fiber-reinforced starch-based composites with high mechanical strength and optimized biodegradability. Starch - Stärke. 2017;69(11–12):1700052.
49. Nasri-Nasrabadi B, Mehrasa M, Rafienia M, Bonakdar S, Behzad T, Gavanji S. Porous starch/cellulose nanofibers composite prepared by salt leaching technique for tissue engineering. Carbohydr Polym. 2014;108:232–8.
50. Susano MA, Leonor IB, Reis RL, Azevedo HS. Elastic biodegradable starch/ethylene-co-vinyl alcohol fibre-mesh scaffolds for tissue engineering applications. J Appl Polym Sci 2014;131 (14):40504.
51. Hosseinzadeh H, Ramin S. Fabrication of starch-graft-poly(acrylamide)/graphene oxide/hydroxyapatite nanocomposite hydrogel adsorbent for removal of malachite green dye from aqueous solution. Int J Biol Macromol. 2018;106:101–15.
52. Patil NV, Rahman MM, Netravali AN. "Green" composites using bioresins from agro-wastes and modified sisal fibers. Polym Compos. 2019;40(1):99–108.
53. Li X, Hou T, Lu Y, Yang B. A method for controlling the surface morphology of centrifugally spun starch-based fibers. J Appl Polym Sci. 2018;135(6):45810.
54. Jaiturong P, Sirithunyalug B, Eitsayeam S, Asawahame C, Tipduangta P, Sirithunyalug J. Preparation of glutinous rice starch/polyvinyl alcohol copolymer electrospun fibers for using as a drug delivery carrier. Asian J Pharm Sci. 2018;13(3):239–47.
55. Farokhi M, Mottaghitalab F, Reis RL, Ramakrishna S, Kundu SC. Functionalized silk fibroin nanofibers as drug carriers: advantages and challenges. J Control Release. 2020;321:324–47.
56. Du Z, Lv H, Wang C, He D, Xu E, Jin Z, et al. Organic solvent-free starch-based green electrospun nanofiber mats for curcumin encapsulation and delivery. Int J Biol Macromol. 2023;232:123497.
57. Cui H, Lu J, Li C, Rashed MMA, Lin L. Antibacterial and physical effects of cationic starch nanofibers containing carvacrol@casein nanoparticles against Bacillus cereus in soy products. Int J Food Microbiol. 2022;364:109530.
58. Pooresmaeil M, Namazi H. Development of the new pH-driven carrier from alginate/carboxymethyl starch bio-coated co-drugs@COF-OH for controlled and concomitant colon cancer treatment. Int J Biol Macromol. 2023;239:124322.
59. Lancuški A, Abu Ammar A, Avrahami R, Vilensky R, Vasilyev G, Zussman E. Design of starch-formate compound fibers as encapsulation platform for biotherapeutics. Carbohydr Polym. 2017;158:68–76.

60. Tang S, Zhao Z, Chen G, Su Y, Lu L, Li B, et al. Fabrication of ampicillin/starch/polymer composite nanofibers with controlled drug release properties by electrospinning. J Sol-Gel Sci Technol. 2016;77(3):594–603.
61. Zhu W, Zhang D, Liu X, Ma T, He J, Dong Q, et al. Improving the hydrophobicity and mechanical properties of starch nanofibrous films by electrospinning and cross-linking for food packaging applications. LWT. 2022;169:114005.
62. Liang Q, Gao Q. Effect of amylose content on the preparation for carboxymethyl starch/pullulan electrospun nanofibers and their properties as encapsulants of thymol. Food Hydrocoll. 2023;136:108250.
63. Alidadi-Shamsabadi M, Behzad T, Bagheri R, Nari-Nasrabadi B. Preparation and characterization of low-density polyethylene/thermoplastic starch composites reinforced by cellulose nanofibers. Polym Compos. 2015;36(12):2309–16.
64. Pan W, Liang Q, Gao Q. Preparation of hydroxypropyl starch/polyvinyl alcohol composite nanofibers films and improvement of hydrophobic properties. Int J Biol Macromol. 2022;223:1297–307.
65. Li Z, Weng W, Ren Z, Zhang Y, Li S, Shi L. Electrospun octenylsuccinylated starch-pullulan nanofiber mats: adsorption for the odor of oyster peptides and structural characterization. Food Hydrocoll. 2022;133:107992.

Chapter 5
Starch Nanovesicles and Nanomicelles

Lina Maria Rayo-Mendez and Jaiber Rodriguez-Llanos

5.1 Introduction

Nanotechnology has revolutionized the field of medicine and food area by enabling the manipulation of materials at the nanoscale level. Among the diverse starch-based nanostructures that have garnered attention for their potential in healthcare, food, and biotechnology applications, starch nanoparticles (SNPs), such as nanovesicles (SNVs) and nanomicelles (SNMs), have emerged as promising players [1, 2]. Both SNVs and SNMs are smaller structures below 200 nm and they are used as nanocarriers [3] for drugs/active ingredients. Furthermore, these nanomaterials can penetrate biological barriers and exhibit low immunogenicity [4–6]. However, nanomicelles and nanovesicles differ in their composition, structure, and applications due to their stability. Nanomicelles are composed with of a core-shell structure, primarily to encapsulate hydrophobic substances, whereas nanovesicles, have a bilayer structure (membrane-bound) that allows them to encapsulate both hydrophilic and hydrophobic compounds. They are known for their high stability and versatility in drug delivery and other applications [2, 7–9]. Nanovesicles can encapsulate drugs [10], nucleic acids [11], proteins [12], and mRNAs [13], or even they can be used in agriculture to boost plant development, protect crops from disease, and increase crop yields [14, 15], transporting bioactive substances such as plant hormones, insecticides [16], seed coatings, or foliar sprays [17]. In contrast, nanomicelles are typically nanocarriers of non-water-soluble drugs or ingredients [18] such as curcumin [18], β-carotene [19], insulin delivery [20], and lutein

L. M. Rayo-Mendez (✉)
Food, Bioprocessing and Nutrition Department, North Carolina State University, Raleigh, NC, USA
e-mail: lrayome@ncsu.edu

J. Rodriguez-Llanos
Engineering Department, Anhanguera University, São Paulo, Brazil

G. Ayala Valencia (ed.), *Starch Nanomaterials and Food Applications*,
https://doi.org/10.1007/978-3-031-60086-9_5

[21]. Because of their smaller size, nanomicelles enhanced stability more than conventional micelles in physiological conditions [7]. In recent years, SNVs and SNMs have attracted attention due to starch is an abundant and renewable macromolecule, that can be isolated mainly from roots, tubers, and grains. Morphological characteristics such as size, shapes, and its chemical characteristics derived from its polymer chains such as amylose (a α-1,4-linked glucose units linear polymer) and amylopectin (both α-1,4-linked glucose units highly branched structure and α-1,6-linkages at the branch points) offer a diversity of modifications that can range from trapping bioactive compounds to later being released in vivo or providing resistance to food packaging [3, 9]. This chapter focuses on the fundamentals and literature review regarding the production and potential food applications of SNVs and SNMs.

5.2 Starch Nanovesicles

Nanovesicles are small lipid bilayer membrane-bound structures, also known as extracellular vesicles, that are released by cells into the extracellular eStarch Nanovesiclesnvironment. They are involved in cell-to-cell communication and have gained significant attention in scientific research due to their potential applications in various fields, including medicine, biotechnology, and drug delivery [1]. Nanovesicles possess inherent biocompatibility and biodegradability, making them suitable for in vivo applications. They can be easily modified with various surface functionalizations, allowing for the attachment of targeting ligands or other functional molecules to achieve specific therapeutic goals [22]. For example, surface modification of nanovesicles with targeting ligands can facilitate their specific binding to receptors on the surface of target cells, leading to enhanced cellular uptake and therapeutic effects [23].

According to Marote et al. [24], nanovesicles can be classified into different types based on their biogenesis, size, and content. The most well-known types are exosomes, microvesicles, and apoptotic bodies. Exosomes are small vesicles (30–150 nm in diameter) that are within multivesicular bodies (MVBs) and released by the fusion of MVBs with the cell membrane. Microvesicles, also known as ectosomes, are larger vesicles (100–1000 nm in diameter) that bud directly from the plasma membrane. Apoptotic bodies are vesicles (1–5 μm in diameter) that are released during programmed cell death or apoptosis [1].

Nanovesicles have shown great potential as therapeutic tools. They can be isolated from various sources, such as blood, urine, saliva, and cell culture media, and can be engineered to carry specific cargo molecules, making them attractive for drug delivery [25]. In contrast, nanovesicles can also be isolated from plants and vegetable sources [23]. They can protect their cargo from degradation, bypass biological barriers, and selectively target specific cells or tissues, making them promising for delivering therapeutic agents, including drugs, proteins, and nucleic acids, with high precision and minimal side effects [26]. Additionally, nanovesicles can be engineered to exhibit controlled release of their cargo, enabling sustained drug release over time, which can be particularly useful in chronic disease management or in some specific applications. The carrier may contain certain analytes or various types of bioactive molecules, including proteins, lipids, nucleic acids (such as DNA, RNA, and mRNA), and metabolites, which can be transferred to recipient cells upon uptake [25]. This transfer of bioactive molecules allows nanovesicles to play crucial roles in intercellular communication, immune response modulation, tissue repair, and disease pathogenesis.

One of the most significant advantages of nanovesicles is their ability to encapsulate and protect cargo molecules, such as drugs, from degradation and clearance in the body. This property makes them ideal for drug delivery applications, as they can enhance the bioavailability and therapeutic efficacy of drugs by improving their stability, solubility, and targeted delivery to specific tissues or cells [5]. Moreover, nanovesicles can overcome biological barriers, such as the blood-brain barrier, which limits the delivery of many drugs to the brain [5, 22]. This makes them particularly attractive for treating Alzheimer's and Parkinson's diseases [27].

Furthermore, nanovesicles have shown great potential in other areas of nanomedicine beyond drug delivery. For instance, they can be used for diagnostic purposes because they can encapsulate and protect diagnostic agents, such as contrast agents for imaging, and enable their targeted delivery to disease sites [28]. By analyzing the content of nanovesicles, including proteins, nucleic acids, and metabolites, researchers can obtain valuable information for disease diagnosis, monitoring treatment response, and personalized medicine [29]. Nanovesicles can also be used as theragnostic agents that combine diagnostic and therapeutic functions in a single platform [30]. Moreover, nanovesicles have shown promise in regenerative medicine because they can be loaded with growth factors or other regenerative agents to promote tissue repair and regeneration [23].

SNVs are nano-sized structures composed of starch molecules arranged in a vesicular or capsule-like structure [31–33]. They are a type of nanomaterial that has attracted significant attention in the field of nanotechnology [34]. Starch, a naturally occurring carbohydrate polymer composed of glucose units, is widely available, biocompatible, and biodegradable, making it an ideal candidate for various applications in fields such as food, pharmaceuticals, and biomedicine [35]. SNVs have unique properties that make them suitable for a range of applications. Their small size, typically in the range of 50–500 nm, and ability to self-assemble into vesicular structures provide high surface area and large internal volume, which can be exploited for the encapsulation and delivery of bioactive compounds such as drugs, nutrients, and functional ingredients [32, 36, 37]. The starch matrix within the nanovesicles provides structural integrity, stability, and biocompatibility, while the lipid coating, if present, can further modulate their properties, such as drug loading capacity and release behavior [37]. SNVs have shown great potential in drug delivery, where they can protect drugs from degradation, improve their solubility, and enable targeted delivery to specific sites in the body. They have also been explored for the encapsulation and delivery of other bioactive compounds in areas such as nutraceuticals, cosmetics, and agriculture

[38]. Additionally, SNVs have been investigated for their potential in functional food ingredients, where they can enhance properties such as texture, stability, and bioavailability [35].

5.2.1 Preparation of Starch Nanovesicles

The synthesis of SNVs can be achieved through several methods such as nanoprecipitation, emulsification/evaporation, self-assembly, and coating with lipids, which provide precise control over their size, shape, and properties. Therefore, each method depends on the application of the SNVs.

The nanoprecipitation or "solvent displacement" method is performed by two dropping-in techniques, where a starch solution (dropping) is mixed with a non-solvent or a surfactant to induce the precipitation of starch nanoparticles, which then self-assemble into nanovesicles. The size, shape, and properties of the nanovesicles can be controlled by adjusting the composition and concentration of the starch solution, non-solvent, and surfactant [39, 40]. The velocity of non-solvent evaporation (ethanol) could be crucial with respect to loading efficiency in non-crosslinked starch or starch with relatively low cross-linkage [41].

Emulsification/evaporation is one of the most common methods used to prepare nanodispersions [42]. The starch solution is emulsified in an organic solvent, and the solvent is then evaporated to form SNPs that self-assemble into nanovesicles [43]. To optimize the physicochemical properties of a nanodispersion system, all the control parameters should thus be optimized. The emulsification and stabilization of the particles are critical factors; thus, the amount of emulsifier plays an important role in the emulsification process and in the protection and stabilization of the droplets, because it can prevent the coalescence of particles [44].

Self-assembly is a process facilitated by weak physical interactions between molecules, including hydrogen bonds, ionic bonds, hydrophobic interactions, and van der Waals interactions [3]. In the case of starch molecules, self-organization into nanovesicles occurs through various mechanisms, including solvent-induced self-assembly or temperature-induced self-assembly [3, 45]. Starch polymeric vesicles are commonly made from amphiphilic chains that can form a bilayer structure, The hydrophilic head is attracted to water/solvent molecules, while the hydrophobic tail repels the aqueous environment; therefore, the bilayer structure is finally formed. Several factors, such as starch concentration, temperature, pH, and the presence of other additives, can influence the self-assembly of starch molecules [46, 47].

SNVs can also be coated with lipids to improve their properties. This can be achieved by incorporating lipids during the preparation of SNVs or by post-coating the nanovesicles with lipids using methods such as lipid film hydration or lipid injection [38, 48, 49]. Furthermore, high-pressure homogenization, ultrasonication, and mechanical shear are highly effective in breaking down larger starch particles, resulting in the formation of smaller nanoparticles and the subsequent generation of the desired SNVs [49]. According to Sintov [32], the formation of amylolipid

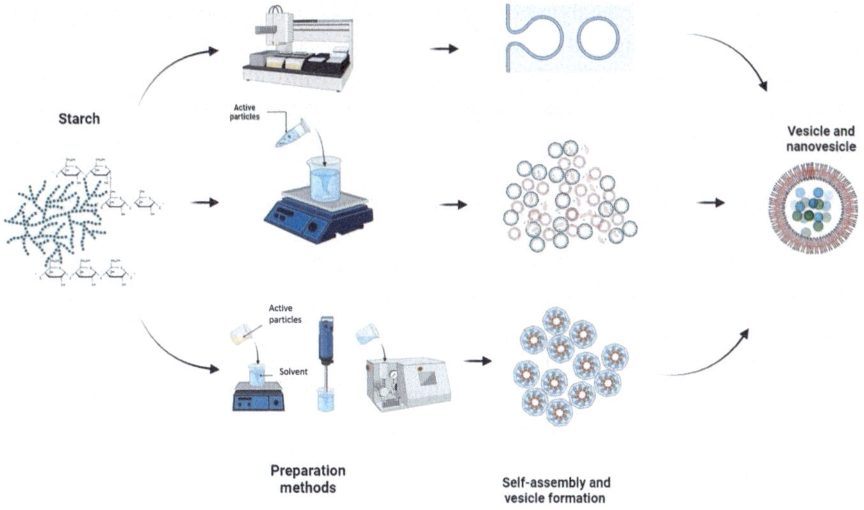

Fig. 5.1 Process for obtaining starch-based vesicles and nanovesicles

nano-vesicles (ALNs) can be achieved through a heated water-in-oil (W/O) microemulsion at 40 °C, incorporating a solid lipid above its melting point and crosslinked starch. The crosslinked starch undergoes modification by reacting with divanillin during the production process. Upon dilution and subsequent cooling to room temperature, the resulting amylolipid nanovesicles, loaded with curcumin, exhibit exceptional stability under refrigeration for a minimum of 3 months, without any observed aggregation. The researchers concluded that the results support the claim that the ALN formulation has great potential as an efficient carrier for targeted delivery of specific analytes.

In general, the synthesis of SNVs is directly linked to the synthesis methods, which normally exhibit characteristics such as variability in size, shape, and application potential. Thus, SNVs are promising for several applications, including drug delivery [50], encapsulation of bioactive compounds [51] and functional food additives [34], among others [52], because of their biocompatibility, biodegradability, and versatility. Figure 5.1 illustrates the process of obtaining vesicles and nanovesicles, highlighting the key steps involved in the development of starch bio-carriers.

5.2.2 *Morphology of Starch Nanovesicles*

The rapid advancement of nanotechnology has opened countless opportunities for developing and obtaining nanoparticles with different formats and applications [53]. Nanoparticles, in addition to being used as matrix reinforcers [54], can also be used as carriers for analytes of interest [50]. One of the main advantages of nanotechnology, highlighted by Ansari et al. [55], is its ability to overcome the challenge of frequently dosing drugs with a short half-life, allowing for prolonged or controlled release. In recent years, there has been a growing interest in exploring the impact of different nanocarrier shapes and architectures on drug transport efficiency [56].

According to Ansari et al. [55], vesicular carriers could have spherical structures composed of lipid molecules arranged in unilamellar or multilamellar bilayers. Thus, these vesicular systems, including liposomes and niosomes, have shown great potential as it is to increase the oral bioavailability of therapeutic agents and increase immunogenic responses [57]. Nanovesicles exhibit a variety of morphologies, predominantly spherical or cup-shaped, although other shapes such as oval, tubular or irregular can also be observed depending on their origin and charge [46]. The nanovesicles typically have diameters ranging from approximately 30 to 150 nm, although they can also be larger or smaller depending on the specific type and origin [4]. Their small size allows them to be easily absorbed by the recipient cells when the focus is on the transport of analytes of interest or the release of drugs [57].

SNVs normally have diameters ranging from 50 to 500 nm. In addition, they can have various shapes, including spherical, ellipsoidal, or irregular, depending on the specific conditions employed during their production [3, 4]. The starch matrix inside the nanovesicles provides structural integrity, stability, and biocompatibility, while the lipid coating, if present, can modulate their properties, such as drug loading capacity and release behavior. The properties of SNVs, including their biocompatibility, biodegradability, stability, and versatility, make them highly attractive for various applications [39].

The performance of SNVs is directly influenced by their morphological attributes. In relation to their size and large surface area, SNVs exhibit efficient encapsulation and delivery capabilities of bioactive compounds [3, 50, 53]. In addition, factors such as stability, release kinetics, and interactions with target systems must be considered [4]. Therefore, understanding the morphological characteristics of starch nanovesicles is vital for optimizing their synthesis, taking advantage of their potential in several applications.

5.2.3 Rheology of the Starch Nanovesicles

Rheology is the study of the flow and deformation behavior of materials under the action of applied forces or stresses [58]. SNVs exhibit viscoelastic behavior, simultaneously implying viscous (dissipative) and elastic (recoverable) properties [59, 60]. This behavior is attributed to the organization and structural arrangement, as well as the intermolecular interaction between the nanovesicles [31, 61]. Thus, under low stresses or low strain frequencies, suspensions containing SNPs behave more like a viscous fluid. However, at high frequencies, they exhibit more elastic and solid behavior. The viscoelastic response of suspensions containing SNVs can be characterized using dynamic oscillatory rheological measurements, such as the storage modulus (G′) and loss modulus (G″), which represent the elastic and viscous components, respectively [62]. This property is relevant in applications that require materials with fluid and solid characteristics, such as in the formulation of viscoelastic gels, hydrogels, or films, because this behavior influences their stability, deformation and response to external forces [36, 63].

According to Ahmad and Gani [62], foods with high storage or elastic nature may take longer to be digested when compared with foods with more complex character or loss modules. The same researchers developed functional snacks with the addition of nanoencapsulated resveratrol, focusing on nutraceutical and physical properties. The rheological behavior was studied, and it was established that the resveratrol release behavior from the snacks will therefore be faster in samples with less elasticity. However, the elastic nature of the snacks was diminished in the samples with fortification of encapsulated resveratrol, which was influenced by molecular interactions in the sample.

SNVs often exhibit shear thinning behavior [64, 65]. This property is desirable in many applications such as coatings, inks, and gels because it allows for easy flow and spread during applications [66]. On the other hand, SNPs can also exhibit shear thickening behavior, where the viscosity increases with increasing shear rate [67, 68]. This is useful in applications that require greater stability and resistance to deformation [69, 70], such as protective coatings [71] or impact-absorbing materials [72].

Nanoparticle suspensions and SNVs generally exhibit a yield stress, which is the minimum stress required to initiate flow or deformation [70]. Above the yield stress, nanoparticles flow and deform, whereas below it, they behave like a solid [73]. The presence of a yield stress in suspensions containing SNPs is advantageous in applications where controlled release or stability is desired, such as in encapsulation systems or drug delivery formulations [74]. The yield stress ensures that the nanovesicles maintain their structure and functionality until the applied stress exceeds the yield value. This behavior can be quantified using techniques such as oscillatory rheology or stress ramp experiments, which provide information about their structural integrity and flow characteristics [73]. Understanding the yield point allows the design of formulations with the desired rheological properties for different applications, such as encapsulation systems or controlled release matrices [75, 76].

5.2.4 Thixotropy of Starch Nanovesicles

Suspensions containing SNVs frequently display thixotropic behavior, which is defined as the time-dependent viscosity reduction under constant shear stress and the slow viscosity recovery after stress removal [77]. When the nanovesicle network is subjected to shear stresses, it structurally reorganizes and is disrupted, resulting in thixotropy [78]. It is possible to characterize the structural recovery kinetics by measuring the hysteresis loop observed in thixotropic starch nanovesicle systems using time sweep studies [42, 79]. Applications that require for simple processing flow and storage stability, such as in creams, gels, or emulsions, depend on understanding thixotropic behavior [80, 81].

The interactions between the nanovesicles, including interparticle forces and the development of a network structure, affect the rheological properties of SNVs [3]. Nanovesicle concentration, size, surface changes, and processing conditions are only a few of the variables that affect the formation of networks [33]. The configuration of nanovesicles and network development can be examined using methods such as small-angle X-ray scattering [82] or cryogenic transmission electron microscopy [34]. Rheological parameters, such as viscosity, elasticity, and stability, can be optimized in starch nanovesicle systems for purposes by comprehending and directing particle interactions and network development.

Therefore, it is important to emphasize that understanding and controlling the rheological properties of SNVs are essential for their successful use in different areas of industries [83]. By manipulating particle size, concentration, surface modifications, and processing conditions, researchers can tailor the rheological behavior of nanoparticle systems to meet the specific requirements of intended applications [32]. Thus, this knowledge allows the development of innovative products and materials with improved performance, stability, and functionality [84].

5.3 Starch Nanomicelles

SNMs are colloidal structures with a nanoscale dimensions below 200 nm [3, 85]. They possess a core-shell structure, composed of amphiphilic polymeric blocks, which represents an advantage as a bioactive carrier of lipophilic compounds [19]. The hydrophobic starch chains form the core where water-insoluble components are deposited, whereas the hydrophilic starch chains form the component-protecting shell to release it in a prolonged manner [85]. The formation of the SNMs is facilitated by the amphiphilic nature of starch, which allows it to simultaneously interact with water and hydrophobic substances [86, 87].

In general, SNMs are versatile materials offering advantages not only for pharmaceutical purposes as drug carriers but also for the food industry, due to their lower toxicity, higher stability in water, enhanced drug/ingredient solubility, and prolonged circulation in vivo [2, 6, 86, 88]. SNMs can be considered a cost-effective nanocarriers because starch is an abundant and renewable material, derived primarily from corn [19, 21, 89], potatoes [3], cassava/tapioca [86], rice, and wheat [3], besides their functional groups can lead to multiple modifications. These characteristics are desirable for food industry applications because nanomicelles are edible and biodegradable [90].

5.3.1 Preparation of Starch Nanomicelles

SNMs can be formed with bottom-up processes using a combination of self-assembly and grafting methods in an aqueous environment, starting from a modified starch with amphiphilic groups [3]. Self-assembly occurs when the starch molecules organize spontaneously into micelles at the nanoscale. As the starch molecules begin to dissolve in the solvent, they undergo a self-assembly process driven by hydrophobic and hydrophilic interactions. In other words, the hydrophobic regions (non-polar) of the starch molecules aggregate together to form the core of nanomicelles, while the hydrophilic head regions interact with the surrounding solvent (Fig. 5.2) [85, 87, 89, 90].

In contrast, grafting involves attaching either hydrophobic or hydrophilic fractions, such as fatty acids or polymers [18, 86], onto the hydrophilic starch molecules. This modification imparts amphiphilic properties to the starch, enabling it's self-assembly in aqueous solutions. The hydrophobic grafts drive the starch molecules to organize into micelles, with the hydrophobic regions shielded from the aqueous environment and the hydrophilic portions exposed to the solvent [9]. Graft polymers offer advantages over starch-based micellar systems because the side chains can stabilize micelles by increasing the overall hydrophilic-hydrophobic balance [2]. This enhanced stability prevents micelle dissociation and maintains its structural integrity in various environments, including biological systems, which is essential for drug/ingredient delivery [87]. Both self-assembly and grafted methods require the selection of the appropriate type of native starch. Due to the inherent absence of starch molecules hydrophobicity, a modification such as acid [19], alkaline

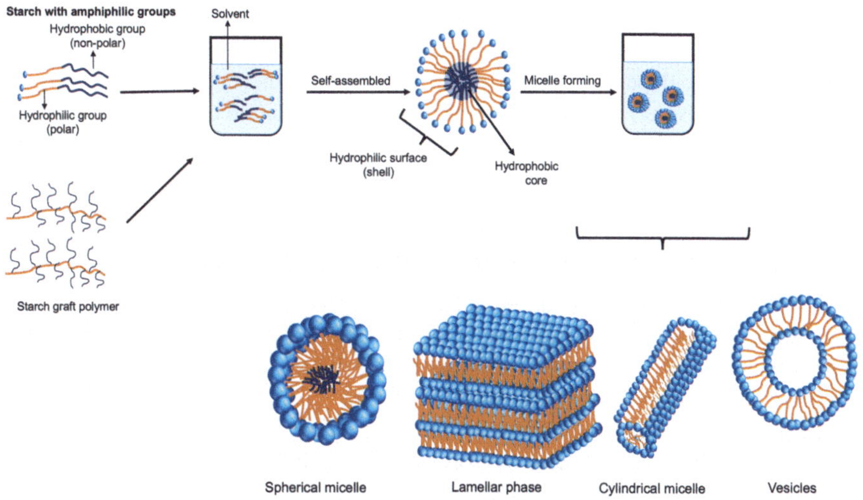

Fig. 5.2 Preparation of self-assembly and grafted starch-based nanomicelles and the types of nanomicelles formed. (Adapted from Tawfik et al. [7])

hydrolysis [91], gelatinization [18, 92], and even a dual modification such as enzymatic debranching [21] are needed to increase their stability for nano micellar formation [19, 93].

5.3.2 Grafting and Self-Assembly of Starch Nanomicelles

Self-assembly of SNMs can be achieved through various methods in an aqueous environment, including direct dissolution, stirring and heating, solvent evaporation, dropwise, dialysis, film hydration, ultrasound-assisted, and oil-in-water emulsion [90]. However, factors such as the type of solvent (polar and non-polar), temperature (low and high), solution pH, hydrophobic domain size, and amphiphile amount can influence noticeably the micellar formation. Another factor is the critical micelle concentration (CMC), which is directly related to micellar formation. Below the CMC, individual molecules exist as monomers, but as the concentration surpasses the CMC, micelles start to form spontaneously (self-assembly) [7]. Micellar concentrations can affect the encapsulation capacity of drugs/ingredients because low micelle stability affects their structure [19]. Table 5.1 summarizes the preparation methods, conditions of the starch-based nanomicelles and their characteristics.

SNMs have been studied using various techniques; however, the most common method to start micelle formation is the direct dissolution [7, 19, 21, 91, 93–95]. Because the characteristics of micelles can be controlled by changes in the conditions of the aqueous system, solvent selection is crucial because starch is insoluble in water at room temperature. Therefore, a proper solvent or a solvent mixture is required [93]. Water is often chosen as a solvent due to its availability and polarity [19], making it effective in solubilizing components. However, higher temperatures are commonly applied [7], and modified starches develop a negative or positive charge depending on the pH of the solution, increasing their polarity to solubilize in a solvent [19]. Organic solvents or a combination of water and organic solvents can also be employed to control self-assembly [90]. In the direct dissolution, the amphiphilic components are dissolved in organic solvents compatible with the hydrophobic groups. The solvent is then evaporated under reduced pressure, leaving a thin film or residue of the starch polymer on the container surface [90]. Usually, water or other organic solvent/buffer is added to the container, causing the starch-polymer film to hydrate and form micelles as the hydrophobic blocks come together to minimize exposure to water [19]. Stirring and heating are also applied to aid dissolution, ensure a uniform distribution of starches, and enhance self-assembly to increase starch solubility [21, 90].

Lin et al. [19], prepared nanomicelles by self-assembly based on octenyl succinic anhydride (OSA) and an acid-hydrolyzed waxy maize starch; therefore, the OSA-modified starch had both hydrophilic hydroxyl groups and hydrophobic octenyl side chains that provided the amphiphilic properties to form micelles in water. Mili-Q water was used as the solvent to form nanomicelles, stirred, and heated to boiling point for 20 min. Nanomicelles were self-assembled above the CMC, with

Table 5.1 Preparation conditions of starch-based nanomicelles, their aims, and characterization techniques

Starch sources and other components	Solvent	Preparation method	Micelle aim	Morphology and size	Reference
Waxy maize starch, octenyl succinic anhydride (OSA)	Milli-Q water	Synthesis of Starch: Native waxy maize starch was acid hydrolyzed (by HCl for 1, 6, and 12 h) and esterified by OSA Self-assembled micelles: The OSA-modified starch (0.01–10%, w/w) was gelatinized in Mili-Q water at boiling point water bath for 20 min, cooled and centrifuged at 4000 rpm for 2 min. β-carotene powder was added to the aqueous starch (1, 2 and 5% w/v) and homogenized at high-speed for 1 min, then magnetic stirred at 1500 rpm, room temperature for 16 h, and filtered through 0.45 μm filters	To enhance the solubility of β-carotene	Micelle structures: Spherical Sizes: 5–20 nm	[19]
Tapioca starch, Monomethoxy Polyethylene glycol polymer (mPEG, Mn = 2000) with a carboxyl-terminal COOH	DMSO, N-(3-dimethylaminopropyl)-N-ethylcarbodiimide (EDC), 1-bromohexadecane, and 4-dimethylamino pyridine (DMAP) Phosphate buffer saline (PBS, 100 mL)	Synthesis of Starch: Tapioca starch (5 g) was dissolved in isopropanol (30 mL), heated at 40 °C on water bath, stirred and mixed with a solution of NaOH (0.75 g/2 mL water) for 10 min, then incubated for 30 min	To curcumin delivery, as a drug release	Micelle structure: Spherical Size: 90–110 nm	[86]

(continued)

Table 5.1 (continued)

Starch sources and other components	Solvent	Preparation method	Micelle aim	Morphology and size	Reference
		For hydrophobic modification (hexadecyl-starch (R-St) polymers): 1-bromohexadecane was dissolved in isopropanol (3.0 g/15 mL), then added to starch, heated to 60 °C for 12 h, cooled, filtered, washed with 75% ethanol, and dried under vacuum at 60 °C The hexadecyl-starch grafted PEG copolymer: Starch (1 g) was mixed with carboxyl-terminated PEG (9.4 g), DMSO (60 mL), EDC (0.72 g), and DMAP (0.023 g), stirred at room temperature, and then dialyzed and finally freeze dried The micellar solution was prepared with 100 mg of polymer, PBS (100 mL), shaken at room temperature for 48 h, then ultrasonicated at 100 W (W) for 10 min, and filtered			
Waxy corn starch (2% amylose and 98% amylopectin), Octenyl succinic anhydride	Water, Phosphate buffered solution (PBS) HCl	Synthesis of Starch: Waxy corn starch (2% amylose and 98%) aqueous was	To encapsulated lutein	Micelle structure: Spherical Size: 10–20 nm,	[21]

(OSA) modified short glucan chains (SGC)		gelatinized, cooled, and debranched by pullulanase enzyme for 6 h, centrifuged at 3000 rpm for 2 min, and supernatant heated to fully inactivate the enzyme. The SGC was washed with alcohol, and distilled water until neutral, and then freeze dried Synthesis of OSA-SGC: SGC solution (5% w/v) was incubated at 121 °C for 30 min, at pH 8 and 9 by adding 0.1 mol/L of NaOH. Then OSA (25%, 50%, and 100%, weight of SGC) were added, stirred for 6 and 10 h at 50 °C. The pH was adjusted to 6.5 using 0.1 mol/L HCl aqueous solution to stop the reaction. OSA-SGC was washed with ethanol and freeze-dried Amphiphilic OSA-SGC and self-assembly nanomicelles: OSA-SGC (10, 50, and 100 mg) were dissolved in 10 mL PBS at pH values 5.5, 7.4, and 9.2, and salt concentrations (0.1, 0.8 M NaCl) stirred at room temperature, heated to 37 °C for		30–40 nm, and 200–400 nm	

(continued)

Table 5.1 (continued)

Starch sources and other components	Solvent	Preparation method	Micelle aim	Morphology and size	Reference
		6 h and cooled Lutein was added to ethanol (0.5 mg/mL) and sonicated for 1 min, then added to 10 mL of 0.1% OSA-SGC solution, heated at 37 °C and stirred for 6 h and then ethanol evaporated. Lutein OSA-SGC micelles were cooled at 4 °C			
Waxy corn starch (98% amylopectin)	Deuterated dimethyl sulphoxide (DMSO~d6) DMSO, Acetic anhydride, Water	Synthesis of Starch: Gelatinized waxy corn starch was debranched by pullulanase enzyme at 58 °C for 6 h and precipitated with ethanol. Then freeze-dried Acetylated debranched starch (ADBS): DBS (1 g) was dissolved in water (6 mL), stirred at 300 rpm and 30 °C for 5 min. Dropwise acetic anhydride (0.4 mL) was added, at constant pH 8.5 by using 5 M NaOH solution for 2 h. Then pH was adjusted to 7.0, washed with ethanol, and freeze dried Self-assembled ADBS	To encapsulated curcumin	Structure: Spherical Size: 50–100 nm	[18]

		micelles: ADBS (30 mg) were dissolved in 2 mL of DMSO, then dropwise into 10 mL of ultrapure water, and stirred at 25 °C for 3 h. Solution was dialyzed for 24 h Curcumin was added (3 mg) to 30 mg of ADBS dissolved in DMSO and dropwise to ultrapure water. Ethanol (20 mL) was added and centrifuged at 12.000 for 30 min, and then washed again with ethanol			
Partially oxidized starch derivatives ~100 kDa, Polydispersity (PD) 1.5 (named St-100 k) ~ 45 kDa PD 1.4 (named St-45 k)	Ethylenediamine, Water and Ethanol (1:1 v/v), Chloroform.	Synthesis of Starch: 1,1′,2-trisnorsqualenic aldehyde (Sq-CHO) (1 mol eq) was dropwise in ethylenediamine (5 mol eq solution) and ethanol (EtOH)/water (9:1 v/v) as solvent, at room temperature for 3 h Sodium cyanoborohydride ($NaBH_3CN$) (5 molar eq) was added for 20 h washed with water then extracted with chloroform and dried with $MgSO_4$ St-100k or St-45k were dissolved in distilled water at a 4 mg/mL concentration,	To drug (Nile red (MW 318.4 Da, log P 4.62), curcumin (MW 368.4 Da, log P ~ 3.048), minocycline (MW 457.5 Da, log P − 3.3) and tobramycin (MW 467.5 Da, log P − 5.8) delivery	Structure: Spherical Size: 200–300 nm	[94]

(continued)

Table 5.1 (continued)

Starch sources and other components	Solvent	Preparation method	Micelle aim	Morphology and size	Reference
		added ethanol to ratio 1:1 (v/v) of water, and DMTMM was added Squalenyl ethylenediamine (Sq-NH_2) was dissolved in 1 mL of ethanol, added into the activated starch solution for 24 h reaction at room temperature. Water was added to the solution to reach a 1:3 ratio (v/v) of EtOH/ water and chloroform to extract unreacted Sq- NH_2. The amphiphilic starch was centrifuge at room temperature 24 400 rpm, for 30 min and freeze-dried Self-assembled amphiphilic starch with Nile red, curcumin, minocycline and tobramycin loaded: Starch was dissolved in water/EtOH (1:1 v/v), centrifuge at 260 rpm, 40 °C, and 40 mbar. Then dialyzed and cooled at 4 °C Nile red or curcumin (0.15 mg) and St-100k-Sq-1% (0.85 mg) were dissolved			

		in EtOH:water (2 mL, 1:1 v/v). Ethanol was removed and 1 mL of ethyl acetate was added. The slurry was washed with ethanol, centrifuged, and washed with water St-100k- Sq-1% (2 mL) was dropwise in 2 mL of minocycline or tobramycin solution at a concentration of 0.25 mg/mL, stirred at room temperature, 200 rpm for 2 h. The slurry was centrifuged for 10 min at 15,000 rpm and washed with water			
Waxy maize starch (95% amylopectin and 5% amylose), Propylene oxide and pyrene	Dimethyl sulfoxide (DMSO)	Synthesis of Starch: Waxy maize starch was modified twice. First, in a 20% starch slurry, 10% of the starch quantity with sodium sulfate was introduced. The pH was adjusted to 11 with 5% NaOH and 10% of the starch amount in propylene oxide was mixed in at room temperature for 30 min. The slurry was shaken at 130 rpm, 37 °C for 22 h. The slurry pH was adjusted to 5.5 with 10% HCl, centrifuge at 3500 rpm, for 5 min, washed	To improve the water solubility and stability of curcumin	Structure: Amorphous Size: 50–100 nm	[91]

(continued)

Table 5.1 (continued)

Starch sources and other components	Solvent	Preparation method	Micelle aim	Morphology and size	Reference
		with ultrapure water and freeze-dried Second, 20% of the hydroxypropylated starch (0.1 M PB buffer at pH 4.66 as solvent) was heated to 95 °C, stirred for 30 min and cooled to 55 °C. Then enzymatically (pullunase) hydrolyzed for 1.5, 4.0 and 8.5 h, cooled, washed with ethanol and centrifuge at 3500 rpm for 5 min, then freeze-dried Self-assembled micelles with curcumin load: the modified starch (30 mg) was dissolved in 2 ml of DMSO, dropwise to 10 ml of ultrapure water, stirred at room temperature for 3 h, dialyzed for 24 h and freeze-dried Curcumin (2.5 mg) and starch (30 mg) were dissolved in DMSO (2 mL) and dropwise to ultrapure water (10 mL), stirred at room temperature for 3 h in dark conditions. The solution was centrifuge at 3500 for			

		5 min, dialyzed and freeze-dried			
Hydrophobically modified starch (HMS, Hi-Cap 100)	Deionized water, pyrene, chloroform	Synthesis of Starch: Starch was already modified (HMS, Hi-Cap 100) Self-assembled micelles with curcumin load: HMS Hi-Cap 100 solution (1%) was mixed with abundant amount of curcumin, homogenized at 24,000 rpm for 10 min, and magnetic stirred overnight at room temperature. Solution was centrifuge and filtered through 0.45 μm filter	To enhanced solubility of curcumin	Structure: Ni Size: 14 nm	[95]
Corn starch, Octenyl succinic anhydride (OSA)	Water NaOH HCl Ethanol	Corn starch (50 g) was dissolved in water, stirred, and pH adjusted to 8–9 with 3% NaOH. OSA was added (3–5% based on the weight of starch), stirred at room temperature for 14 h. pH was adjusted to 6.5 with HCl. Slurry was filtered, washed with distilled water and 95% ethanol, oven-dried at 40 °C for 48 h to 8% moisture OSA-starch dried was dissolved in 25 mL phosphate buffer solution (0.1–4% concentration) at	To predict the OSA-starch micelles stability	Structure: bell-shaped, long rod, flattened prolate ellipsoid Size: 46–80 nm	[93]

(continued)

Table 5.1 (continued)

Starch sources and other components	Solvent	Preparation method	Micelle aim	Morphology and size	Reference
		pH 4.0, orbital shacked at 30 °C and 200 rpm for 4 h			
Native potato starch (NPS)	Sulfuric Acid Ethanol anhydrous Cetyl trimethylammonium bromide (CTAB) Sodium dodecyl sulfate (SDS) Tween 20	Synthesis of Starch: NSP was modified twice, potato starch (36.725 g) was dissolved with 250 mL of 3.26 M of sulfuric acid at 40 °C for 5 days, incubated and shaken at 100 rpm, then washed with water, centrifuged at 7025 rpm for 20 min, and drops of chloroform were added. Slurry was freeze-dried and ethanol was used to precipitate the nanoparticles NPS (2 g) was dissolved in 100 mL of deionized water, heated, and stirred at 60 °C for 2 h. Dropwise of ethanol (100 mL) was added at 30 °C under stirring. The slurry was centrifuged at 7025 rpm for 20 min and washed with ethanol Dried starch (100 mg) was again modified with CTAB, SDS, and Tween 20 (5, 10, and 15 mM) with and	To increase solubility of Paclitaxel (PTX), an anti-cancer drug.	Structure: flake and spherical Size: 85–165 nm	[96]

		without the addition of 10 mM NaCl. Nanoparticles suspension was incubated and shacked at 200 rpm, 30 °C for 24 h, centrifuge at 9503 rpm, for 5 min Self-assembled micelles with paclitaxel (PTX): starch (2 mg) was mixed with PTX and ethanol at 20%, 40%, 60%, 80%, and 100%, agitated at 200 rpm, 20 °C for 30 min.Then centrifuge at 9503 rpm for 10 min and freeze dried			
Soluble starch (Mw 8.8 kDa), Zwitterionic dialdehyde, 1, 3-propanesultone, 1-chloro-3-dimethylaminopropane hydrochloride (CDMAP·HCl) (99%), Monomethyl ether (mPEG) (Mn 1.9 kDa), 1-ethyl-3-(3-dimethylaminopropyl) carbodiimide hydrochloride (EDC·HCl)	Water, phosphate buffered solution (PBS), dimethyl sulfoxide (DMSO)	Synthesis of Starch: A type of sulfobetaine (SB) named as 3-dimethyl (chloropropyl) ammonium propanesulfonate (DCAPS) was chosen to prepare the zwitterionic etherifying agent DCAPS was synthesized by conjugating 1, 3-propane sultone with 1-chloro-3-(dimethylamino) propyl. The DCAPS modified dialdehyde starch (SB-DAS) was conveniently synthesized by conjugating DCAPS with DAS. SB-DAS was	To insulin delivery	Structure: Spherical Size: 88–245 nm	[20]

(continued)

Table 5.1 (continued)

Starch sources and other components	Solvent	Preparation method	Micelle aim	Morphology and size	Reference
		synthesized using a Williams etherification reaction Dialdehyde starch (1.43 g) was dispersed in 10 mL of NaOH (11 wt % solution), then DCAPS (5 mL) was added, under stirring at 60 °C for 6 h and then cooled. The slurry was washed with methanol and dried in a vacuum oven overnight SB-DAS (0.1 g) and APBA (0.1 g) were mixed with dimethyl sulfoxide (DMSO) under stirring at nitrogen atmosphere, 45 °C for 12 h, then dialyzed and freeze-dried DAS (0.2 g), mPEG-COOH (0.38 g), EDC HCl (0.23 g) and DMAP (0.05 g) were dissolved in DMSO, under stirring at 25 °C for 48 h, dialyzed, and lyophilized mPEG-DAS (0.1 g) and APBA (0.1 g) were dissolved in DMSO at nitrogen atmosphere, 60 °C for 6 h, dialyzed and freeze-			

		dried Self-assembly micelles of SB-DAS-APBA and mPEG-DAS-APBA and insulin loaded: Phosphate buffered solution (PBS) (20 mL) was dropwise in SB-DAS-APBA or mPEG-DAS- APBA polymers (20 mg), shacked for 3 h, sonicated for 10 min and filtered through 0.45 nm membrane filters SB-DAS-APBA polymers (10 mg) were added to 10 mL of insulin solution (0.25 mg/mL), stirred at 37 °C and dialyzed			
Corn starch (0.2 mmol of anhydroglucose units (AGU)) Starch-graft-poly(methyl methacrylate)	Ionic liquid [AMIM]Cl 2-bromoisobutyryl bromide (BIBB) N,N-dimethyl formamide (DMF) Tetrahydrofuran (THF) Acetone 1,4- dioxane Acetonitrile Methanol Potassium chloride (KCl) Water	Synthesis of Starch: Corn starch (1.0 g,) was dissolved in [AMIM]Cl (20.0 g), stirred at 80 °C, cooled and BIBB (4.8 g,) was added, stirred at room temperature for 3 h, and washed with deionized water, filtered, and dissolved in DMF, washed, and finally dried in vacuum at 60 °C for 24 h Starch-graft-poly (methyl methacrylate) (S-g- PMMA): Strach-Br (40 mg) was	To observe the influence factors of the self-assembly of starch-graft-poly (methyl methacrylate) (S-g- PMMA)	Structure: Spherical Size: 100–200 nm	[89]

(continued)

Table 5.1 (continued)

Starch sources and other components	Solvent	Preparation method	Micelle aim	Morphology and size	Reference
		dissolved in DMF (10 mL), then PMDETA and MMA were added, and degassed with freeze-evacuate-thaw cycles, then Cu power was added at 30 °C. Dropwise in methanol, filtered, washed with more methanol, and dried in vacuum at 50 °C for 24 h			

Ni not informed

spherical shapes and less than 20 nm in size. Once nanomicelles were formed, β-carotene (a poor water-soluble component) was incorporated at room temperature under a high-speed homogenizer and stirred at 1500 rpm for 16 h.

Self-assembly micelles of acetylated debranched starch were made from waxy corn starch (98% amylopectin) [18], but first, native tapioca starch was gelatinized and then debranched through pullulanase enzyme. Subsequently, samples were dissolved in deuterated dimethyl sulfoxide (DMSO) solvent and slowly dropwise into 10 mL of ultra-purified water, stirred for 3 h at room temperature to induce micelle formation, and then dialyzed for 24 h to remove the solvent. The starch solution was added drop by drop to the aqueous phase, which acted as the continuous phase in which the hydrophobic regions of starch molecules aggregated to minimize contact with water, forming micelles. Therefore, dropwise addition ensured that small, uniform-sized micelles were created, which is critical for achieving the desired properties and functionalities of starch-based micelles. The authors reported that this method provided an excellent control over micelle size, stability, and encapsulation efficiency. Spherical shapes and micelles with good dispersibility were observed by the same authors.

Recently, Phan et al. [94] dissolved partially oxidized starch derivatives (~100 kDa) in a ratio concentration of 1:1 v/v mix of ethanol and water, followed by an ethanol evaporation at 40 mbar and 40 °C, and therefore the self-assembly of amphiphilic molecules occurred at the remained finally with water. The authors reported that amphiphilic starch had a high solubility in the solvent mix when compared to water. The influence of pH on the self-assembly of octenyl succinic anhydride modified short glucan chains (OSA-SGC) nanomicelles was studied by Ranran et al. [21]. The modified starch was dissolved under constant stirring and 37 ° C of temperature into the phosphate buffered solution (PBS) and the authors successfully encapsulated lutein, a water insoluble antioxidant. The smallest micelles (40 nm) resulted at neutral pH 7.4, whereas at alkaline pH 9.2, micelles increased in size to 100 nm due to the deprotonation and swelling of the micelles. In contrast the largest micelles were obtained at a pH value of 5.5, ranged from 200 to 400 nm because of the highly protonated carboxyl groups of the OSA, which cause the aggregation of nanoparticles.

The grafting method includes the attachment of amphiphilic polymer chains to the surface of pre-existing nanoparticles or micelle cores. Amphiphilic polymer chains consist of hydrophilic and hydrophobic segments, which enable them to self-assemble into micelles when grafted onto core materials [9, 86]. The synthesis begins with the preparing of nanoparticle or micelle cores, followed by the creating amphiphilic polymer chains. These polymer chains are then covalently or electrostatically grafted onto the core surface. The hydrophobic tails of the polymer anchor onto the core, whereas the hydrophilic heads extend outward into the surrounding solution, forming a stabilizing polymer shell. This grafting process enhances the stability and dispersibility of the nanomicelles in solution, particularly in aqueous environments [2]. Wang et al. [89] prepared starch-graft-poly (methyl methacrylate) (S-g-PMMA) copolymers and starch-graft-polystyrene as macroinitiators for a single electron transfer living radical polymerization (SET-LRP), and found that both the number average molecule weights and grafting ratio increased when the molar ratio of monomer to macroinitiator and reaction time were higher. The size of the micelles was larger as the polymer concentration, water amount, and the length of the grafted chains were increased. Therefore, selecting polymer structures, and grafting conditions are basic for tailoring nanomicelles to specific requirements, such as targeted drug delivery and controlled release systems [9]. Kou et al. [86] grafted both hydrophobic hexadecane fractions and polyethylene glycol (PEG), a hydrophilic group, onto the native tapioca starch structure to improve the micelle shell-core stability and control the release of encapsulated curcumin, a poorly water-soluble ingredient.

5.3.3 Morphology and Physicochemical Properties of the Starch Nanomicelles

Morphological analysis of SNMs is carried out to understand their structure and function. Techniques such as the transmission electron microscopy (TEM), atomic force microscopy (AFM), and scanning electron microscopy (SEM) are commonly employed to visualize and determine the shape, size, and dispersion of nanomicelles through images. Studies have revealed that nanomicelles can not only have a spherical structure, but also ellipsoidal, lamellar, and vesicles [7]. Micelles are capable of reaching sizes from 200 nm and below [3]; however, some micelles have reached further sizes because of nanoparticle aggregation [21]. Kou et al. [86] made tapioca starch-based nanomicelles grafted onto hydrophilic carboxyl-terminated obtained nanomicelles with a diameter of 200 nm measured via TEM. Spherical micelles were obtained, however, the authors found that nanomicelles prepared at pH 5.5 (acidic) were unable to be dispersed, displaying an agglomeration with each other, instead, at a neutral pH 7.4, where micelles were well dispersed with individual shapes. Liu et al. [18] observed that nanomicelles made by self-assembly from acetylated debranched starch (ADBS) had a good dispersibility and a spherical structure, with particle size diameters ranging from 50 to 100 nm, as observed by TEM and AFM.

In another research, Wang et al. [89] reported by SEM images an increase in the size of the micelles with the increasing length of the grafted chains of starch-graft-poly(methyl methacrylate) copolymers, but also observed micelles with irregular shapes when long chains were grafted, affecting the thermodynamic equilibrium being unstable. Dynamic light scattering (DLS) measurements provided the CMC concentration of OSA-modified starch assembled into spherical micelles with sizes below of 20 nm [18]. The hydrodynamic diameter determined by DLS of octenyl succinic anhydride modified short glucan chains based amphiphilic biopolymer (OSA-SGC) nanoparticles ranged from 108.5 to 339.8 nm for concentrations of 0.1%, 0.5% and 1% OSA-SGC polymer; however. 0.5% of OSA-SGC polymer formed ultrasmall micelles, whereas the lowest concentration of 0.1% OSA-SGC polymers formed large compound vesicles, and the 1% OSA-SGC polymers formed monodispersed ultrasmall micelles but also aggregated micelles observed by TEM [21].

The internal structure of SNMs can be probed using cryogenic transmission electron microscopy (cryo-TEM), which provides insight into the arrangement of hydrophobic and hydrophilic regions within the micelles [97]. The precise control of the morphological features allows the adaptation of nanomicelles for specific applications in drug carriers and nanocomposite materials. Phan et al. [94] observed the successful formation of the amphiphilic starch compounds by proton nuclear magnetic resonance spectroscopy (^{1}H NMR) and the spherical morphology of starch-based nanomicelles made from partially oxidized starch derivates by TEM. The micelle diameter is directly proportional to the length of the grafted chains; the larger the chains, the larger the micelles will increase. However, the size can cause a thermodynamic imbalance that makes it difficult to form micelles of regular size.

SNMs are versatile materials, and because of their amphiphilic nature, they can encapsulate hydrophobic compounds, such as drugs or nutraceuticals, within their hydrophobic cores [3]. The stability of SNMs is governed by factors such as the grafting density, the size of the hydrophilic and hydrophobic regions, and the pH of the surrounding environment [9]. SNMs have been shown to be stable against aggregation and coalescence, making them valuable as emulsifiers in food and as drug carriers in biological environments. Pourjavadi et al. [98] showed that the surface charge, determined by the degree of grafting, can be manipulated to control micelle formation and their interactions with other molecules, such as graft copolymers, allowing for diverse applications in materials science and biotechnology.

OSA-modified starch nanomicelles exhibited high stability properties that make them versatile and ingredient carriers; however, it was noticed by authors that β-carotene encapsulation to the hydrophobic core of nanomicelles increased their size from 100 to 10,000 nm, which caused an increase in size but also an aggregation of the starch micelles. Their amphiphilic nature allows them to efficiently encapsulate hydrophobic compounds within their hydrophobic cores, but their stability depends on grafting, ingredient density, and the relative size of the hydrophilic and hydrophobic regions. Trivedi et al. [87] found that SNMs are renowned for their exceptional stability against aggregation and coalescence, making them highly sought after as emulsifiers in the food and cosmetic industries. Furthermore, their surface charge, which is influenced by the degree of grafting, can be precisely tailored to control interactions with other molecules, facilitating applications in materials science and biotechnology.

The characteristics and properties of SNMs dispersions are crucial for understanding their behavior in various applications [9]. Nanomicelles can significantly influence the solubility of water-insoluble components such as curcumin [18], the shear-thinning behavior, and the thixotropy of solutions in which they are dispersed [40]. The rheological behavior of starch nanomicelles is depends on their concentration, size, and surface charge. At low concentrations, they often act as Newtonian fluids, whereas at higher concentrations, their shear-thinning behavior are not affected by micellar interactions. The nature of starch nanomicelles dispersions means that their viscosity decreases under shear stress and recovers over time when the stress is removed [9, 85, 99]. Understanding the properties of starch nanomicelles aids in their effective use in various industrial and biomedical applications [21, 100].

5.4 Conclusions

Nanovesicles can advance in several industrial areas, but they present a possible significant development in nanomedicine due to their inference with application in humans (patient), despite some issues that need to be resolved. To realize the full potential of nanovesicles and apply them in clinical settings, further studies and development in this field are required. New perspectives and uses for starch nanovesicles are expected to be discovered through further studies and development in this field. Despite the potential of starch nanovesicles, there are still issues that need to be resolved. This includes aspects of further research, such as manufacturing scalability, standardization of characterization techniques, and safety concerns, such as possible toxicity and immunogenicity. This is necessary for its practical implementation, and other factors such as regulatory issues and commercial feasibility must be considered. Starch nanomicelles represent an attractive class of colloidal structures with a wide array of applications. Their preparation through grafting and self-assembly, coupled with their diverse morphological, physicochemical, and rheological properties, positions them as valuable materials in fields ranging from pharmaceuticals to food science. Controlling these properties effectively requires a deep understanding of their fundamental characteristics, which is important for future applications.

Acknowledgments We would also like to thank God and all those who have assisted us with our careers. The authors gratefully acknowledge the North Carolina State and Anhanguera universities for their support.

Conflicts of Interest The authors declare no conflict of interest.

References

1. Dad HA, Gu T-W, Zhu A-Q, Huang L-Q, Peng L-H. Plant exosome-like nanovesicles: emerging therapeutics and drug delivery nanoplatforms. Biog Physiol Actions PELNVs Mol Ther. 2021;29(1):13–31.
2. Yu M, Ji N, Wang Y, Dai L, Xiong L, Sun Q. Starch-based nanoparticles: stimuli responsiveness, toxicity, and interactions with food components. Compr Rev Food Sci Food Saf. 2021;20(1):1075–100.
3. Alves MJ dos S, Chacon WDC, Gagliardi TR, Henao ACA, Monteiro AR, Valencia GA. Food applications of starch nanomaterials: a review. Starch/Stärke. 2021;73:2100046.
4. Chavda VP, Pandya A, Kumar L, Raval N, Vora LK, Pulakkat S, et al. Exosome nanovesicles: a potential carrier for therapeutic delivery. Nano Today. 2023;49:2–24. Elsevier B.V.

5. Ding Y, Wang L, Li H, Miao F, Zhang Z, Hu C, et al. Application of lipid nanovesicle drug delivery system in cancer immunotherapy. J Nanobiotechnol. 2022;20:2–22. BioMed Central Ltd.
6. Weng YH, Ma XW, Che J, Li C, Liu J, Chen SZ, et al. Nanomicelle-assisted targeted ocular delivery with enhanced antiinflammatory efficacy in vivo. Adv Sci. 2018;5(1):1–11.
7. Tawfik SM, Azizov S, Elmasry MR, Sharipov M, Lee Y-I. Recent advances in nanomicelles delivery systems salah. Nano. 2021;11(70):1–34.
8. Das Bidla P, Panda PK, Verma A, Raikwar S, Jain SK. Nanovesicles for colon-targeted drug delivery. In: Applications of nanovesicular drug delivery. Elsevier; 2022. 253–266 p.
9. Atanase LI, Desbrieres J, Riess G. Micellization of synthetic and polysaccharides-based graft copolymers in aqueous media. Prog Polym Sci. 2017;73:32–60.
10. Martinelli C, Gabriele F, Dini E, Carriero F, Bresciani G, Slivinschi B, Dei Giudici M, Zanoletti L, Manai F, Paolillo M, et al. Development of artificial plasma membranes derived nanovesicles suitable for drugs encapsulation. Cells. 2020;9:1–17.
11. Umezu T, Takanashi M, Murakami Y, Ohno S-I, Kanekura K, Sudo K, et al. Acerola exosome-like nanovesicles to systemically deliver nucleic acid medicine via oral administration. Mol Ther Methods Clin Dev. 2021;21:199–208.
12. Dehnad D, Emadzadeh B, Ghorani B, Rajabzadeh G, Tucker N, Jafari SM. Bioactive-loaded nanovesicles embedded within electrospun plant protein nanofibers; a double encapsulation technique. Food Hydrocoll. 2023;141:108683.
13. Küpeli Akkol E, Bardakci H, Yücel Ç, Şeker Karatoprak G, Karpuz B, Khan H. A new perspective on the treatment of Alzheimer's disease and sleep deprivation-related consequences: can curcumin help? Oxid Med Cell Longev; 2022. p. 2–23. Hindawi Limited.
14. Kalia A, Sharma SP, Kaur H, Kaur H. Novel nanocomposite-based controlled-release fertilizer and pesticide formulations: Prospects and challenges. Multifunctional Hybrid Nanomaterials for Sustainable Agri-food and Ecosystems; 2020. p. 99–134.
15. Bin Z, Shawon Z, Hoque E, Chowdhury SR. Nanosensors and nanobiosensors: agricultural and food technology aspects.
16. Chaud M, Souto EB, Zielinska A, Severino P, Batain F, Oliveira-Junior J, et al. Nanopesticides in agriculture: benefits and challenge in agricultural productivity, toxicological risks to human health and environment. Toxics. 2021;9:2–19. MDPI.
17. Lugovitskaya TN. Sulfite lignin nanoparticles and nanovesicles as biologically active crop growth stimulants. ACS Appl Nano Mater. 2022;5(6):8048–58.
18. Liu Q, Li F, Ji N, Dai L, Xiong L, Sun Q. Acetylated debranched starch micelles as a promising nanocarrier for curcumin. Food Hydrocoll. 2021;111:1–8.
19. Lin Q, Liang R, Zhong F, Ye A, Hemar Y, Yang Z, et al. Self-assembled micelles based on OSA-modified starches for enhancing solubility of β-carotene: effect of starch macromolecular architecture. J Agric Food Chem. 2019;67(23):6614–24.
20. Wen N, Lü S, Gao C, Xu X, Bai X, Wu C, et al. Glucose-responsive zwitterionic dialdehyde starch-based micelles with potential anti-phagocytic behavior for insulin delivery. Chem Eng J. 2018;335(October 2017):52–62.
21. Chang R, Yang J, Ge S, Zhao M, Liang C, Xiong L, et al. Synthesis and self-assembly of octenyl succinic anhydride modified short glucan chains based amphiphilic biopolymer: Micelles, ultrasmall micelles, vesicles, and lutein encapsulation/release. Food Hydrocoll. 2017;67:14–26.
22. Omidi Y, Omidian H, Kwon Y, Castejon A. Blood–brain barrier and nanovesicles for brain-targeting drug delivery. In: Applications of nanovesicular drug delivery. Elsevier; 2022. p. 167–99.
23. Gioia S Di, Hossain MN, Conese M. Biological properties and therapeutic effects of plant-derived nanovesicles. Open Med. 2020;15(1):1096–122.
24. Marote A, Teixeira FG, Mendes-Pinheiro B, Salgado AJ. MSCs-derived exosomes: cell-secreted nanovesicles with regenerative potential. Front Pharmacol. 2016;7:1–8. Frontiers Media S.A.

25. Elkhoury K, Koçak P, Kang A, Arab-Tehrany E, Ward JE, Shin SR. Engineering smart targeting nanovesicles and their combination with hydrogels for controlled drug delivery.
26. Molinaro R, Evangelopoulos M, Hoffman JR, Corbo C, Taraballi F, Martinez JO, et al. Design and development of biomimetic nanovesicles using a microfluidic approach. Adv Mater. 2018;30(15):1–9.
27. Haney MJ, Klyachko NL, Zhao Y, Gupta R, Plotnikova EG, He Z, et al. Exosomes as drug delivery vehicles for Parkinson's disease therapy. J Control Release. 2015;207:18–30.
28. Cardador CM, Muehlmann LA, Coelho CM, Silva LP, Garay AV, Carvalho AM dos S, et al. Nucleotides entrapped in liposome nanovesicles as tools for therapeutic and diagnostic use in biomedical applications. Pharmaceutics. 2023;15:1–25. MDPI.
29. Gao J, Su Y, Wang Z. Engineering bacterial membrane nanovesicles for improved therapies in infectious diseases and cancer. Adv Drug Deliv Rev. 2022;186:1–16.
30. Li X, Yu S, Lee Y, Guo T, Kwon N, Lee D, et al. In vivo albumin traps photosensitizer monomers from self-assembled phthalocyanine nanovesicles: a facile and switchable Theranostic approach. J Am Chem Soc. 2019;141(3):1366–72.
31. Falcão L de S, Coelho DB, Veggi PC, Campelo PH, Albuquerque PM, de Moraes MA. Starch as a matrix for incorporation and release of bioactive compounds: fundamentals and applications. Polymers. 2022;14:2361.
32. Sintov AC. AmyloLipid Nanovesicles: A self-assembled lipid-modified starch hybrid system constructed for direct nose-to-brain delivery of curcumin. 2020;588:1–8.
33. Sun D, Mou S, Chen L, Yang J, Wang R, Zhong A, Wang W, Tong J, Wang Z, Sun J. High yield engineered nanovesicles from ADSC with enriched miR-21-5p promote angiogenesis in adipose tissue regeneration. Biomater Res. 2022;26:1–20.
34. Rostamabadi H, Falsafi SR, Jafari SM. Nanostructures of starch for encapsulation of food ingredients. In: Biopolymer nanostructures for food encapsulation purposes. Elsevier; 2019. p. 419–62.
35. Hassan NA, Darwesh OM, Smuda SS, Altemimi AB, Hu A, Cacciola F, et al. Recent trends in the preparation of nano-starch particles. Molecules. 2022;27(17):5497.
36. Jacob JK, Paliyath G. Physico-chemical characteristics of nanovesicle-carbohydrate complexes in grape juice concentrate. J Agric Food Chem. 2008;56(4):1305–15.
37. Li D, Cao G, Yao X, Yang Y, Yang D, Liu N, et al. Tartary buckwheat-derived exosome-like nanovesicles against starch digestion and their interaction mechanism. Food Hydrocoll. 2023;141:108739.
38. Carrion CC, Nasrollahzadeh M, Sajjadi M, Jaleh B, Soufi GJ, Iravani S. Lignin, lipid, protein, hyaluronic acid, starch, cellulose, gum, pectin, alginate and chitosan-based nanomaterials for cancer nanotherapy: Challenges and opportunities. Int J Biol Macromol. 2021;178:193–228.
39. Chin SF, Yazid SNAM, Pang SC. Preparation and characterization of starch nanoparticles for controlled release of curcumin. Int J Polym Sci. 2014;2014:1–8.
40. Fu Y, Yang J, Jiang L, Ren L, Zhou J. Encapsulation of lutein into starch nanoparticles to improve its dispersity in water and enhance stability of chemical oxidation. Starch - Stärke. 2018;71(5–6):1800248.
41. Sintov AC, Yariv D. A new nanoparticulate system based on divanillin-crosslinked starch: mode of manufacturing and in-vitro evaluation of skin penetration. Starch/Staerke. 2022;74:1–8.
42. Cheng M, Kong R, Zhang R, Wang X, Wang J, Chen M. Effect of glyoxal concentration on the properties of corn starch/poly(vinyl alcohol)/carvacrol nanoemulsion active films. Ind Crops Prod. 2021;171:1–12.
43. Pisoschi AM, Pop A, Cimpeanu C, Turcuş V, Predoi G, Iordache F. Nanoencapsulation techniques for compounds and products with antioxidant and antimicrobial activity – A critical view. Eur J Med Chem. 2018;157:1326–45.
44. Anarjan N, Nehdi IA, Tan CP. Influence of astaxanthin, emulsifier and organic phase concentration on physicochemical properties of astaxanthin nanodispersions. Chem Cent J. 2013;7(1):1–11.
45. Liu R, Hudalla GA. Using self-assembling peptides to integrate biomolecules into functional supramolecular biomaterials. Molecules. 2019;24:1–21. MDPI AG.

46. Li D, Zhang X, Chen X, Li W. Research progress and prospects for polymeric nanovesicles in anticancer drug delivery. Front Bioeng Biotechnol; 2022. p. 1–10. Frontiers Media S.A.
47. Wang Q, Sun Y, Zhang Z, Duan Y. Targeted polymeric therapeutic nanoparticles: Design and interactions with hepatocellular carcinoma. Biomaterials 2015;56:229–40.
48. Cañas MA, Giménez R, Fábrega MJ, Toloza L, Baldomà L, Badia J. Outer membrane vesicles from the probiotic Escherichia coli Nissle 1917 and the commensal ECOR12 enter intestinal epithelial cells via clathrin-dependent endocytosis and elicit differential effects on DNA damage. PLoS One. 2016;11:1–22.
49. Lang C, Mission EG, Ahmad Fuaad AAH, Shaalan M. Nanoparticle tools to improve and advance precision practices in the Agrifoods Sector towards sustainability—a review. J Clean Prod. 2021;293:1–24. Elsevier Ltd.
50. Joseph J. Nanovesicles for delivery of anticancer drugs. Applications of Nanovesicular Drug Delivery; 2022. p 267–83.
51. Rezaul M, Shishir I, Xie L, Sun C, Zheng X, Chen W. Advances in micro and nano-encapsulation of bioactive compounds using biopolymer and lipid-based transporters. 2018.
52. Li A, Li D, Gu Y, Liu R, Tang X, Zhao Y, Qi F, Wei J, Liu J. Plant-derived nanovesicles: Further exploration of biomedical function and application potential. Acta Pharmaceutica Sinica B; 2023. p. 3300–20.
53. Modi S, Prajapati R, Inwati GK, Deepa N, Tirth V, Yadav VK, et al. Recent trends in fascinating applications of nanotechnology in allied health sciences. Crystals. 2022;12:1–16.
54. Llanos JHR, Tadini CC. Preparation and characterization of bio-nanocomposite films based on cassava starch or chitosan, reinforced with montmorillonite or bamboo nanofibers. Int J Biol Macromol. 2018;107:371–82.
55. Danish Ansari M, Saifi Z, Pandit J, Khan I, Solanki P, Sultana Y, et al. Spanlastics a novel nanovesicular carrier: its potential application and emerging trends in therapeutic delivery. AAPS PharmSciTech. 2022;23:112.
56. De R, Mahata MK, Kim KT. Structure-based varieties of polymeric nanocarriers and influences of their physicochemical properties on drug delivery profiles. Advanced Science. John Wiley and Sons Inc; 2022. p. 1–32.
57. Rizwanullah ZS, Rizwanullah M, Mir SR, Amin S. Bilosomes nanocarriers for improved oral bioavailability of acyclovir: A complete characterization through in vitro, ex-vivo and in vivo assessment. J Drug Deliv Sci Technol Editions de Sante. 2020;57:1–13.
58. Dong H, Zhang Q, Gao J, Chen L, Vasanthan T. Comparison of morphology and rheology of starch nanoparticles prepared from pulse and cereal starches by rapid antisolvent nanoprecipitation. Food Hydrocoll. 2021;119:1–13.
59. Fuongfuchat A, Seetapan N, Makmoon T, Pongjaruwat W, Methacanon P, Gamonpilas C. Linear and non-linear viscoelastic behaviors of crosslinked tapioca starch/polysaccharide systems. J Food Eng. 2012;109:571–8.
60. Hu Y, Li C, Tan Y, McClements DJ, Wang L. Insight of rheology, water distribution and in vitro digestive behavior of starch based-emulsion gel: Impact of potato starch concentration. Food Hydrocoll Elsevier B.V. 2022;132:1–9.
61. Marto J, Gouveia LF, Gonç Alves LM, Gaspar DP, Pinto P, Carvalho FA, et al. A quality by design (QbD) approach on starch-based nanocapsules: a promising platform for topical drug delivery. Colloids Surf B Biointerfaces. 2016;143:177–85.
62. Ahmad M, Gani A. Development of novel functional snacks containing nano-encapsulated resveratrol with anti-diabetic, anti-obesity and antioxidant properties. Food Chem. 2021;352:1–8.
63. Sundaram Sivamaruthi B, Kumar Nallasamy P, Suganthy N, Kesika P, Chaiyasut C. Pharmaceutical and biomedical applications of starch-based drug delivery system: a review. J Drug Deliv Sci Technol. 2022;77:1773–2247.
64. Charles APR, Mu R, Jin TZ, Li D, Pan Z, Rakshit S, Cui SW, Wu Y. Application of yellow mustard mucilage and starch in nanoencapsulation of thymol and carvacrol by emulsion electrospray. Carbohydr Polym. 2022;298:1–10.

65. Sana S, Manson J, Zivkovic V, Boodhoo K. Solvent-antisolvent precipitation in a spinning disc reactor: Precipitation kinetics and mathematical modelling of starch nanoparticles. Chem. Eng. Process – Process Intensif. 2022;176:1–17.
66. Hategekimana J, George Masamba K, Ma J, Zhong F. Encapsulation of vitamin E: effect of physicochemical properties of wall material on retention and stability. Carbohydr Polym. 2015;124:172–9.
67. Rehman A, Jafari SM, Tong Q, Karim A, Mahdi AA, Iqbal MW, Aadil RM, Ali A, Manzoor MF. Role of peppermint oil in improving the oxidative stability and antioxidant capacity of borage seed oil-loaded nanoemulsions fabricated by modified starch. Int J Biol Macromol. 2020;153:697–707.
68. Sharif HR, Abbas S, Majeed H, Safdar W, Shamoon M, Khan MA, et al. Formulation, characterization and antimicrobial properties of black cumin essential oil nanoemulsions stabilized by OSA starch. J Food Sci Technol. 2017;54(10):3358–65.
69. Kumar Y, Roy S, Devra A, Dhiman A, Prabhakar PK. Ultrasonication of mayonnaise formulated with xanthan and guar gums: rheological modeling, effects on optical properties and emulsion stability. LWT-Food Sci Technol. 2021;149:111632.
70. Tavares L, Pelayo C, Noreña Z, Barros HL, Smaoui S, Lima PS, et al. Rheological and structural trends on encapsulation of bioactive compounds of essential oils: a global systematic review of recent research. Food Hydrocoll. 2022;129:107628.
71. Wanyi W, Lu L, Zehan H, Xinan X. Comparison of emulsifying characteristics of different macromolecule emulsifiers and their effects on the physical properties of lycopene nanoemulsions. J Dispers Sci Technol. 2020 Mar;41(4):618–27.
72. Ozogul Y, Karsli GT, Durmuş M, Yazgan H, Oztop HM, McClements DJ, Ozogul F. Recent developments in industrial applications of nanoemulsions. Adv Colloid Interface Sci. 2022;304:1–15.
73. Mandala I, Apostolidis E. Rheological characterization of liquid nanoencapsulated food ingredients by viscometers. In: Characterization of nanoencapsulated food ingredients. Elsevier; 2020. p. 529–45.
74. Kumari S, Yadav BS, Yadav R. Characterization of acid hydrolysis based nano-converted mung bean (Vigna radiata L.) starch for morphological, rheological and thermal properties. Int J Biol Macromol. 2022;211:450–9.
75. Farnad N, Farhadi K. Simple and complex coacervation methods for the nanoencapsulation of Rosa damascena mill L. anthocyanin in zein/potato starch: a new approach to enhance antioxidant and thermal properties. J Food Sci. 2023 Mar;88(3):1019–32.
76. Montoya-Yepes DF, Jiménez-Rodríguez AA, Aldana-Porras AE, Velásquez-Holguin LF, Méndez-Arteaga JJ, Murillo-Arango W. Starches in the encapsulation of plant active ingredients: state of the art and research trends. Polymer Bulletin. Springer Science and Business Media Deutschland GmbH; 2024. p. 135–63.
77. Dolz M, Hernández MJ, Delegido J, Alfaro MC, Muñoz J. Influence of xanthan gum and locust bean gum upon flow and thixotropic behaviour of food emulsions containing modified starch. J Food Eng. 2007;81:179–86.
78. Zhang Q, Gao B, Xiao Y, Yang H, Wang Y, Du L, et al. Purification and characterization of a novel β-glucuronidase precisely converts glycyrrhizin to glycyrrhetinic acid 3-O-mono-β-D-glucuronide from plant endophytic Chaetomium globosum DX-THS3. Int J Biol Macromol. 2020;159:782–92.
79. Roman L, Walker MR, Detlor N, Best J, Martinez MM. Pregelatinized Drum-Dried Wheat Starch of Different Swelling Behavior as Clean Label Oil Replacer in Oil-in-Water Emulsions. Foods MDPI. 2022;11:1–18.
80. Kost B, Brzezinski M, Socka M, Basko M, Biela T. Biocompatible polymers combined with cyclodextrins: Fascinating materials for drug delivery applications. Molecules. 2020;25:1–44.
81. Shaikh MS, Kale MA. Formulation and molecular docking simulation study of luliconazole nanosuspension–based nanogel for transdermal drug delivery using modified polymer. Mater Today Chem Elsevier Ltd. 2020;18:1–18.
82. Falsafi SR, Rostamabadi H, Jafari SM. X-ray diffraction (XRD) of nanoencapsulated food ingredients. Characterization of Nanoencapsulated Food Ingredients; 2020. p. 271–93.

83. Das AK, Nanda PK, Bandyopadhyay S, Banerjee R, Biswas S, McClements DJ. Application of nanoemulsion-based approaches for improving the quality and safety of muscle foods: A comprehensive review. Compr Rev Food Sci Food Saf. 2020;19:2677–700.
84. Zhu Y, Liu Y, Jin K, Pang Z. Polysaccharide nanoparticles for cancer drug targeting. Polysaccharide Carriers for Drug Delivery; 2019. p. 365–96.
85. Adams ML, Lavasanifar A, Kwon GS. Amphiphilic block copolymers for drug delivery. J Pharm Sci. 2003;92(7):1343–55.
86. Kou Z, Dou D, Lan L, Zhang J, Lan P, Yu Q, et al. Preparation, characterization, and performance analysis of starch-based nanomicelles. Int J Biol Macromol. 2020;145:655–62.
87. Trivedi R, Kompella UB. Nanomicellar formulations for sustained drug delivery: strategies and underlying principles. Nanomedicine. 2010;5(3):485–505.
88. Qu Q, Wang Y, Zhang L, Zhang X, Zhou S. A nanoplatform with precise control over release of cargo for enhanced cancer therapy. Small. 2016;12(10):1378–90.
89. Wang X, Yang R, Huang L, Li J, Liu Z. Preparation of starch-graft-poly(methyl methacrylate) via SET-LRP at molecular level and its self-assembly. Polymer. 2019;173:11–9.
90. Bu X, Ji N, Dai L, Dong X, Chen M, Xiong L, Sun Q. Self-assembled micelles based on amphiphilic biopolymers for delivery of functional ingredients. Trends Food Sci. 2021;114:386–98.
91. Zhi K, Wang R, Wei J, Shan Z, Shi C, Xia X. Self-assembled micelles of dual-modified starch via hydroxypropylation and subsequent debranching with improved solubility and stability of curcumin. Food Hydrocoll. 2021;118:1–11.
92. Chang R, Ji N, Li M, Qiu L, Sun C, Bian X, Qiu H, Xiong L, Sun Q. Green preparation and characterization of starch nanoparticles using a vacuum cold plasma process combined with ultrasonication treatment. Ultrason Sonochem. 2019;58:1–13.
93. Zhu J, Li L, Chen L, Li X. Nano-structure of octenyl succinic anhydride modified starch micelle. Food Hydrocoll. 2013;32(1):1–8.
94. Phan VHG, Trang Duong HT, Tran PT, Thambi T, Ho DK, Murgia X. Self-assembled amphiphilic starch based drug delivery platform: synthesis, preparation, and interactions with biological barriers. Biomacromolecules. 2021;22(2):572–85.
95. Yu H, Huang Q. Enhanced in vitro anti-cancer activity of curcumin encapsulated in hydrophobically modified starch. Food Chem. 2010;119(2):669–74.
96. Putro JN, Ismadji S, Gunarto C, Soetaredjo FE, Ju YH. A study of anionic, cationic, and nonionic surfactants modified starch nanoparticles for hydrophobic drug loading and release. J Mol Liq. 2020;298:1–9.
97. Wang D, Xiao Z, He J, Xu W, Wang J. Strong synergistic molecular interaction in catanionic surfactant mixtures: unravelling the role of the benzene ring. Langmuir. 2023;39 (36):12649–61.
98. Pourjavadi A, Kashani FB, Doroudian M, Amin SS. Synthesis and characterization of stimuli responsive micelles from chitosan, starch, and alginate based on graft copolymers with polylactide-poly(methacrylic acid) and polylactide- poly[2(dimethyl amino)ethyl methacrylate] side chains. International Journal of Biological Macromolecules Elsevier B.V. 2023;253:1–12.
99. Honda C, Abe Y, Nose T. Relaxation kinetics of micellization in micelle-forming block copolymer in selective solvent. Macromolecules. 1996;29(21):6778–85.
100. Hong Y, Liu G, Gu Z. Recent advances of starch-based excipients used in extended-release tablets: a review. Drug Deliv. 2016;23(1):12–20.

Chapter 6
Starch Nanomaterials for Adsorption and Encapsulation

Wilson Daniel Caicedo Chacon, Silvani Verruck, Alcilene Rodriges Monteiro, and Germán Ayala Valencia

6.1 Introduction

Bioactive compounds are substances present in foods or plants that exert beneficial effects on living organisms, such as antioxidant, anti-inflammatory, and antitumor properties [1]. These compounds have been used in both medicine and as nutritional ingredients, they are mostly unstable or poorly soluble in aqueous media, limiting their efficacy and applicability [2]. As one of the promising solutions that have emerged in recent years, the use of starch nanoparticles (SNPs) has been proposed as an encapsulant and adsorptive compound due to their large specific surface area and ability to interact with different types of compounds [3]. Adsorption is a process that involves the attraction between the molecules of a fluid and the matrix of a solid, this mechanism has been used to encapsulate bioactive compounds [4]. Studying the adsorption mechanisms is essential to understand how the interaction between SNPs and bioactive compounds occurs and how SNPs can be used to encapsulate bioactive compounds. It has been observed that the adsorption capacity of SNPs is mainly due to hydrogen bonding and electrostatic interactions [5]. Electrostatic interactions are attractive forces between the electric charges of the involved components, and the SNPs can acquire a positive or negative net charge depending on the preparation conditions. On the other hand, hydrogen bonding interactions allow the attraction of complementary functional groups of the adsorbed compounds, which expands the

W. D. C. Chacon · A. R. Monteiro · G. Ayala Valencia (✉)
Department of Chemical and Food Engineering, Federal University of Santa Catarina, Florianópolis, SC, Brazil
e-mail: alcilene.fritz@ufsc.br; g.ayala.valencia@ufsc.br

S. Verruck
Department of Food Science and Technology, Agricultural Sciences Center, Federal University of Santa Catarina, Florianópolis, SC, Brazil
e-mail: silvani.verruck@ufsc.br

G. Ayala Valencia (ed.), *Starch Nanomaterials and Food Applications*,
https://doi.org/10.1007/978-3-031-60086-9_6

ability of SNPs to interact with a wide range of bioactive molecules [5]. Although SNPs have a higher adsorption capacity and encapsulating efficiency than unmodified starch, these nanoparticles show limited adsorption or no capacity to adsorb and encapsulate hydrophobic compounds [6]. To enhance the adsorption capacity of SNPs, several chemical modification techniques have been employed, such as esterification, carboxylation, hydroxypropylation, amination, and oxidation [7–11]. These chemical modifications can alter the properties of SNPs and make them more suitable for adsorbing bioactive compounds with specific characteristics, both hydrophilic and hydrophobic, thus allowing greater efficiency in adsorption and retention.

Because of various physicochemical characterizations, clear insights into adsorption mechanisms, interaction types with the carrier material, and efficiency in the stability of bioactive compounds have been achieved. Analytical techniques such as infrared spectroscopy, thermogravimetry, calorimetry, and X-ray diffraction play a fundamental role in studying these properties [6, 12–14]. In addition to their adsorptive properties, the cellular viability, antioxidant, and antimicrobial effects of SNPs have been investigated. These functional properties are also of paramount importance because they ultimately provide approval for consumption and define potential applications for such nanomaterials [11, 15–17].

This chapter focuses on the fundamentals and literature review regarding the use of SNPs as an adsorbent material for encapsulating bioactive compounds by means of adsorption, covering three key aspects: adsorption mechanisms, chemical modifications to enhance adsorption capacity, physicochemical properties, and functional properties.

6.2 Adsorption and Encapsulation of Bioactive Compounds Using Starch Nanomaterials

6.2.1 Mechanisms of Adsorption

Adsorption occurs due to the attractive forces between fluid molecules and the solid surface. When Van der Waals forces are involved, it leads to physisorption on the adsorbent surface results from weak intermolecular interactions between the solid and the fluid. Activated adsorption or chemisorption occurs when chemical bonds form between the fluid molecules and the adsorbent surface [4]. In particular, solid nanoparticles have drawn attention for this purpose because they have a higher adsorption capacity due to their large specific surface area, allowing for greater loading capacity of compounds of interest compared with larger sized particles.

Nanometer-scale polymers have been of special focus for adsorbing bioactive compounds, and understanding their adsorption mechanisms has been essential to develop and optimize new nanomaterials for releasing compounds [18]. SNPs are a type of nanomaterial obtained from native starch through physical, chemical, or enzymatic methods (Table 6.1) [34–36]. The adsorption mechanism of SNPs depends on various factors, such as size, shape, charge, and composition of the particles, as well as the properties of the adsorbed compound and the medium [37, 38]. Smaller sizes lead to higher adsorption capacity, and more irregular and porous materials result in increased adsorption [39, 40]. The charge and composition of the SNPs play a crucial role as they primarily determine the adsorption mechanisms. The adsorption mechanism of SNPs, depending on their chemical structure under specific conditions or modifications, is based on electrostatic, hydrogen bonding, and hydrophobic interactions established between the starch molecules and the surface of the adsorbed compound (Fig. 6.1) [41–43].

Electrostatic interactions occur between the electric charges of the involved components [43]. In the case of SNPs, they can have positive or negative charges on their surface, whereas the adhering compounds may have opposite charges [44, 45]. These charges attract SNPs to other compounds and contribute to their adsorption. SNPs are polysaccharides composed of glucose molecules linked in long chains. In its natural form, starch does not have a net electric charge, meaning that is neither inherently positive nor negative [46]. Therefore, it cannot generate electrostatic interactions naturally. However, starch can acquire a positive or negative net charge depending on the preparation and processing conditions. The properties of SNPs can be modified to have a positive or negative charge through chemical treatments or surface modifications, such as amination. Depending on the charge of the starch nanoparticles, they can have electrostatic interactions with proteins with positive ($-NH_3^+$) or negative ($-COO^-$) charges.

Hydrophobic interactions are attractive forces that occur between nonpolar or weakly polar regions of molecules. SNPs have hydrophobic regions in their structure, allowing them to adhere to hydrophobic compounds [47]. The hydrophobic regions of SNPs can attract and adsorb nonpolar molecules due to intermolecular dispersion forces, also known as London forces. Amylose and amylopectin contain hydrophobic regions due to interactions between glycosidic bonds and spatial configurations of glucose chains [47]. Generally, starch has a moderate affinity for adsorbing hydrophobic compounds; therefore, chemical modifications are often performed to improve its efficiency. Based on this type of interaction, SNPs can have an affinity for aromatic compounds, hydrophobic dyes, pharmaceutical compounds, etc., in aqueous media. For instance, Acevedo-Guevara et al. [6] achieved high adsorption of curcumin in banana SNPs, promoting hydrophobic interaction through starch acetylation. Acetylated starch can enhance curcumin adsorption by introducing acetyl groups, increasing the hydrophobicity of starch. These acetyl groups can interact with curcumin through hydrophobic interactions, thereby enhancing their affinity and increasing the adsorption efficiency of acetylated starch. In other studies, propylated starch particles were loaded with flufenamic acid, testosterone, and caffeine, which showed higher effectiveness in human skin penetration studies [48]. The presence of propyl groups in starch facilitates hydrophobic interactions with these compounds, resulting in greater efficiency in their adsorption and retention.

Table 6.1 Production of starch nanoparticles (SNPs) for the adsorption and encapsulation of compounds

Adsorbed compound	Starch source	SNPs synthesis method	Adsorption process	Main particle size (nm)	PDI	Encapsulation efficiency (%)	Release capacity	ξ-potential (mV)	Reference
Curcumin	Banana starch, acetylated banana starch	Nanoprecipitation	Magnetic stirring with dropwise addition	147–201	0.02–0.391	85.23–90.63	Ni	(−) 3.67–4.49	[6]
Curcumin	Acetylated waxy corn starch	Alkaline hydrolysis	Stirring with dropwise addition	100	Ni	43.38–61.60	45–55% in 7 h	Ni	[8]
Loteonin	Oxided lotus starch	Nanoprecipitation	Simultaneous adsorption to nanoprecipitation processes	305	0.173	82.2	85% in 360 min	(−) 20.8	[11]
Curcumin	Soluble starch	Nanoprecipitation	Simultaneous adsorption to nanoprecipitation processes	182–255	0.09–0.18	45.27–66.57	Ni	(−) 17.89–26.26	[13]
Sodium tripolyphosphate	Maize starchB7:B11	Nanoprecipitation	Stirring	21.04	0.2	95.1	20% in 4 h	(−) 35.3	[14]
Menthone	Waxy corn starch	Enzymatic debranching with pullulanase and nanoprecipitacion	Stirring with dropwise addition	84–104	0.156–0.262	72.4–86.6	55–70% in 72 h	Ni	[15]
Resveratrol	Horse chestnut, lotus stem and water chestnut starch	Planetary milling	Magnetic stirring with dropwise addition	419.59–797.32	Ni	79.37–81.46	54.54–64.51% in 180 min	14.77–24.28	[16]
Rosemary Essential Oil	High amylose corn starch	Alkaline hydrolysis	Stirring	133.7 ± 35.8	Ni	94.63 ± 0.762	9.65 ± 0.08%	Ni	[17]
Rutin	Quino and maize starch	Ultra-sonication	Simultaneous adsorption to nanoprecipitation processes	21.5–106.9	0.36; 0.40	26.56–67.4	Ni	(−) 18.0 ± 0.3	[19]

Myricetin	Tapioca starch	Vibrating super-fine mill	Stirring with dropwise addition	55.27–120.1	Ni	453 ± 8.07 mg/g (LC)	70% in 1000 min	Ni	[20]
Catechin	Horse chest-nut, water chestnut and lotus stem starch	Nanoprecipitation	Simultaneous adsorption to nanoprecipitation processes	322.2–615.6	0.20; 0.62; 0.59 (R)	48.30–59.09	3.39–5.35 mg in 4 h	(−) 18.05–20.1	[21]
Ciprofloxacin	Corn starch	Nanoprecipitation	Simultaneous adsorption to nanoprecipitation processes	221–324	0.11–0.51	67.7–89.1	Ni	(−) 5.1–32.3	[22]
Piroxicam	Quinoa starch	Nanoprecipitation	Simultaneous adsorption to nanoprecipitation processes	282–870	0.339–0.772	78	98.8% in 2 h	Ni	[23]
Ardisia compressa anthocyanins	Normal, high-amylose, and high-amylopectin succinylated corn starch	Extrusion and nanoprecipitation	Simultaneous adsorption to nanoprecipitation processes	60–370	0.38–0.73	49.4–52.5	Ni	(−) 15–34	[24]
Vitamin D3	High amylose corn starch and potato starch	Alkaline hydrolysis	Simultaneous adsorption to alkaline hydroly-sis processes	38.3–99.2	0.16–0.38	1.09–2.98	Ni	(−) 8.49–33.39	[25]
Cordycepin	Cassava starch	Physical and acid treatment	Centrifugation and ultrasonication	103–116	Ni	3.89–15.18	Ni	Ni	[26]
Apogossypolone and doxorubicin hyaluronic acid	Potato starch	Grafting stearic acid	Magnetic stirring with dropwise addition	14.7–68.5	Ni	94 ± 1.7	30% in 80 h	(+) 27.6 ± 3.6; (−) 22.3 ± 2.1	[27]

(continued)

Table 6.1 (continued)

Adsorbed compound	Starch source	SNPs synthesis method	Adsorption process	Main particle size (nm)	PDI	Encapsulation efficiency (%)	Release capacity	ξ-potential (mV)	Reference
(+)catechin, (−)-epicatechin, (−)-epigallocatechin-3-gallate, and proanthocyanidins	Waxy maize starch	Nanoprecipitation	Centrifugation	222–398; 145–171; 128–206; 89–171 (R)	Ni	65	80% in 10 h	(−) 20.6–9.3	[28]
Cisplatin	Cassava starch acetate	Nanoprecipitation	Simultaneous adsorption to nanoprecipitation processes	311.7–350.1	Ni	66.9–94.1	70–80% in 700 min	(−) 5.5–10.0	[29]
β-carotene	Banana starch	Nanoprecipitation	Stirring with dropwise addition	231 ± 0.3	0.25 ± 0.04	95.32 ± 1.23	40% in 20 h	(−) 15.2 ± 0.23	[30]
Ciprofloxacin	Soluble starch	Emulsion cross-linking	Nanoparticles dispersed in ciprofloxacin solution	Ni	Ni	7.67–9.44	35% in 120 min	Ni	[31]
Capsaicin	Rice starch	Enzymatic debranching with pullulanase	Simultaneous adsorption to nanoprecipitation processes	617.84 ± 6.38	0.24 ± 0.07	70.05 ± 1.78	80% in 36 h	Ni	[32]
Ginkgo biloba extract	Dafozhi, damaling, daguo, and, corn starch	Nanoprecipitation	Simultaneous adsorption to nanoprecipitation processes	201.2–250.9	0.428–0.478	600–900 μg/mg (LC)	80% in 12 h	Ni	[33]

LC load capacity, *PDI* polydispersity index, *SNPs* starch nanoparticles, *Ni* not informed

Finally, hydrogen bonding is a chemical interaction between a hydrogen atom and a more electronegative atom (such as oxygen, nitrogen, or fluorine) in another molecule. SNPs have functional groups, such as hydroxyl groups (−OH), that can form hydrogen bonds with complementary functional groups of the adsorbed compounds that adhere to them [19, 20]. Some of the compounds that can be adsorbed through hydrogen bonding with their respective functional groups include sugars (−OH), polysaccharides (−OH), proteins ($-NH_2$) (−COOH) (−C=O), phenolic compounds (−OH), essential oils (−OH), etc. Various studies have reported that SNPs have been used mainly to adsorb phenolic compounds due to the ease of forming hydrogen bonds with these types of functional molecules. Ahmad and Gani [16] achieved high efficiency in stabilizing resveratrol in various types of SNPs, such as horse chestnut, lotus stem, and water chestnut starch. Ahmad et al. [21] stabilized catechin in high-amylose corn SNPs. On the other hand, Misken et al. [49] stabilized citric acid in corn SNPs, an active compound that is also adsorbed onto the starch molecule through hydrogen bonding.

Considering this information, it is evident that SNPs exhibit versatility in adsorbing bioactive and pharmaceutical compounds due to their ability to have electrostatic, hydrophobic, and hydrogen bonding interactions with various types of compounds. In addition, the possibility of modifying the functional groups of this material allows for changes or improvements in its adsorption capacity through different interaction mechanisms.

6.2.2 Chemical Modifications to Enhance Adsorption Capacity and Encapsulation Efficiency

It has been observed that SNPs are versatile for the adsorption of different types of compounds. However, natural SNPs naturally exhibit low affinity toward hydrophobic compounds, and in some cases, the adsorption of polar compounds is not efficient. Therefore, various methods of chemical modification of starch have been developed to alter or enhance its characteristics, such as esterification, carboxylation, hydroxypropylation, amination, and oxidation (Fig. 6.1). It should be noted that some of these chemical modifications have not been performed on nanoscale starch but on native starch. Nevertheless, because of their similar nature, these modifications can also be applied to SNPs.

6.2.2.1 Esterification

Esterification involves of introducing ester groups into the glucose chains through reactions with organic acids or their anhydrides. These reactions can occur in an aqueous medium or in organic solvents, with or without catalysts, and can affect both amylose and amylopectin in SNPs [7]. This modification can improve the

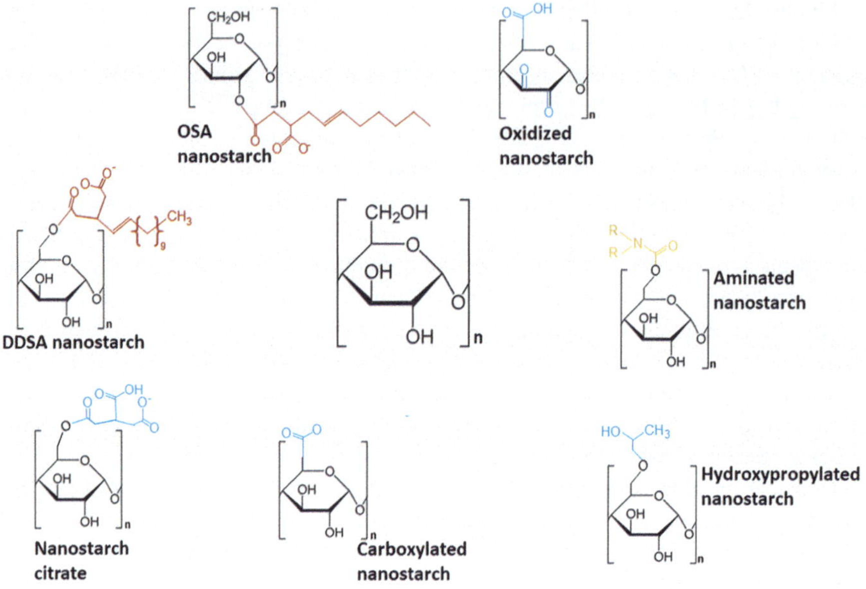

Fig. 6.1 Chemical modifications performed on starch nanoparticles (or nanostarch) to enhance the adsorption capacity of bioactive compounds. OSA: octenylsuccinic acid; DDSA: dodecenylsuccinic acid

adsorption capacity for hydrophobic compounds because ester groups can interact with these molecules through hydrophobic interactions, thus modifying their swelling capacity, solubility, viscosity, thermal stability, and ability to form gels and emulsions [7, 50, 51].

Among the organic acids used to modify starch are acetic acid, octenyl succinic acid (OSA), dodecenylsuccinic acid (DDSA), and citric acid [52–54]. Each of these acids imparts specific properties to starch, making it suitable for different applications. Acetic acid is the simplest and most commonly used organic acid to modify starch. The esterification reaction with acetic acid or acetic anhydride produces acetylated starch, which has a degree of substitution (DS) between 0.01 and 3, where DS is the average number of ester groups per glucose unit [55]. Acetylated SNPs have demonstrated higher hydrophobicity, thermoplasticity, and thermal stability than unmodified SNPs, as well as reduced tendency to retrogradation and increased capacity to form thermoreversible gels [52–54]. Acetylated starch has great potential for use as a stabilizer, emulsifier, and encapsulating agent. For instance, Acevedo-Guevara et al. [6] synthesized acetylated banana nanostarch using acetic anhydride to enhance the adsorption capacity of curcumin, resulting in increased encapsulation efficiency (EE) and loading capacity (LC) compared with unmodified starch nanoparticles, with EE values of 90.63 ± 1.5% and LC values of 3.2 ± 0.01 mg/mg, and EE values of 85.23 ± 2% and LC values of 2.03 ± 0.02 mg/mg, respectively. This suggests that the acetylation of banana starch leads to more hydrogen bonding sites in the starch molecule, allowing interaction with more curcumin molecules. Similarly, Najafi and Ashori [22] acetylated corn starch with acetic acid to improve the adsorption of ciprofloxacin as a model compound, finding that increasing the DS in the starch acetate led to an increase in EE from 67.7% to 89.1%. Moreover, increasing the water/acetone ratio from 1:1 (DS: 0.33) to 3:1 (DS: 2.66) increased the EE from 48.5% to 89.1%. The presence of more hydrogen bonds and layers of ciprofloxacin retained by starch molecules leads to the formation of larger spheres of acetylated starch nanoparticles loaded with ciprofloxacin.

On the other hand, octenylsuccinic acid (OSA) is an unsaturated dicarboxylic acid obtained by reacting maleic anhydride with 1-octene. The esterification reaction with OSA or octenylsuccinic anhydride produces octenylsuccinate starch (OSAS), which has a DS between 0.01 and 0.1. The FAO and WHO have evaluated octenyl succinate as a starch modifier that can be used in food, and there are no restrictions on the daily requirements of the modified starch obtained [43]. Modified starch with OSA is used in the food industry as an emulsifier, coating agent (e.g., microencapsulated materials), fat substitute, and slow-digesting starch [56–61]. For example, Wang et al. [62] investigated the effect of starch modification with OSA on the formation of maize starch-lipid complexes. According to the complexity index (CI) in this study, native corn starch formed more complexes with monopalmitoylglycerol (MPG) than with palmitic acid (PA), whereas dipalmitoylglycerol (DPG) was not effective in forming complexes with native maize starch. After modification with OSA, the complexation between OSA-starch and lipids increased considerably, especially for PA and DPG, with CI values increasing from 79.6% to 93.3% for OSA-starch-PA complexes, and from 80.3% to 93.2% for OSA-starch-DPG complexes, with an increase in the DS of OSA-starch.

Furthermore, DDSA is an unsaturated dicarboxylic acid obtained by reacting maleic anhydride with 1-dodecene. The esterification reaction with DDSA or dodecenylsuccinic anhydride produces dodecenylsuccinate starch (DDSAS), which has a DS between 0.01 and 0.05 [63]. DDSAS is an important safe additive that possesses good emulsification and thickening abilities and it is biodegradable. This starch sugar ester is an important modified starch in the food industry [64, 65]. DDSAS has higher hydrophobicity and emulsifying capacity than native starch. It has been used to stabilize flavor concentrates in beverages, oil in salad dressings, and to encapsulate flavors, fragrances, and vitamins [63]. Ren et al. [66] observed that starch nanoparticles modified with DDSA exhibited higher hydrophobicity at lower DS compared with those modified with OSA or acetic anhydride, tested with chloroform as a model for hydrophobic compound adsorption.

Lastly, among the modifications made by esterification, citric acid was used. It is a tricarboxylic acid naturally found in many citrus fruits. The esterification reaction with citric acid or citric anhydride produces citrate starch, which has a DS between 0.01 and 0.5. Citrate starch has higher solubility in cold water, greater swelling capacity, and a greater ability to form thermoreversible gels than native starch [67, 68]. Furthermore, it exhibits greater heat and acid resistance and higher biodegradability [49]. Citrate starch has been used as a gelling agent, thickener, stabilizer, and antioxidant [67]. Ma et al. [68] and Misken et al. [49] produced citric acid-modified starch nanoparticles, which show potential for adsorbing and stabilizing hydrophilic compounds.

6.2.2.2 Carboxylation

Carboxylation of starch is a process of chemical modification in which carboxyl groups (-COOH) are introduced into starch molecules. This process is achieved by reacting starch with carboxylating agents such as acrylic acid, maleic anhydride, or enzymes [8]. One of the main advantages of this modification is that the introduced carboxyl groups make the starch more hydrophilic. This modification of starch mainly occurs in the amorphous region within the surface layer of the granules [8]. Carboxylated starch has several applications in various industries. Because of its strong hydrophilicity and polarity, carboxylated starch has been widely used as a drug carrier, adsorbent, flocculant, and in other applications [69, 70].

6.2.2.3 Hydroxypropylation

This chemical modification involves the introduction of hydroxypropyl groups into the starch structure. This modification is carried out by reacting starch with a hydroxypropylation agent, such as propylene oxide [9, 71]. Hydroxypropylation can modify the physical and chemical properties of starch, such as solubility, viscosity, and adsorption capacity [9, 72]. This modification increases the hydrophilicity of starch, which enhances its adsorption capacity for hydrophilic compounds. In addition, hydroxypropylation can alter the starch structure, making it more flexible and with a greater availability of functional groups, facilitating interactions with the compounds to be adsorbed [9]. The literature on hydroxypropylation using organic solvents to achieve higher molar substitution is limited. Kshirsagar [73] observed that hydroxypropylated and hydrolyzed corn and waxy amaranth starches were good encapsulating agents for lemon oil but not for orange oil.

6.2.2.4 Amination

Amination of starch is a chemical modification in which amino groups ($-NH_2$) are introduced into the starch structure. This modification is performed through chemical reactions that may involve the substitution of hydroxyl groups with amino groups. The introduction of amino groups into starch can have several beneficial effects in terms of adsorption capacity. Starch molecules modified with amino groups contain many active hydroxyl and amino groups, that can chelate metal ions [10]. Aminated starches have been primarily used for the adsorption of metal ions. Dong et al. [74] studied the adsorption capacity of amino-modified starch for

Cu^{2+} and Cr^{6+}, and the results showed that the adsorption capacity of amino-modified starch for Cu2+ and Cr^{6+} can reach 29.4 mg/g and 12.5 mg/g, respectively. Ren et al. [75] prepared an adsorbent based on starch functionalized with amino groups through a three-step reaction of esterification, crosslinking, and aminolysis, with the aim of adsorbing uranium ions. They achieved satisfactory results under conditions of pH 6, mass ratio of 35 mg/g, temperature of 25 °C, concentration of 0.25 g/L, and time of 5 h, with the adsorbent's adsorption capacity for uranium ions being 118.92 mg/g.

6.2.2.5 Oxidation

In addition, oxidation reactions have also been employed to modify starch. The purpose of oxidizing starch is to alter its physical and chemical properties to enhance its use in the food industry. Oxidized starch is an additive obtained by treating starch with oxidizing agents such as hydrogen peroxide or sodium hypochlorite [76, 77]. Oxidized starch has a lower viscosity, higher heat and acid stability, increased clarity, and reduced tendency to retrograde compared with native starch. Previous studies have reported that oxidation can improve the solubility of starch and affect its encapsulation capacity [76–78]. Chen et al. [11] developed oxidized lotus root starch nanoparticles loaded with luteolin to enhance the stability and antioxidant activity of luteolin. In this research, it was observed that luteolin was successfully encapsulated in the nanoparticles, with an encapsulation efficiency of 87.2%. Furthermore, this study demonstrated the controlled release of this active compound in simulated gastric fluid.

6.3 Physicochemical Properties for Understanding the Adsorption of Compounds on Starch Nanoparticles

The physicochemical characterization of SNPs and the study of their capacity to adsorb bioactive or pharmaceutical compounds are essential for understanding their behavior and optimizing their application in various fields. In this regard, analytical techniques such as Fourier-transform infrared spectroscopy (FTIR), thermogravimetry (TG), differential scanning calorimetry (DSC), and X-ray diffraction (XRD) play a crucial role.

6.3.1 Fourier-Transform Infrared Spectroscopy (FTIR)

FTIR spectroscopy has allowed the identification and analysis of functional groups present in starch nanoparticles and adsorbed bioactive compounds. This technique provides information about different types of molecular interactions and allows for the identification of functional groups on the surface, such as those adsorbed compounds on starch nanoparticles. In the study conducted by Acevedo-Guevara et al. [6], acetylated and non-acetylated starch nanoparticles loaded with curcumin were compared with simple mixtures of curcumin with these nanoparticles. These mixtures showed a peak at 3520 cm^{-1} attributed to the stretching of the O–H group in the benzene ring of curcumin. However, this band disappeared in curcumin-loaded nanoparticles, indicating that little to no free curcumin molecules were present. Ahmad and Gani [16] studied the stabilization of resveratrol in starch nanoparticles and observed that the characteristic peaks of free resveratrol in the FTIR spectrum were not very evident in the resveratrol encapsulated in the nanoparticles. This suggests that most of the loaded resveratrol is trapped inside the encapsulating agent's cavity, with a smaller amount possibly present on the surface as free resveratrol. Similar results were also observed in studies where luteolin [11], catechin [21], rosemary essential oil [17], piroxicam [23], *Ardisia compressa* anthocyanins [24], cisplatin [29], rutin [19], β-carotene [30], and capsaicin [32] were stabilized in starch nanoparticles. In these cases, the adsorption was mainly evidenced by the decrease or disappearance of the characteristic bands of the bioactive compounds in the FTIR spectra.

6.3.2 Thermogravimetry (TG)

Thermogravimetry has been used to determine the thermal stability of starch nanoparticles and their capacity to retain adsorbed compounds. By subjecting the samples to a controlled temperature increase, the loss of mass over time is recorded. This technique allows for the evaluation of the amount of adsorbed compound and its thermal stability, which is essential to ensure its effectiveness in specific applications. In the study conducted by Amani et al. [17], the thermal stability of rosemary essential oil was increased when encapsulated in starch nanoparticles. Pure rosemary essential oil exhibited a degradation range of 45–160 °C, whereas when encapsulated in starch nanoparticles, the degradation range shifted to 230–300 °C. It is possible that the thermal stability of essential oils increases when adsorbed onto starch nanoparticles due to the formation of inclusion complexes between the phenolic compounds and SNPs. Phenolic compounds can form hydrogen bonds with the starch and, therefore, be trapped within the starch nanoparticle matrix. Other researchers have reported that the thermal stability of polyunsaturated fatty acids increased after their incorporation into amylose, and the amylose-fatty acid inclusion complex was well protected against oxidation and thermal degradation [12, 79].

6.3.3 Differential Scanning Calorimetry (DSC)

DSC can help understand the adsorption of bioactive compounds by providing information about changes in thermal properties during the adsorption process. This characterization can reveal the decomposition temperature, phase transitions, enthalpy changes, and other thermal events associated with the interaction between the bioactive compound and the adsorbent substrate. These thermal data have been instrumental in understanding the stability, compatibility, and adsorption affinity of the compound and starch nanoparticles. Previous research has shown that both the DSC thermogram of native curcumin and the physical mixture of curcumin and starch exhibit a sharp endothermic peak around 178 °C, corresponding to the melting point of curcumin crystals. However, in all samples of SNPs with adsorbed curcumin, the endothermic peak corresponding to curcumin crystal melting disappeared, suggesting that after being loaded into the starch nanoparticle (SNP) matrix, curcumin loses its crystalline state [13]. In a similar study, the melting point of vitamin D3 was identified at 87.8 °C, but when stabilized in potato starch and high-amylose corn starch nanoparticles, the melting peak of vitamin D3 was not observed, indicating successful encapsulation of the vitamin [25]. It has also been observed that nanoparticles of starch from normal, high-amylose, and high-amylopectin corn starch sources increased the thermal stability of *Ardisia compressa* anthocyanins, shifting the peak fusion temperature from approximately 130 °C in its pure state to approximately 160 °C in its adsorbed state. This suggests enhanced thermal stability, possibly due to a recrystallization phenomenon experienced by the nanoparticle-anthocyanin interaction [80, 81]. In addition, a decrease in the enthalpy of fusion (ΔH) was observed, reflecting the perfect packing of double helices in the crystalline and amorphous regions of the starch particles [82]. The decrease in ΔH suggests that the anthocyanins positioned themselves in the space between helices and altered the structural order of the double helix, resulting in semi-crystalline single-helix structures [23, 24]. The results of these investigations suggest that, from a thermal properties perspective, starch nanoparticles with stabilized bioactive compounds can be applied in functional beverages that undergo thermal processes such as pasteurization, as well as in cereal-based flours processed at high temperatures, such as the extrusion process.

6.3.4 X-Ray Diffraction (XRD)

XRD has been employed to determine the crystalline structure of starch nanoparticles and to investigate the presence of crystalline or amorphous phases. Additionally, this technique has provided information about the characteristic peaks of both starch nanoparticles and the adsorbed compounds, which is relevant for understanding and confirming the stability of bioactive compounds during the adsorption process. XRD analysis conducted by Bhatia and Rohilla [13, 23] revealed the characteristic peaks of native starch and piroxicam, observing that starch nanoparticles produced by adsorbing piroxicam exhibit properties of both types of nature, i.e., both amorphous (due to the nanoprecipitation of starch) and crystalline (from piroxicam), thus confirming the adsorption of piroxicam. On the other hand, Li et al. [13] observed that the physical mixture of curcumin and starch exhibited both curcumin and raw starch diffraction peaks. However, curcumin-loaded starch nanoparticles did not show obvious diffraction peaks, displaying a diffractogram more similar to that of starch after gelatinization and freeze-drying. When curcumin is encapsulated in starch nanoparticles, the particle size can become so small that X-ray diffraction cannot detect the crystalline structure. As the particle size decreases, the X-ray diffraction becomes less pronounced. This same behavior was observed in a research by Miskeen et al. [39] when stabilizing curcumin in starch nanoparticles. Other researchers have also characterized starch nanoparticles using XRD [6, 11, 14, 16, 17].

6.4 Functional Properties and Cytotoxicity

Functional properties play a crucial role in the application of starch nanoparticles in the adsorption of bioactive compounds, as these properties determine the efficiency of the production of these materials. Among the functional properties commonly evaluated for starch nanoparticles loaded with bioactive compounds are encapsulation efficiency (EE), loading capacity (LC), release capacity, as well as antioxidant, antimicrobial, cytotoxic, and cell viability effects.

6.4.1 Encapsulation Efficiency (EE) and Loading Capacity (LC)

Retention capacity is essential to ensure the encapsulation efficiency and loading capacity of bioactive compounds in starch nanoparticles. Encapsulation efficiency (EE) refers to the number of bioactive compounds that have been encapsulated relative to the total amount of bioactive compounds added in the process. On the other hand, loading capacity (LC) refers to the number of encapsulated compounds relative to the total amount of particles (encapsulated compounds plus encapsulating compound). A high retention capacity ensures that bioactive compounds are effectively adsorbed and retained in the starch matrix, thereby preventing losses and maximizing their activity. This is particularly important in pharmaceutical and food applications, where precise and sustained delivery of compounds is desired. Research has shown that SNPs have a high potential for adsorbing bioactive compounds (Table 6.1). Qiu et al. [15] observed that nanoparticles of waxy corn starch achieved EE ranging from 78.6% to 83.2% and LC from 0.14 to 0.19 ml/g for absorbed menthone using the nanoprecipitation process. In this study, it was evident that temperature played an important role, with higher EE observed at elevated temperatures. This could be attributed to slower nucleation and distinct crystal growth at higher complexation temperatures, resulting in a more ordered crystalline structure, indicating increased complexation capacity with essential oils [29, 83]. Raj and Prabha [29] studied the EE and LC of acetylated cassava starch nanocomposites and complexes of these nanoparticles with polyethylene glycol and polyethylene glycol with gelatin, for the stabilization of the drug cisplatin, obtaining EE of 85.9%, 91.6%, and 94.1%, and LC of 50.6%, 53.9%, and 57.0%, respectively, indicating that this increase can be attributed to the fact that polyethylene glycol and gelatin have a higher attraction or affinity toward this drug.

6.4.2 Release Capacity

The release capacity of the adsorbed compounds is another crucial aspect. Controlled release allows for regulation of the delivery of bioactive compounds over time, optimizing their therapeutic or functional activity. Starch nanoparticles can be designed and have shown potential for sustained release of compounds (Table 6.1). This property is particularly relevant in the field of bioactive compound release, where controlled and maximum release at the site of action is desired. Ahmad and Gani [16] studied the release of resveratrol from horse chestnut, lotus stem, and water chestnut SNPs under simulated gastrointestinal conditions. The first stage simulating gastric conditions showed resveratrol release between 16.30% and 19.24%, which could be due to the release of compounds located on the surface or compounds not encapsulated within the nanoparticles. The second stage, referring to release in simulated intestinal fluid, showed a significant increase in release, with release between 54.54% and 64.51%. This could be attributed to the rupture of the starch nanocapsule surrounding resveratrol due to swelling of the capsules by absorption of the simulated intestinal fluid solution or due to the action of enzymes on the outer layer. The release mechanism of encapsulated particles was also reported by Arifin et al. [84], who suggested similar pathways of release behavior, such as diffusion, swelling of the capsules, or degradation of the wall materials. Another reason for the higher release of resveratrol could be the high surface area of the nanoparticles, which could have increased their percentage of release under simulated intestinal conditions [85]. A higher percentage of bioactive compounds was also observed in the simulated intestinal fluid in previous studies by the same authors [86–88]. The higher release of resveratrol in the simulated intestinal fluid is important because it can exert maximum health benefits for the host.

6.4.3 Antioxidant Capacity

The adsorption and release of bioactive compounds have enabled SNPs to have various active functionalities, such as acting as an antioxidant agent, neutralizing free radicals, and protecting against oxidative stress. This is particularly relevant in applications related to health and food, where antioxidant compounds can play an important role in disease prevention and product quality maintenance. Oxidized lotus starch nanoparticles have been used to stabilize luteolin, demonstrating the capacity to scavenge DPPH radicals (85% radical scavenging), and even showing higher antioxidant capacity than free luteolin in ethanol and water (80% and 60% radical scavenging, respectively) [11]. Free luteolin can form aggregates because of its low solubility in water, limiting the interaction between luteolin and the DPPH free radical [89]. Li et al. [13], on the other hand, studied the radical scavenging capacity of both free and adsorbed curcumin in soluble starch nanoparticles. In this case, curcumin-loaded starch nanoparticles showed radical scavenging percentages of up to 90% and 70% for DPPH and ABTS radicals, respectively, compared to 100% for both radicals in free curcumin. The antioxidant capacities of myricetin [20], rutin [19], catechin, epicatechin, epigallocatechin-3-gallate, and proanthocyanidins [28] have also been investigated in starch nanoparticles, showing similar efficacy in free radical scavenging, thus demonstrating that nanoscale starch is an efficient carrier material to fulfill this function.

6.4.4 Antimicrobial Capacity

The antimicrobial properties of starch nanoparticles are of interest in applications related to food preservation and infection prevention. Bioactive compounds adsorbed on starch nanoparticles have been shown to inhibit the growth and proliferation of pathogenic microorganisms. Amani et al. [17] evaluated the antimicrobial activity of both free rosemary essential oil and rosemary essential oil adsorbed on high-amylopectin corn starch nanoparticles against Gram-positive bacteria (*Staphylococcus aureus* and *Listeria monocytogenes*) and Gram-negative bacteria (*Shigella flexneri* and *Escherichia coli*). For all strains, both free essential oil and starch nanoparticles with essential oil showed minimum inhibitory and minimum bactericidal concentrations of 3.52 mg/mL and 2.5 mg/mL, respectively. Although a decrease in antimicrobial capacity was observed for the adsorbed essential oil, it still caused significant inhibition of bacterial growth. Qiu et al. [90] indicated that different essential oils loaded into starch nanoparticles had varying effects on microorganisms. In another study, antibiotics (penicillin and streptomycin) were encapsulated in starch nanoparticles, and the results demonstrated that encapsulation could enhance the inhibitory properties of the antibiotics [91, 92].

6.4.5 *Cytotoxicity/Cell Viability*

Evaluating the cytotoxicity of new products developed for use in the food and pharmaceutical industries is essential to ensure their safety and quality for human consumption. Cytotoxicity is a crucial indicator of the potential harmful effects of certain components may have on human cells. In the context of food, it is important to ensure that additives and contaminants do not have a negative impact on health. In pharmaceuticals, it is vital to verify that medications are not toxic to the body's cells and do not cause undesirable side effects. Assessing this parameter is crucial to certify the safety of consumption, and thus, starch nanoparticles with bioactive compounds have been subjected to such studies. Starch nanoparticles loaded with luteolin showed low toxicity in RAW264.7 cells at concentrations of 100 μg/mL, with over 81% of the cells exhibiting good viability [11]. Similar results were also found for desramified starch nanoparticles [93]. The cytotoxicity of acetylated starch nanoparticles loaded with hydrophobic functional ingredients (catechin, epicatechin, epigallocatechin-3-gallate, and proanthocyanidins) was also evaluated in Caco-2 cells at concentrations ranging from 0 to 1000 μg/mL, maintaining a viability of >90%, indicating low toxicity and high biocompatibility [8]. Furthermore, the cytotoxicity of tapioca starch nanoparticles loaded with myricetin was studied in HepG2 cells at concentrations from 0 to 250 mg/L, and it was observed that over 90% of the cells survived the incubation [20]. These findings were consistent with previous results where acylated starch nanoparticles did not have toxic effects on mouse embryonic fibroblasts (3T3L1) [94]. These studies underscore the importance of cytotoxicity evaluation to ensure the safety and biocompatibility of starch nanoparticles and their loaded bioactive compounds for potential use in food and pharmaceutical applications.

6.5 Conclusions

Starch nanoparticles (SNPs) have shown great promise as powerful tools for the adsorption, encapsulation, stabilization, and controlled release of bioactive compounds in various pharmaceutical and food applications. Their large specific surface area and ability to interact with different types of compounds make them highly efficient adsorbents. Chemical modification techniques have further enhanced their adsorption capacity, allowing the retention of hydrophobic compounds and improving the retention of hydrophilic compounds. Physicochemical characterization is crucial for understanding the adsorption mechanisms and evaluating the stability efficiency of bioactive compounds in SNPs.

In addition to their effectiveness as adsorbents, it has been demonstrated that SNPs are biocompatible and exhibit low cytotoxicity, making them a safe option for medical and nutritional applications. Furthermore, their antioxidant and antimicrobial capabilities have been verified, expanding their potential use in food preservation and infection prevention.

Research in the field of SNPs is expected to continue to advance to further improve their functional properties and applications. Chemical modification and optimization of encapsulation techniques could enable even greater efficiency in the controlled release of bioactive compounds, potentially increasing their therapeutic and functional effectiveness in medical and nutritional applications. Furthermore, exploring new combinations of bioactive compounds with starch nanoparticles could lead to the development of innovative formulations with synergistic and improved properties. Another crucial aspect for the future is the scalability of starch nanoparticle production for commercial applications. Research into more efficient and cost-effective synthesis methods would allow for large-scale production, facilitating their incorporation into products and treatments available in the market. With continued progress in research and the development of new technologies, these nanostructures could revolutionize the way beneficial compounds are administered and used in medicine and nutrition, opening up new possibilities for human well-being and the industry.

Acknowledgments W.D.C. Chacon gratefully acknowledges the Fundação de Amparo à Pesquisa e Inovação do Estado de Santa Catarina (FAPESC) for the PhD fellowship (3003/2021), as well as the Hermes 57983 project of the Universidad Nacional de Colombia sede Palmira, the opportunity to learn skills for academic writing. G. A. Valencia would like to thank the National Council for Scientific and Technological Development (CNPq) (grant 302434/2022-4) for financial support. The authors gratefully acknowledge the Federal University of Santa Catarina for the support.

Conflicts of Interest The authors declare no conflicts of interest.

References

1. Remali J, Aizat WM. A review on plant bioactive compounds and their modes of action against coronavirus infection. Front Pharmacol. 2021;11:589044. Frontiers Media S.A.
2. Dincheva I, Badjakov I, Galunska B. New insights into the research of bioactive compounds from plant origins with nutraceutical and pharmaceutical potential. Plants. 2023;12:258.
3. Lu H, Tian Y. Nanostarch: preparation, modification, and application in Pickering emulsions. J Agric Food Chem. 2021;69:6929–42. American Chemical Society.
4. Ruthven DM. Principles of adsorption and adsorption processes, vol. 1. 1st ed. New York: Wley-Interscience; 1984. p. 1–464.
5. Yu M, Ji N, Wang Y, Dai L, Xiong L, Sun Q. Starch-based nanoparticles: stimuli responsiveness, toxicity, and interactions with food components. Compr Rev Food Sci Food Saf. 2021;20 (1):1075–100.
6. Acevedo-Guevara L, Nieto-Suaza L, Sanchez LT, Pinzon MI, Villa CC. Development of native and modified banana starch nanoparticles as vehicles for curcumin. Int J Biol Macromol. 2018;111:498–504.
7. Hong J, Zeng XA, Brennan CS, Brennan M, Han Z. Recent advances in techniques for starch esters and the applications: a review. Food Secur. 2016;5(3):1–15.

8. Liu Q, Li F, Ji N, Dai L, Xiong L, Sun Q. Acetylated debranched starch micelles as a promising nanocarrier for curcumin. Food Hydrocoll. 2021;111:106253.
9. Fu Z, Zhang L, Ren MH, BeMiller JN. Developments in hydroxypropylation of starch: a review. Starch/Staerke. 2019;71:1800167. Wiley-VCH Verlag.
10. Bayramoglu G, Arica MY. MCM-41 silica particles grafted with polyacrylonitrile: modification in to amidoxime and carboxyl groups for enhanced uranium removal from aqueous medium. Microporous Mesoporous Mater. 2016;226:117–24.
11. Chen YY, Liu K, Zha XQ, Li QM, Pan LH, Luo JP. Encapsulation of luteolin using oxidized lotus root starch nanoparticles prepared by anti-solvent precipitation. Carbohydr Polym. 2021;273:118552.
12. Marinopoulou A, Papastergiadis E, Raphaelides SN, Kontominas MG. Morphological characteristics, oxidative stability and enzymic hydrolysis of amylose-fatty acid complexes. Carbohydr Polym. 2016;141:106–15.
13. Li J, Shin GH, Lee IW, Chen X, Park HJ. Soluble starch formulated nanocomposite increases water solubility and stability of curcumin. Food Hydrocoll. 2016;56:41–9.
14. El-Naggar ME, El-Rafie MH, El-sheikh MA, El-Feky GS, Hebeish A. Synthesis, characterization, release kinetics and toxicity profile of drug-loaded starch nanoparticles. Int J Biol Macromol. 2015;81:718–29.
15. Qiu C, Chang R, Yang J, Ge S, Xiong L, Zhao M, et al. Preparation and characterization of essential oil-loaded starch nanoparticles formed by short glucan chains. Food Chem. 2017;221: 1426–33.
16. Ahmad M, Gani A. Ultrasonicated resveratrol loaded starch nanocapsules: characterization, bioactivity and release behaviour under in-vitro digestion. Carbohydr Polym. 2021;251: 117111.
17. Amani F, Sami M, Rezaei A. Characterization and antibacterial activity of encapsulated rosemary essential oil within amylose nanostructures as a natural antimicrobial in food applications. Starch/Staerke. 2021;73:2100021.
18. Caicedo Chacon WD, Verruck S, Monteiro AR, Valencia GA. The mechanism, biopolymers and active compounds for the production of nanoparticles by anti-solvent precipitation: a review. Food Res Int. 2023;168:112728. Elsevier Ltd.
19. Remanan MK, Zhu F. Encapsulation of rutin using quinoa and maize starch nanoparticles. Food Chem. 2021;353.
20. Xia W, Zheng B, Li T, Lian F, Lin Y, Liu R. Fabrication, characterization and evaluation of myricetin adsorption onto starch nanoparticles. Carbohydr Polym. 2020;250:116848.
21. Ahmad M, Mudgil P, Gani A, Hamed F, Masoodi FA, Maqsood S. Nano-encapsulation of catechin in starch nanoparticles: characterization, release behavior and bioactivity retention during simulated in-vitro digestion. Food Chem. 2019;270:95–104.
22. Mahmoudi Najafi SH, Baghaie M, Ashori A. Preparation and characterization of acetylated starch nanoparticles as drug carrier: ciprofloxacin as a model. Int J Biol Macromol. 2016;87:48–54.
23. Bhatia M, Rohilla S. Formulation and optimization of quinoa starch nanoparticles: quality by design approach for solubility enhancement of piroxicam. Saudi Pharm J. 2020;28(8):927–35.
24. Escobar-Puentes AA, García-Gurrola A, Rincón S, Zepeda A, Martínez-Bustos F. Effect of amylose/amylopectin content and succinylation on properties of corn starch nanoparticles as encapsulants of anthocyanins. Carbohydr Polym. 2020;250:116972.
25. Hasanvand E, Fathi M, Bassiri A. Production and characterization of vitamin D3 loaded starch nanoparticles: effect of amylose to amylopectin ratio and sonication parameters. J Food Sci Technol. 2018;55(4):1314–24.
26. Kaokaen P, Jaiboonma A, Chaicharoenaudomrung N, Kunhorm P, Janebodin K, Noisa P, et al. Cordycepin-loaded nanoparticles from cassava starch promote the proliferation of submandibular gland cells and inhibit the growth of oral squamous carcinoma cells. Nutr Cancer. 2021;73 (10):2014–29.

27. Li K, Liu H, Gao W, Chen M, Zeng Y, Liu J, et al. Mulberry-like dual-drug complicated nanocarriers assembled with apogossypolone amphiphilic starch micelles and doxorubicin hyaluronic acid nanoparticles for tumor combination and targeted therapy. Biomaterials. 2015;39:131–44.
28. Liu C, Ge S, Yang J, Xu Y, Zhao M, Xiong L, et al. Adsorption mechanism of polyphenols onto starch nanoparticles and enhanced antioxidant activity under adverse conditions. J Funct Foods. 2016;26:632–44.
29. Raj V, Prabha G. Synthesis, characterization and in vitro drug release of cisplatin loaded Cassava starch acetate–PEG/gelatin nanocomposites. J Assoc Arab Univ Basic Appl Sci. 2016;21:10–16.
30. Santoyo-Aleman D, Sanchez LT, Villa CC. Citric-acid modified banana starch nanoparticles as a novel vehicle for β-carotene delivery. J Sci Food Agric. 2019;99(14):6392–9.
31. Shi A, Li D, Liu H, Adhikari B, Wang Q. Effect of drying and loading methods on the release behavior of ciprofloxacin from starch nanoparticles. Int J Biol Macromol. 2016;87:55–61.
32. Tao X, Zhan L, Huang Y, Li P, Liu B, Chen P. Preparation, characterization and evaluation of capsaicin-loaded indica rice starch nanoparticles. Food Chem. 2022;386:132692.
33. Wang T, Wu C, Fan G, Li T, Gong H, Cao F. Ginkgo biloba extracts-loaded starch nanospheres: preparation, characterization, and in vitro release kinetics. Int J Biol Macromol. 2018;106:148–57.
34. Hakke VS, Bagale UD, Boufi S, Uday Bhaskar Babu G, Sonawane SH. Ultrasound assisted synthesis of starch nanocrystals and it's applications with polyurethane for packaging film. J Renew Mater. 2020;8(3):239–50.
35. Dukare AS, Arputharaj A, Bharimalla AK, Saxena S, Vigneshwaran N. Nanostarch production by enzymatic hydrolysis of cereal and tuber starches. Carbohydr Polym Technol Appl. 2021;2: 100121.
36. Harun Z, Arsad A, Pang AL, Zaini MAA, Abdurrahman M, Awang N, et al. Acid hydrolysis and optimization techniques for nanoparticles preparation: current review. Appl Biochem Biotechnol. 2022;194:3779–801. Springer.
37. Odeniyi M, Omoteso O, Adepoju A, Jaiyeoba K. Starch nanoparticles in drug delivery: a review. Polym Med. 2019;48(1):41–5.
38. Hassan NA, Darwesh OM, Smuda SS, Altemimi AB, Hu A, Cacciola F, et al. Recent trends in the preparation of nano-starch particles. Molecules. 2022;27(17):5497. MDPI.
39. Miskeen S, An YS, Kim JY. Application of starch nanoparticles as host materials for encapsulation of curcumin: effect of citric acid modification. Int J Biol Macromol. 2021;183:1–11.
40. Zhu F. Encapsulation and delivery of food ingredients using starch based systems. Food Chem. 2017;229:542–52. Elsevier Ltd.
41. Gad E, Owda M, Yahia R. A novel starch nanoparticle citrate based adsorbent for removing of crystal violet dye from aqueous solution. Egypt J Chem. 2020;63(6):2075–95.
42. Caldonazo A, Almeida SL, Bonetti AF, Lazo REL, Mengarda M, Murakami FS. Pharmaceutical applications of starch nanoparticles: a scoping review. Int J Biol Macromol. 2021;181:697–704. Elsevier B.V.
43. Zhang J, Ran C, Jiang X, Dou J. Impact of octenyl succinic anhydride (OSA) esterification on microstructure and physicochemical properties of sorghum starch. LWT. 2021;152:112320.
44. Bobacka V, Eklund D. The influence of charge density of cationic starch on dissolved and colloidal material from peroxide bleached thermomechanical pulp. Colloids Surf A Physicochem Eng Asp. 1999;152:285–91.
45. Lekniute E, Peciulyte L, Klimaviciute R, Bendoraitiene J, Zemaitaitis A. Structural characteristics and flocculation properties of amphoteric starch. Colloids Surf A Physicochem Eng Asp. 2013;430:95–102.
46. Pérez S, Bertoft E. The molecular structures of starch components and their contribution to the architecture of starch granules: a comprehensive review. Starch/Staerke. 2010;62:389–420.
47. Nowak T, Mazela B, Olejnik K, Peplińska B, Perdoch W. Starch-silane structure and its influence on the hydrophobic properties of paper. Molecules. 2022;27:3136.

48. Santander-Ortega MJ, Stauner T, Loretz B, Ortega-Vinuesa JL, Bastos-González D, Wenz G, et al. Nanoparticles made from novel starch derivatives for transdermal drug delivery. J Control Release. 2010;141(1):85–92.
49. Miskeen S, Hong JS, Choi HD, Kim JY. Fabrication of citric acid-modified starch nanoparticles to improve their thermal stability and hydrophobicity. Carbohydr Polym. 2021;253:117242.
50. Tian S, Chen Y, Chen Z, Yang Y, Wang Y. Preparation and characteristics of starch esters and its effects on dough physicochemical properties. J Food Qual. 2018;2018. Hindawi Limited.
51. Otache MA, Duru RU, Achugasim O, Abayeh OJ. Advances in the modification of starch via esterification for enhanced properties. J Polym Environ. 2021;29:1365–79. Springer.
52. Couto Pereira A, Bubolz L, Colussi R. Modificação de amido de castanha portuguesa (catanea sativa mill) com anidrido octenil succinico e seus efeitos na capacidade de absorção de água e de óleo.
53. Li G, Xu X, Zhu F. Physicochemical properties of dodecenyl succinic anhydride (DDSA) modified quinoa starch. Food Chem. 2019;300:125201.
54. Golachowski A, Drożdz W, Golachowska M, Kapelko-Zeberska M, Raszewski B. Production and properties of starch citrates—current research. Foods. 2020;9:1311. MDPI AG.
55. Imre B, Vilaplana F. Organocatalytic esterification of corn starches towards enhanced thermal stability and moisture resistance. Green Chem. 2020;22(15):5017–31.
56. Marefati A, Wiege B, Haase NU, Matos M, Rayner M. Pickering emulsifiers based on hydrophobically modified small granular starches—part I: manufacturing and physico-chemical characterization. Carbohydr Polym. 2017;175:473–83.
57. No J, Shin M. Preparation and characteristics of octenyl succinic anhydride-modified partial waxy rice starches and encapsulated paprika pigment powder. Food Chem. 2019;295:466–74.
58. Punia S, Siroha AK, Sandhu KS, Kaur M. Rheological and pasting behavior of OSA modified mungbean starches and its utilization in cake formulation as fat replacer. Int J Biol Macromol. 2019;128:230–6.
59. Quintero-Castaño VD, Castellanos-Galeano FJ, Álvarez-Barreto CI, Bello-Pérez LA, Alvarez-Ramirez J. In vitro digestibility of octenyl succinic anhydride-starch from the fruit of three Colombian Musa. Food Hydrocoll. 2020;101:105566.
60. Sharma M, Singh AK, Yadav DN, Arora S, Vishwakarma RK. Impact of octenyl succinylation on rheological, pasting, thermal and physicochemical properties of pearl millet (Pennisetum typhoides) starch. LWT. 2016;73:52–9.
61. Yu ZY, Jiang SW, Zheng Z, Cao XM, Hou ZG, Xu JJ, et al. Preparation and properties of OSA-modified taro starches and their application for stabilizing Pickering emulsions. Int J Biol Macromol. 2019;137:277–85.
62. Wang J, Ren F, Yu J, Copeland L, Wang S. Octenyl succinate modification of starch enhances the formation of starch-lipid complexes. J Agric Food Chem. 2021;69(49):14938–50.
63. Zhongdong L, Boxiang L, Guohua W, Xin Z, Huabin W. Studies on the property and application of starch sugar ester dodecenylsuccinic. In: Chemical properties of starch. London: IntechOpen; 2020.
64. Ntawukulilyayo JD, De Smedt SC, Demeester J, Remon JP. Stabilisation of suspensions using sucrose esters and low substituted n-octenylsuccinate starch-xanthan gum associations. Int J Pharm. 1996;128(1–2):73–9.
65. Trubiano PC. The role of specialty food starches in flavor encapsulation. 1997;244–53.
66. Ren L, Dong Z, Jiang M, Tong J, Zhou J. Hydrophobization of starch nanocrystals through esterification in green media. Ind Crop Prod. 2014;59:115–8.
67. Zhang M, Jia H, Wang B, Ma C, He F, Fan Q, et al. A prospective review on the research progress of citric acid modified starch. Food Secur. 2023;12(3):458.
68. Ma X, Jian R, Chang PR, Yu J. Fabrication and characterization of citric acid-modified starch nanoparticles/plasticized-starch composites. Biomacromolecules. 2008;9(11):3314–20.
69. Shahriarpanah S, Nourmohammadi J, Amoabediny G. Fabrication and characterization of carboxylated starch-chitosan bioactive scaffold for bone regeneration. Int J Biol Macromol. 2016;93:1069–78.

70. Sarmah D, Karak N. Double network hydrophobic starch based amphoteric hydrogel as an effective adsorbent for both cationic and anionic dyes. Carbohydr Polym. 2020;242:116320.
71. Tang H, Qu Y, Li Y, Dong S. Synthesis of hydroxypropylated debranched pea starch with high substitution degree in an ionic liquid, and its characterization and properties. J Polym Res. 2018;25:235.
72. Shen Y, Yao Y, Wang Z, Wu H. Hydroxypropylation reduces gelatinization temperature of corn starch for textile sizing. Cellulose. 2021;28(8):5123–34.
73. Kshirsagar AC, Singhal RS. Preparation of hydroxypropyl corn and amaranth starch hydrolyzate and its evaluation as wall material in microencapsulation. Food Chem. 2008;108(3): 958–64.
74. Dong A, Xie J, Wang W, Yu L, Liu Q, Yin Y. A novel method for amino starch preparation and its adsorption for Cu(II) and Cr(VI). J Hazard Mater. 2010;181(1–3):448–54.
75. Ren Z, Liu C, Zhang B, Wu M, Tan Y, Fang X, et al. Preparation of amino-functionalized starch-based adsorbent and its adsorption behavior for uranyl ions. J Radioanal Nucl Chem. 2021;328(3):1253–63.
76. Wang S, Chen Y, Liang H, Chen Y, Shi M, Wu J, et al. Intestine-specific delivery of hydrophobic bioactives from oxidized starch microspheres with an enhanced stability. J Agric Food Chem. 2015;63(39):8669–75.
77. Zhang ZH, Li MF, Peng F, Zhong SR, Huang Z, Zong MH, et al. Oxidized high-amylose starch macrogel as a novel delivery vehicle for probiotic and bioactive substances. Food Hydrocoll. 2021;114:106578.
78. Zhou Y, Li X, Lv Y, Shi Y, Zeng Y, Li D, et al. Effect of oxidation level on the inclusion capacity and solution stability of oxidized amylose in aqueous solution. Carbohydr Polym. 2016;138:41–8.
79. Di Marco AE, Ixtaina VY, Tomás MC. Inclusion complexes of high amylose corn starch with essential fatty acids from chia seed oil as potential delivery systems in food. Food Hydrocoll. 2020;108:106030.
80. Pereira Souza AC, Deyse Gurak P, Damasceno Ferreira Marczak L. Maltodextrin, pectin and soy protein isolate as carrier agents in the encapsulation of anthocyanins-rich extract from jaboticaba pomace. Food Bioprod Process. 2017;102:186–94.
81. Zhao Y, Khalid N, Shu G, Neves MA, Kobayashi I, Nakajima M. Complex coacervates from gelatin and octenyl succinic anhydride modified kudzu starch: insights of formulation and characterization. Food Hydrocoll. 2019;86:70–77.
82. Liu R, Xu C, Cong X, Wu T, Song Y, Zhang M. Effects of oligomeric procyanidins on the retrogradation properties of maize starch with different amylose/amylopectin ratios. Food Chem. 2017;221:2010–17.
83. Biliaderis CG, Galloway G. Crystallization behavior of amylose-v complexes: structure-property relationships. Carbahydr Res. 1989;189:31–48. Elsevier Science Publishers B.V.
84. Arifin DY, Lee LY, Wang CH. Mathematical modeling and simulation of drug release from microspheres: implications to drug delivery systems. Adv Drug Deliv Rev. 2006;58:1274–1325.
85. Gigliobianco MR, Casadidio C, Censi R, Di Martino P. Nanocrystals of poorly soluble drugs: Drug bioavailability and physicochemical stability. Pharmaceutics. 2018;10:134. MDPI AG.
86. Ahmad M, Qureshi S, Maqsood S, Gani A, Masoodi FA. Micro-encapsulation of folic acid using horse chestnut starch and β-cyclodextrin: microcapsule characterization, release behavior & antioxidant potential during GI tract conditions. Food Hydrocoll. 2017;66:154–60.
87. Ahmad M, Ashraf B, Gani A, Gani A. Microencapsulation of saffron anthocyanins using β glucan and β cyclodextrin: microcapsule characterization, release behaviour & antioxidant potential during in-vitro digestion. Int J Biol Macromol. 2018;109:435–42.
88. Ahmad M, Gani A, Hamed F, Maqsood S. Comparative study on utilization of micro and nano sized starch particles for encapsulation of camel milk derived probiotics (Pediococcus acidolactici). LWT. 2019;110:231–8.

89. Yi J, Fan Y, Zhang Y, Wen Z, Zhao L, Lu Y. Glycosylated α-lactalbumin-based nanocomplex for curcumin: physicochemical stability and DPPH-scavenging activity. Food Hydrocoll. 2016;61:369–77.
90. Qiu C, Wang Y, Teng Y, Zhao M. Influence of glycosylation of deamidated wheat gliadin on its interaction mechanism with resveratrol. Food Chem. 2017;221:431–8.
91. Ismail NS, Gopinath SCB. Enhanced antibacterial effect by antibiotic loaded starch nanoparticle. J Assoc Arab Univ Basic Appl Sci. 2017;24:136–40.
92. Ricci D, Fraternale D, Giamperi L, Bucchini A, Epifano F, Burini G, et al. Chemical composition, antimicrobial and antioxidant activity of the essential oil of Teucrium marum (Lamiaceae). J Ethnopharmacol. 2005;98(1–2):195–200.
93. Qin Y, Wang J, Qiu C, Hu Y, Xu X, Jin Z. Effects of degree of polymerization on size, crystal structure, and digestibility of debranched starch nanoparticles and their enhanced antioxidant and antibacterial activities of curcumin. ACS Sustain Chem Eng. 2019;7(9):8499–511.
94. Thakore S, Valodkar M, Soni JY, Vyas K, Jadeja RN, Devkar RV, et al. Synthesis and cytotoxicity evaluation of novel acylated starch nanoparticles. Bioorg Chem. 2013;46(1):26–30.

Chapter 7
Starch Nanomaterials as Food Emulsifiers

Edith Agama-Acevedo, Maria A. Hernandez-Aguirre, and Luis A. Bello-Perez

7.1 Introduction

Emulsions are colloidal systems in which two fluids are dispersed, and one is broken up into small particles for dissolution into the other. The emulsification method is based on the use of surfactants that decrease the surface tension between the two immiscible fluids. Surfactants reduce the repellent force between the two fluids and the adhesion between the components of the same fluid. Food emulsions are mainly oil-in-water (O/W) and water-in-oil (W/O), such as salad dressing, mayonnaise, sauces, butter, margarine, and ice-cream. The main problem in emulsions is phase separation, where flocculation and coalescence occur. When these phenomena are present (e.g., oil and water are then apart from the emulsion), the emulsion loses its functionality and sensory characteristics. Flocculation and coalescence are present while the emulsion is stored; therefore, surfactants are necessary to prevent or slow these phenomena and keep the emulsion stable. Conventional emulsion stabilization uses low molecular weight components with amphiphilic characteristics to reduce interfacial tension, and the emulsion is stabilized by steric hindrance and/or electrostatic repulsion between the droplets [1]. Solid particles are an alternative to surfactant agents; they can be adsorbed into the oil-water interface and produce a mechanical barrier, preventing droplet destabilization. This concept was introduced in the early 1900s by Pickering and, therefore, those emulsions stabilized by solid particles are named after him [2]. The adsorption of solid particles in the Pickering emulsion is irreversible unlike low molecular weight emulsifiers, and shows better stabilization against coalescence and Ostwald ripening than low molecular weight emulsifiers [3].

E. Agama-Acevedo (✉) · M. A. Hernandez-Aguirre · L. A. Bello-Perez
Instituto Politécnico Nacional, CEPROBI, Yautepec, Mexico
e-mail: eagama@ipn.mx; labellop@ipn.mx

G. Ayala Valencia (ed.), *Starch Nanomaterials and Food Applications*,
https://doi.org/10.1007/978-3-031-60086-9_7

Food-grade polysaccharides as starches have been used to stabilize emulsions. Small-sized starches such as amaranth, quinoa, taro, and rice are good candidates for stabilizing Pickering emulsions [4]. Starches esterified with octenyl succinic anhydride (OSA) have superior amphiphilic characteristics, proving that esterification increases the emulsifier properties of starch [5, 6]. It has been suggested that small OSA starch granules are better candidates for stabilizing emulsions [7]. It has also been reported that OSA esterification is affected by the amylose content. Starch with a higher amylose content shows a higher degree of substitution than its counterpart with lower amylose concentrations [8]. Small starch granules (quinoa, amaranth, taro, rice) have low amylose content (around 1–15%); therefore, the challenge is to find starch with small granule size and higher amylose content. Furthermore, the molar mass distribution in native starch before the OSA esterification must be considered to determine whether more OH groups are available for substitution.

In recent years, starch nanoparticles (SNPs) and nanocrystals (SNCs) produced by acid hydrolysis, mechanical treatment (ball milling and ultrasound), and enzymatic processes have been alternatively used to stabilize emulsions [9]. They are particularly effective when esterified with OSA. The present chapter aims to review the use of starch nanomaterials as stabilizers of food emulsions.

7.2 Starch Nanocrystals and Nanoparticles Stabilizing Emulsions

7.2.1 Starch Nanocrystals

The production of modified nanomaterials to improve functionality, their size, and yield are described. SNCs are amphiphilic because they can adsorb at the interface of oil and water due to steric hindrances that prevent contact between adjacent oil droplets. The incorporation of substituent groups in SNCs changes the functionality and physicochemical properties as solubility (Table 7.1). The use of SNCs in foods is limited by their solubility and stability, which represents an opportunity to modify them and prepare hydrophobic materials with improved functionality. There are two strategies: (1) to introduce substituents in surface OH-groups through reagents, such as alkenyl succinic anhydride, phenyl isocyanate, citric acid, and acetic anhydride [10–12]; and (2) surface copolymerization by grafting with polymers such as polyethylene glycol, polytetrahydrofuran, polycaprolactone, polystyrene [13–16]. OSA has been used to esterify starch granules and produce an amphiphilic character. SNCs are hydrophilic and, when modified with OSA, become hydrophobic. Their amphiphilicity improves oil solubility to stabilize O/W emulsions or be used as fillers in foods. Identifying the degree of substitution (DS) of OSA-modified SNCs remains a challenge because the two determination methods, alkaline titration, and nuclear magnetic resonance (^{1}H NMR), require extensive training to obtain good results. SNCs have been prepared from waxy potato starch and modified with

Table 7.1 Production of modified starch nanocrystals from diverse botanical sources and their amylose content

Starch source and amylose content	Production method	Size (nm)	Yield (%)	Esterification	DS	Size (nm)	Yield (%)	Reference
Amaranth 1.5%	AH using sulfuric acid (immersion for 5 days)	374	9.3	OSA: 3% Reaction time: 4 h Temperature: 35 °C pH: 8.5	0.009	Ni	90	[17, 18]
Waxy maize 3.3%		462	21.8		0.012	Ni	92	
Waxy potato 1%	Enzymatic pretreatment (glucoamylase) and AH using sulfuric acid (immersion for 5 days)	Ni	Ni	Reaction time: 1 h OSA: 12.5–75%	0.0575–0.1271	200–800 SNCs form agglomerates	Ni	[19]
Maize flour Ni	AH using sulfuric acid (immersion for 5 days)	125	Ni	Reaction time: 5 h OSA : SNCs (w/w) 1 : 5–2 : 5 Temperature: 40 °C pH: 8.5	0.013–0.045	125–82.6	Ni	[20]
Mize flour Ni	AH using sulfuric acid	Lateral size 100 Thickness 8–10	Ni	Crosslinking Sodium hexametaphosphate: 9%, 18%, and 27% w/w Reaction time: 4 h Temperature: 40 °C	2.5×10^{-3}–7.2×10^{-3}	Size increase as a function of DS 110-370	Ni	[21]
				Acetylation Acetic anhydride Reaction time: 1 h Temperature: 50 °C	0.018	Reduces particle size, but does not the thickness and stiffness	Ni	

(continued)

Table 7.1 (continued)

Starch source and amylose content	Production method	Size (nm)	Yield (%)	Esterification	DS	Size (nm)	Yield (%)	Reference
Sago 24–31%	AH using sulfuric acid (immersion for 5 days)	40–140	25	Ratio: SNCs /lauroyl chloride	Ni	SNCs were disaggregated and distributed randomly	Ni	[22]

AH acid hydrolysis, *DS* degree of substitution, *OSA* octenyl succinic anhydride, *Ni* not informed

OSA at different concentrations. The DS was determined by ^{1}H NMR, and some molecular features are also identified. Higher OSA concentrations increase the DS, which ranges between 0.0575 (12.5% OSA) and 0.1271 (75% OSA). This occurs because the OH groups in glucose monomers are more likely to link to OSA long chains when greater amounts of the anhydride are found in the reaction medium. The source of SNCs is key to the DS; for instance, SNCs obtained from waxy maize and esterified with 100% (w/w) OSA show a DS of 0.034 [23]. This pattern is explained by an increase in the diffusion of the OSA reagent toward waxy potato SNCs.

The OSA esterification of nanocrystals did not produce evident changes in the crystalline structure. In contrast, SNCs from maize starch show a slight decrease in crystallinity when DS is increased [20]. A similar pattern has been found in SNCs from waxy potato starch, where the crystallinity of unmodified SNCs was 45.80% and that of SNCs with different DS ranged between 44.53% and 41.43% [19]. Other eco-friendly reagents, such as alkenyl succinic anhydride (ASAs) and acetic anhydride (AA), have been used to esterify SNCs and produce hydrophobization. SNCs from waxy maize starch were esterified with dodecenylsuccinic acid (DDSA), AA, and OSA. The authors mentioned that the pH of the reaction media is important for producing a high DS in SNCs [19]. The results show that DDSA in a water/ethanol solution (pH $>$ 8.0) yields a higher DS than in water or ethanol. This is because SNCs swell in water, increasing chain mobility, whereas ethanol can dilute DDSA and promote its diffusion into SNC molecules. The esterification of SNCs by DDSA has been evaluated at different pHs, showing that a higher pH of water/ethanol solution increases the DS between 0.005 and 0.021. The use of different AA/SNC ratios (10/1 and 30/1) and reaction temperatures (30, 40, and 50 °C) increases the DS, especially when larger amounts of AA are used at a high temperature (50 °C). The alteration led to reduced crystallinity in unmodified SNCs (54.2%), whereas that in SNCs modified with the three reagents (the highest DS) ranged between 17.4% and 22.6% [23]. It is important to consider the source of SNCs, the reagent used in esterification, and the pH of the reaction media to increase the DS and maintain the crystalline arrangement, since both alter the functionality of SNCs.

7.2.2 *Starch Nanoparticles*

The production of SNPs involves acid treatment (sulfuric acid) at different times and modification by OSA. Starches with different crystalline arrangements, such as those from maize (A- type), potato (B-type), and pea (C-type), have been used to prepare OSA-esterified SNPs. Esterified potato SNPs show the highest DS when compared with esterified pea SNPs. This pattern is attributed to the fact that B- type starches have a less compact structure than A- and C-type starches, and the less compact structure has higher accessibility to OSA reagent toward the amorphous zones of starch granules with increasing DS.

The OSA esterification of SNPs has been reported in starches with small (quinoa and barley), medium (maize), and large (potato) granule sizes. The OSA levels in both the control and acid-hydrolyzed starches ranged between 1.81% and 2.10%. Longer hydrolysis times reduced the surface-weighted mean degree in potato starch from 24.9 μm (control sample) to 3.87 μm (starch hydrolyzed for 5 days). Quinoa starch hydrolyzed for 5 days showed no changes in size (1.59 μm) when compared to unhydrolyzed starch. It is important to mention that nanometric materials (>100–200 nm) have not been obtained after acid hydrolysis [24]; nevertheless, these SNPs are an alternative to stabilize Pickering emulsions. The DS of corn, potato, and pea SNPs esterified with 3% OSA was similar to that observed in starches and ranged between 0.011 and 0.0199 [25]. Quinoa and maize SNPs have been esterified using different OSA concentrations (0.6, 1.2, 1.8, and 2.4%). The DS (0.032–0.168) was larger (122–168 nm) in maize than quinoa (96–132 nm) SNPs [26]. It has been found that amylose content affects the OSA esterification of starch. High-amylose maize starch (60% amylose) esterified with different OSA concentrations (5%, 10%, and 20%) showed increased DS from 0.032 to 0.177, but no differences with respect to the size of SNPs (≈122 nm) [27]. Ultrasound-assisted OSA esterification of SNPs produced a higher DS and reduced size when the power and cycles of ultrasonication increased [28].

SNPs can be classified into soluble (SSNPs) and insoluble (ISNPs). The first are obtained from sugary-1 maize by grounding in a high-speed blender and precipitated them with ethanol. They can be modified with OSA, and their characterization has not been reported (Table 7.2). On the other hand, ISNPs have been obtained from waxy maize starch by acid hydrolysis (H_2SO_4, 3.16 M) under stirring at 40 °C for 1 week [29]. Both starch nanoparticles are candidates for stabilizing Pickering emulsions.

Starch microparticles (SMPs) could be considered starch nanomaterials because, as starch chain aggregates, their size is 1000 nm or less. They are produced as follows: gelatinization of waxy maize, debranching with pullulanase, centrifugation, heating for 15 min in a boiling water bath, and cooling at 4 °C for 12 h. The debranching of starch components leads to the assembly of short linear chains. SMPs can be esterified with OSA (2%, 5%, 8%, and 10%) to produce amphiphilicity and increase DS from 0.0137 to 0.0583, values higher than those of OSA starch. Because SMPs are considered aggregates, their average particle size is 3.78 μm for unmodified and 4.35 μm for OSA SMPs.

The use of sonication before the particle size distribution test can likely provide more information on the real particle size distribution as observed by scanning electron microscopy.

7.3 Pickering Emulsion Stabilized by Starch Nanomaterials

7.3.1 Unmodified Starch Nanocrystals

The use of SNCs to stabilize Pickering emulsions has been widely reported (Fig. 7.1), as summarized in Table 7.3. Native waxy starch (2% amylose) has been

Table 7.2 Production of modified starch nanoparticles from diverse botanical sources and their amylose content

Starch source and amylose content	Production method	Size (nm)	Yield (%)	Modification	DS	Size (nm)	Reference
Potato Ni	AH using sulfuric acid (immersion for 5 days)	3800	9.1	OSA: 3–4% Reaction time: 1 h	1.86	9.1	[24]
Waxy maize Ni		3800	14.8		1.81	14.8	
Waxy barley Ni		5000	10		1.98	10	
Quinoa Ni		1600	6.8		1.86	6.8	
Corn Ni	Starch dispersion (1%) was heated at 95 °C for 2 h followed by stirring at 14,500 rpm for 15 min	Ni	Ni	OSA: 3% Reaction time: 4 h Temperature: 35 °C pH: 8.5	0.0178	Ni	[25]
Potato Ni		Ni	Ni		0.0199	Ni	
Pea Ni		Ni	Ni		0.0112	Ni	
Quinoa 6.3%	Starch dispersion (1%) was heated at 80 °C for 1 h. Ethanol was added to precipitated gelatinized starch. Starch was separated by centrifugation at 2000 × g for 10 min	92	Ni	OSA: 0.6%, 1.2%, 1.8%, 2.4%, and 3% pH: 7.4–7.5	0.0037–0.0163	96–132	[26]
Maize 25.8%		118	Ni		0.0032–0.0168	122–168	
Potato 24.2%		200	Ni		0.0026–0.0161	222–295	
High amylose maize 60%	Starch (10 g) was mixed with ethanol (20 mL) and NaOH (0.5 M). In sequence, starch suspension was neutralized to pH 6.5 with HCl. Particles were separated by centrifugation at 3000 × g for 5 min	342	Ni	OSA: 5–20% Temperature: 35 °C Reaction time: 15 min pH: 8.5	0.032–0.117	122	[27]
Waxy rice 0.16%	Starch was hydrolyzed using pullulanase at 40 °C for 12 h	215.53	Ni	OSA: 1 mL SNPs: 2.5 g			[28]

(continued)

Table 7.2 (continued)

Starch source and amylose content	Production method	Size (nm)	Yield (%)	Modification	DS	Size (nm)	Reference
				Reaction time: 6 h Ultrasonicated conditions			
				200 W power/ 30 cycles	0.034	169	
				200 W power/ 60 cycles	0.038	116	
				400 W power/ 30 cycles	0.045	128	
				400 W power/ 60 cycles	0.051	74	
Sugary-1 maize Ni	Starch was grounded in water using a high-speed blender, The slurry was filtered through 100-mesh sieves and then centrifuged at 10,000 g for 10 min. Three volumes of ethanol were added to precipitate the soluble starch particles	Ni	Ni	OSA: 10% Reaction time: 8 h Temperature: 35 °C pH: 8.5	0.05 (5.04%)	Ni	[29]

AH acid hydrolysis, *DS* degree of substitution, *OSA* octenyl succinic anhydride, *SNPs* starch nanoparticles, *Ni* not informed

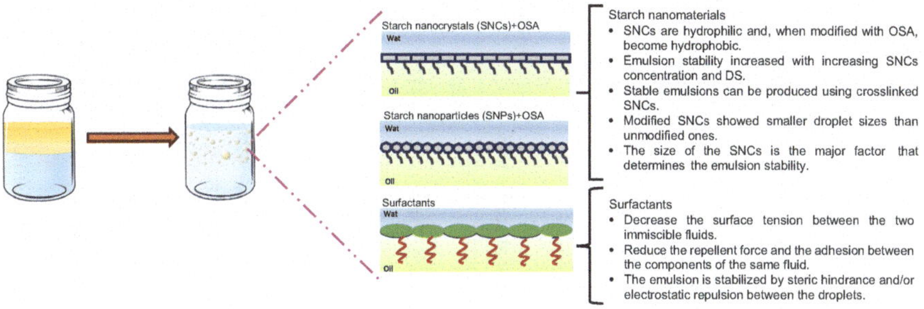

Fig. 7.1 Schematic representation of the starch nanomaterials as emulsion stabilizers

used as a reference to prepare food-grade nanocrystals and stabilize Pickering emulsions by acid hydrolysis. The relative crystallinity of SNCs increased by approximately 12% when compared to native starch. Emulsions with SNCs (1.0 wt%) were prepared with an increased oil phase volume fraction (0.10–0.90) and pH 7.0. The ionic strength increased from 0 to 400 mM and reduced the negative ζ-potential of particles in starch nanocrystal suspensions. The concentration of SNCs in the continuous phase of 1.0 wt% and oil internal phase (ϕ) of 0.75–0.85 led to the formation of gel-like structures with greater stiffness. No creaming occurred even after 30 days of storage, and values of ϕ lower 0.70 phase separation were found, along with a cream layer on top and an aqueous phase at the bottom, where the droplets coated with the SNCs flocculated. The concentration of SNCs in the continuous phase at ϕ value affects the droplet size and properties of high-internal phase emulsions. Emulsions were stable when different starch nanocrystal concentrations (0.5%, 1.0%, 2.0%, and 4.0%) were used, and no changes occurred after 15 days of storage; a minimal concentration of 1.0 wt.% is necessary for gel-like Pickering emulsion.

The structures of SNCs and starch nanomaterials in general are analyzed to determine how structure types alter emulsification properties [30]. Unconventional starches have been used to prepare nanocrystals and stabilize emulsions. Sago (*Metroxylon sago*) was used for starch isolation and the preparation of nanocrystals by acid hydrolysis. Sago SNCs stabilized Pickering emulsions at a 1:1 water-oil ratio; the SNC concentration ranged between 0.5% and 4.0% (w/w) of the total mass. The volume fraction of the emulsified phase decreased slowly after emulsification and remained stable for several weeks. The concentration of 3.5% w/w SNCs showed good stability and no indications of creaming after 2 months of storage at room temperature [31]. The concentration of sago SNCs for the preparation of Pickering emulsions was 1%, 2%, 3%, and 4% w/w. The mean droplet diameter over volume (d_{43}) decreased from 27–40 to 24–36 μm as the nanocrystal concentration increased. Emulsion stability is evaluated by the creaming index, where phase separation occurs due to physical instability caused by gravitational force. The creaming index was higher in the presence of low concentrations of oil and nanocrystals; however, the effect was not observed using the highest amount of

Table 7.3 Emulsion stability properties using unmodified and modified starch nanocrystals

Source of SNCs, amylose content and particle size	Emulsion type	Emulsion stability properties	Reference
Native or unmodified SNCs			
Waxy maize 2% Ni	Pickering O/W emulsion using high internal phase emulsions (HIPEs) Storage time: 15 days SNCs: 0.5–4% Ratio oil/dispersion SNCs: 0.1–0.9 Ionic strengths: 0–400 mM	HIPEs as a gel-like gel network with tunable viscoelastic properties in the system, usually characterized by a minimal internal phase ratio (ø) of 0.74 Concentrations of SNCs between 1% and 4% can form a highly packed and connected state for HIPEs. The higher concentration of SNCs produces a decreased droplet size. The droplet size of all the emulsions was around 50 μm Gel-like Pickering emulsions were formed when ø value was between 0.7–0.9 and they were stable after storage for up to 1 month The droplets became more tightly packed as the ø increased Rheological properties indicated that emulsion gels had an elastic behavior Gel emulsions were stable and strengthening increasing to pH 5–10 To evaluate the ionic strengths on stable HIPE formation was found that stable and gel-like HIPEs values of 0–20 mM. At values of 30–60 mM, the formed Pickering HIPEs were in a liquid state	[30]
Sago 24.3% 20–100 nm	Pickering O/W emulsion (50:50 v:v) SNCs: 0.5 and 4% Storage: 2 months	The emulsification index increased from 0.8 to 1.0 with the presence of SNCs The amount of SNCs is a major factor that determines the quality and characteristics of the Pickering emulsion obtained	[31]
Sago 24% 20–150 nm	Pickering O/W emulsion SNCs: 1–4% Storage: 4 weeks	Various droplet sizes in the range between 0.5 and 50 nm Decreased the average droplet size diameter with the incorporation of SNCs. Dispersion of SNCs was like a skeletal structure that restricts droplet movement Emulsions containing 1%, 2% and 3% of SNCs were stable after 2 h, while the emulsion with 4% SNCs was stable during 4 weeks Creaming was reduced when an extra amount of SNCs was used The higher concentration of SNCs produced higher yield stress, the emulsion was less susceptible to phase separation The flow index was less than 1 indicating that emulsions have shared thinning properties The apparent viscosity increased with the presence of SNCs, improving emulsion	[32]

<table>
<tr><td></td><td></td><td colspan="5">stability because SNCs avoid droplet mobility
The results show that the amount of SNCs is a major factor that determines the quality and characteristics of the Pickering emulsion obtained</td><td></td></tr>
<tr><td rowspan="5">Barley
28.6%
141 nm</td><td>Pickering O/W emulsion (30:70 v/v)
Storage: 24 days</td><td colspan="5">Droplet size and apparent viscosities of the emulsions increased with the content of nanocrystals. ζ-potential values indicate that the surface charge of the droplets prevents their aggregation. Emulsified phase by increasing the content of stabilizing particles in their formulation. The increase in SNC concentration was inversely proportional to the turbiscan stability index and polydispersity index values, thus associated with the higher emulsion stability
All emulsions increased their viscosity as a function of SNCs concentration and exhibited higher values of the storage modulus (G′) than those for the loss modulus (G″) behaviors similar to a gel-like structure</td><td rowspan="5">[33]</td></tr>
<tr><td>Concentration of SNCs (%)</td><td>Droplet diameter (μm)</td><td>Emulsification index</td><td>ζ-potential (mV)</td><td>Turbiscan stability index</td><td>Polydispersity index</td></tr>
<tr><td>1</td><td>8.4</td><td>0.677</td><td>−32.6</td><td>21.2</td><td>4.49</td></tr>
<tr><td>2</td><td>3.2</td><td>0.773</td><td>−30.7</td><td>18.0</td><td>2.14</td></tr>
<tr><td>4</td><td>1.2</td><td>0.976</td><td>−31.9</td><td>4.60</td><td>1.24</td></tr>
<tr><td colspan="8">Modified SNCs</td></tr>
<tr><td>Waxy maize
8%
452 nm
Amaranth
14%
347 nm</td><td>Pickering
O/W (30:70 v/v)
emulsion
Modified SNCs with OSA: 4%
Storage: 7 days
Waxy maize | DS: 0.012
Amaranth | DS: 0.0009</td><td colspan="5">Modified SNCs showed minor droplet sizes than unmodified
Droplet sizes decreased when increased DS (unmodified SNCs stabilized emulsions with droplet size 36 nm until 0.7 nm when OSA SNCs stabilized emulsions with DS 0.045)
OSA amaranth SNCs showed higher ζ-potential (−23 mV) than OSA Waxy maize SNCs, (−9 mV) indicating OSA amaranth SNCs stabilized emulsions had more stables. The size of SNCs is the major factor that determines the emulsion stability</td><td>[18]</td></tr>
</table>

(continued)

Table 7.3 (continued)

Native or unmodified SNCs					
Source of SNCs, amylose content and particle size	Emulsion type		Emulsion stability properties		Reference
Normal Maiz 125 nm	Pickering O/W (30: 70 v/v) emulsion SNCs modified with OSA: 4% Storage: 7 days		Droplet sizes decreased with the increase in DS Emulsion viscosity increased with the modification of SNCs with OSA		[20]
			Droplet size (μm)		
			Unmodified SNCs	36	
			DS: 0.013	1.02	
			DS: 0.018	0.90	
			DS: 0.045	0.7	
Native maize Lateral size: 100 nm Thick ness: 8–10 nm SHP: Sodium hexameta-phos-phate Dex: Dextran PEG: Polyetylene glycol	Pickering O/W emulsion Crosslinking with SHP Storage time: 24 h Aqueous two-phase system Dex : PEG (7:3)		Modified SNCs have more affinity to dextran than polyethylene glycol free energy Emulsion stability increased with the DS in SNCs Crosslinked SNCs had higher viscosity and stronger modulus responses when compared with unmodified SNCs stabilizing emulsions		[21]
	SHP (% w/w)	DS			
	9	0.0023			
	18	0.0040			
	27	0.0065			
	Pickering O/W emulsion Acetylation DS: 0.18 Storage time: 24 h Aqueous two-phase system (Dex : PEG)		Acetylated SNCs had less affinity to aqueous phases		

Sago 24% 40–100 nm	Pickering O/W (1:3) emulsion Lauroyl chloride DS: 0.122 Storage: 24 h	Decreased the droplet size and emulsification index with the incorporation of SNCs. Emulsions were most stable with the incorporation of SNCs		[22]
	Concentration of SNCs (%)	Drople size (μm)	Emulsification index	
	0.5	3–10	0.5	
	1.0	2–8	0.85	
	2.0	1–5	1.0	
	Unmodified SNCs: 2%	2–8	0.8	

DS degree of substitution, *SNCs* starch nanocrystals, *O/W* oil-in-water, *Ni* not informed

SNCs (4% w/w) regardless of oil concentration (5%, 15%, 25%, and 25%) [32]. Sago SNCs are then potential stabilizers, fillers, and emulsifiers for colloidal systems.

Starches of small granule sizes, such as amaranth and barley, have been used to produce nanocrystals by acid hydrolysis. It is well known that smaller granule sizes provide good emulsifying characteristics in Pickering emulsions. These starches, along with their high amylopectin contents, yield large amounts of nanocrystals by acid hydrolysis in short time. Pickering emulsions were prepared with different nanocrystal concentrations (1%, 1.5%, 2%, 3%, and 4%). The dispersed phase was prepared with 14 mL of water, and oil was added to reach a 70:30 ratio at 25 °C. The emulsion with the lowest nanocrystal concentration also showed the lowest emulsification index, which increased with nanocrystal concentration. The turbiscan stability index (TSI), which is related to all the instabilities (changes in droplet size, creaming, sedimentation) detected in the system, decreases in the presence of higher nanocrystal concentrations. No creaming was observed was not observed in the emulsion with the highest nanocrystal concentration after 7 days of storage [33]. The SNCs produced from amaranth starch were used to stabilize Pickering emulsions [17], and nanocrystals from waxy starch were prepared following the same procedure and used as reference. The preparation of both SNCs from different botanical origins is important for comparison. The emulsification indexes of amaranth nanocrystals and waxy maize nanocrystals remained stable throughout the 10 days of the test [18].

7.3.2 Modified Starch Nanocrystals

SNCs have been modified to produce hydrophobic and more stable O/W Pickering emulsions. Chemical modifications of starch, such as crosslinking, acetylation, and esterification, have been reported. Esterification with OSA produces amphiphilicity in starch and nanocrystals, both of which are recommended to stabilize emulsions. The DS characterizes the chemical modifications and is difficult to determine in modified nanocrystals using the traditional titration method used for starch; therefore, ^{1}H NMR is suggested as an alternative. Nanocrystals have been prepared by acid hydrolysis of maize and esterified with OSA. The DS values were 0.013, 0.018, and 0.045, as determined by ^{1}H NMR. The emulsions were elaborated with caprylic/capric triglycerides and an oil/water ratio of 30:70, and the nanocrystal concentration was 0.2–5.0 wt%, relative to the water phase. The emulsifying capacity of nanocrystals increased with the DS, and creaming was reduced as the concentration of nanocrystals increased. Creaming was not present in the emulsion with the highest DS (0.045), the smallest droplet size, and nanocrystal concentrations of 3%, 4%, and 5% [20]. Amaranth SNCs have been esterified with OSA and used to stabilize emulsions. The DS was not determined, and a qualitative test by Fourier-Transformed Infrared Spectroscopy (FTIR) was used to observe the OSA group. The emulsions were prepared with 70:30 water/canola oil and 100 mg nanocrystals/

mL oil, and they were stored for 10 days. The emulsification index of the sample with OSA amaranth nanocrystals was 1.0 and did not change throughout the studied period. The droplet size of the emulsion was smaller than that of the emulsion prepared with OSA maize nanocrystals because it was more stable. The emulsification index was improved by esterification because the non-modified amaranth nanocrystals showed a higher value. The ξ-potential of the emulsion with OSA amaranth nanocrystals presented a higher absolute value (approximately double) than that of the emulsion added with OSA maize nanocrystals. This provides information on the repulsive forces between the oil droplets covered by particles. OSA amaranth nanocrystals show good characteristics for stabilizing emulsions [18]. The cross-linking of starch is a chemical modification that stabilizes starch chains due to intra- and inter-hydrogen bonding between chains by covalent bonds. Esterification with acetic anhydride is a modification by substitution/stabilization where the OH groups of starch chains are substituted, and the esterified starch is dispersed under heating. Both chemical modifications are used in starch nanocrystals to improve the emulsifier properties. Cross-linking is carried out with sodium hexametaphosphate at three different levels of substitution, while acetic anhydride is used in esterification. Emulsions were prepared with three cross-linking nanocrystals and eight concentrations (0.18–5.08 wt%). The increase in nanocrystal concentration and DS produced stable emulsions because no creaming was observed when the highest DS and the six highest concentrations were used. Acetylated nanocrystals produce more stable emulsions than cross-linking nanocrystals at the highest DS. The former did not exhibit coalescence or flocculation, and their droplet size was smaller than that of cross-linking nanocrystals [21]. The type of chemical modification is key to stabilizing Pickering emulsions using SNCs and it deserves further research.

7.3.3 Starch Nanoparticles

Unmodified particles from different starch sources have been used to stabilize Pickering emulsions (Table 7.4). The size of the unmodified maize nanoparticles is a function of the medium pH, and nanoparticle size decreased from 131.9 to 12.6 nm when the pH increased from 3 to 10. It is important to note that transmission electron microscopy showed aggregates under acidic conditions, which is an important fact in emulsions with acidic characteristics. The droplet size of the emulsion is essential for decreasing coalescence and creaming. The addition of starch nanoparticles between 1% and 5%, based on the total weight of the emulsion, rapidly decreased the droplet size, and a high starch nanoparticle concentration (15%) did not produce a significant reduction in the droplet size.

A starch nanoparticle concentration of 5% was used to evaluate the effect of pH on the emulsification properties; the mean size of the droplets with the 5% concentration was approximately 18 nm. The acid pH (3 and 4) produces a homogeneous emulsion immediately after preparation, and at higher pH (up to 4) a dark red layer

Table 7.4 Emulsion stability properties using unmodified and modified starch nanoparticles

Unmodified SNPs				
Source of SNPs, amylose content and particle size	Emulsion type	Emulsion stability properties		Reference
Maize 24% 12.6–131.9 nm in function of pH	O/W emulsion with 10% of oil based on the total weight of emulsion	SNPs: 0–20% pH: 7	Emulsion droplet size decreased from 40 μm to 10 μm with the concentration of SNPs. SNPs > 5% did not impact on the emulsion droplet size	[34]
		pH: 3–10 SNPs: 5%	The droplet size (30–40 μm) was not altered with pH modification Pickering emulsion stability decreased at alkaline conditions when storage time increased The higher values of ζ-potential indicated that pH increases lead to repulsion between SNPs, producing a thin coating layer due to the dissolution of SNPs, creating an unstable emulsion	
Native maize	O/W emulsion SPNs and oil content (%) of the total weight of emulsion Storage time: 24 h	Particle size: 20,000–15 nm	Emulsions were more stable when the particle size decreased The smaller particle size produced higher values of contact angles indicating major emulsion stability	[35]
		SNPs: 0–15% Oil: 4 g	Emulsion stability increased with the incorporation of SNPs When emulsions were subject to a centrifugal force, the phase separation was observed with slight	

(continued)

Table 7.4 (continued)

Unmodified SNPs				
Source of SNPs, amylose content and particle size	Emulsion type	Emulsion stability properties		Reference
			differences as a function of starch nanoparticle content Emulsion stabilized with SNPs $\leq$ 33.5 nm showed an emulsion index close to 1	
		Oil: 30–70% SNPs: 1 g	Emulsion stabilized with SNPs $\leq$ 33.5 nm showed constant values of emulsion index (0.9 approximately) with oil concentration between 30% and 65%	
		Heating and freeze-thawing SNPs: 1 g Oil: 4 g	It was observed an accelerated separation of the oil phase in emulsion stabilized with SNPs > 33.5 nm When the emulsion was stabilized with SNP $\leq$ 30 nm, the heating, and freeze-thawing did not affect its stability	
Tapioca (18.9% and 216 nm) Waxy corn (98.8% and 227 nm) Corn (26.4% and195 nm) Sweet potato (20.6% and 184 nm)	Ni	Storage stability	The low creaming index shows that tapioca, sweet potato, and corn SNPs form effective barriers to stabilized emulsions Spherical droplets were produced with sizes between 20 and 40 μm SNPs from waxy corn showed strong hydrophilicity and then poor emulsion stability SNPs from sweet potatoes produced the most stable emulsion	[36]

(continued)

Table 7.4 (continued)

Unmodified SNPs				
Source of SNPs, amylose content and particle size	Emulsion type	Emulsion stability properties		Reference
Native corn and sweet potato	SNPs: 1% Oil: 0.5%	Corn SNPs had particle size between 59 and 685 nm, whereas sweet potato SNPs had particle size varying between 51 and 712 nm	Emulsion stabilized with SNPs having a particle size between 100 and 220 nm showed a low creaming index	
Native corn	SNPs: 1% Oil fraction: 0.24–0.75	Storage time: 1 month	After 1 month, a remarkable increase in emulsified phase volume was observed The paste volume increased with increasing the oil fraction from 0.25 to 0.5 after a 3 h storage	
	SNPs: 0–2% Oil fraction 0.5	The cream volume increases with the incorporation of SNPs. The creaming index remained at 0% and droplet size was 19 μm when the concentration of SNPs was 2.0% In all cases, emulsion viscosity decreased with the increased shear rate and increased with the concentration of SNPs		
	SNPs: 2% Oil fraction: 0.5 Storage time: 24 h	Heating: 40–90 °C Time: 1–30 min	Temperatures ≥60 °C produce a swelling of SNPs and increase the droplet size emulsion The heating treatment time did not significantly affect the droplet size The emulsion heat-treated were stables for 24 h	
	SNPs: 1% Oil fraction: 0.5 Storage time: 3 days	Ionic strength: 10–90 mM	Stability of emulsion was not affected with addition of NaCl	

(continued)

Table 7.4 (continued)

Unmodified SNPs				
Source of SNPs, amylose content and particle size	Emulsion type	Emulsion stability properties		Reference
Modified SNPs				
Sugary-1 maize	OSA-SNP Storage time: 30 days Medium-chain triacylglycerol (MCT)	SNPs: 0.05–5% MCT: 50% pH: 7	With the increase in concentration of nanoparticles, emulsion index increased from 0.6% to 1.0% With a 3% of SNPs, emulsion index had the higher value [1] Droplet size decreased as increased the percent of SNPs	[29]
		CMT: 5–50% pH: 7 SNPs: 1%	The emulsion index and droplet size of the emulsion increased with the MCT addition. Higher droplet size was obtained with 50% MCT	
		pH: 3.1–11.5 SNPs: 1% MCT: 50%	At pH value below 8.0, the emulsion index and droplet size decreased, and emulsion instability at alkaline conditions could be related with SNPs degradation	
		SNPs: 1.0% MCT: 50% pH: 7	The emulsion display shear thinning behavior, due to breaking of an entangled chains SNPs during shearing Pickering emulsion showed a gel-like structure Droplet size distribution showed a single peak with an average size of 29 μm Emulsion showed high ζ Potential (−	

(continued)

Table 7.4 (continued)

Unmodified SNPs				
Source of SNPs, amylose content and particle size	Emulsion type	Emulsion stability properties		Reference
			37 mV), its stability was improved by a repulsion among droplet of emulsion	
Corn, potato and pea	OSA-SNPs SNPs: 2% Oil: 10% Storage time: 30 days	Different starch sources	The emulsifying activity index and emulsifying stability index values demonstrating that OSA-SNPs resulted in stable emulsion than unmodified SNPs OSA potato SNPs showed major capacity to stabilize emulsions than OSA corn and OSA pea SNPs. These results could be related with higher DS and smaller droplet size of Potato SNPs (0.0199; 2 μm) compared with corn SNPs (0.0178; 3 μm) and pea SNPs (0.0122; 5 μm)	[25]
		pH: 1–11	The OSA-SNPs had improved hydrophobic properties ζ-potential showed an increase with the increasing of pH, and this tendency was a remarkable when SNPs were OSA modified Emulsion stabilized by OSA-SNPs were more stable to pH 5 and more notary OSA potato SNPs. A very acid (pH < 3) and alkaline (pH > 5) condition could be	

(continued)

Table 7.4 (continued)

Unmodified SNPs				
Source of SNPs, amylose content and particle size	Emulsion type	Emulsion stability properties		Reference
			degrading SNPs and decreasing surface coverage oil droplets destabilizing the emulsion	
		Ionic strength: 0.1–0.9 mol/L	ζ-potential decreased with the increasing of ionic strength At low ionic strength droplets remain disaggregated by formation of particle bridges between them Gradually the degree creaming increased as ionic strength increases	
		Temperature: 30–90 °C	Emulsions with OSA-SNPs had small uniform droplets between 30 and 50 °C OSA-SNPs emulsion (heat-treated) have good stability against temperature	
Quinoa 6.3% 132 nm Maize 25.8% 168 nm Potato 24.2% 295 nm	OSA-SNPs SNP: Oil 0.25 w/v % Storage time: 30 days	Higher emulsification index, smaller droplets size and ζ-potential values (−28 to −31 mV), indicate that smaller SNPs produced major emulsion stability, and modification of SNPs with OSA enhancement this property OSA-SNPs from quinoa showed major capacity to stabilize emulsion than OSA-SNP from maize and potato, because form smaller droplets Higher shear rate produced a decrease in the viscosity of emulsion Rheological properties showed that all emulsion form an elastic gel-like, this network packed oil droplets improving its stability at 30 days		[26]

(continued)

Table 7.4 (continued)

Unmodified SNPs				
Source of SNPs, amylose content and particle size	Emulsion type	Emulsion stability properties		Reference
High amylose maize 60% 122 nm	OSA-SNP : Oil 1 : 1 Storage time: 2 months	Concentration of SNPs (1–7%)	Unmodified SNPs could be used to stabilize emulsions at high concentration of SNPs (≥5%) When SNPs were modified, a lower concentration was necessary to stabilize the emulsions (3% OSA-SNPs)	[27]
		Freeze thaw −20 °C/8 h −25 °C/3 h 5 cycles	Freeze-thaw produces phase separation in most emulsions. Emulsion stabilized with 1% of OSA-SNPs (DS 0.117) showed a low freeze-thaw stability due to foaming formation	
		Thermal stability Incubated 30, 40, 50, 60 y 70 °C/4 h 25 °C/ 3 h	Thermal stability of OSA-SNPs at concentration > 1% was high	

DS degree of substitution, *SNPs* starch nanoparticles, *O/W* oil-in-water, *Ni* not informed

appeared in the prepared emulsions. At the lowest pH (3), two layers were observed: one homogeneous and the other a serum layer. Both the droplet size and pH increased during storage for 7 days, indicating that storage stability was reduced as the pH increased. This indicates that alkaline conditions reduce the stability of emulsions containing starch nanoparticles. The use of starch nanoparticles to stabilize emulsions has limitations that may restrict their application in specific emulsions [34]. Regular and waxy starches have been used to prepare nanoparticles and stabilize Pickering emulsions. The mean sizes of the regular and waxy starch nanoparticles were 196 and 228 nm, respectively. The mean droplet diameter of regular maize nanoparticles was 32.2 μm, but waxy nanoparticles were unstable, and their strong hydrophobicity prevented the completion of the analysis. The creaming index of the emulsions stabilized with waxy nanoparticles was higher than that of regular nanoparticles, and it was not determined after 1 day of storage. Emulsions added with regular nanoparticles showed an increase in the creaming index after 1 day of storage, indicating phase separation; however, no changes in the index were observed after storage for 1 month. This shows the stability of the emulsions added with regular SNPs [36].

Unconventional SNPs from tapioca and sweet potato were used to stabilize Pickering emulsions. Both starch nanoparticles showed mean sizes similar to that of maize starch, but the creaming index of the emulsion added with sweet potato nanoparticles was the lowest, followed by that of tapioca starch nanoparticles. Both emulsions were more stable than those containing maize starch nanoparticles. The results show the importance of using starches to prepare nanoparticles with improved functionality to stabilize Pickering emulsions [36].

Starch nanoparticles have been modified to increase the amphiphilicity and improve the stability of Pickering emulsions. Esterification with OSA is widely used in starches and starch nanoparticles. The emulsifying activity index (EAI) and emulsifying stability index (ESI) of the emulsions added with esterified nanoparticles were higher than those of their non-esterified counterparts, and esterified potato starch nanoparticles showed the highest values that were associated with the highest DS. The stability of the emulsions was assessed after 5 weeks by determining the average droplet size. In general, the droplet size of the emulsions increased together with the storage time. The esterified potato starch nanoparticles showed a minor increase, as the esterified pea starch nanoparticles exhibited the highest increase. This pattern was associated with minor creaming (phase separation) in the emulsion containing with esterified potato starch nanoparticles. The emulsion stability with esterified nanoparticles occurs because the hydrophobic sites in OSA nanoparticles penetrate the oil phase more, and the hydrophilic sites of nanoparticles are oriented toward the water phase, producing steric stabilization and long-term emulsion stability. The increase in the DS produces more stable emulsions during storage [25]. The OSA-modified SNPs measuring 132 nm (quinoa), 168 nm (maize), and 295 nm (potato) were used to stabilize Pickering emulsions. The smaller SNP size produced more stable emulsions than the larger SNPs because the smaller droplet size reduced creaming.

7.4 Future Trends

Several studies have investigated the use of traditional and conventional starch sources, such as maize, potato, wheat, and rice, in the production and modification of nanomaterials. The use of these materials to stabilize Pickering emulsions has also been analyzed. The effects of granule size, amylose/amylopectin ratio, and crystalline arrangement are yet to be systematically evaluated, representing an opportunity for further research in this field. Similarly, more studies should be dedicated to the use of unconventional starches from plantain, taro, and sweet potato to produce nanomaterials and increase emulsion stability. The structure–emulsifier relationship, which identifies the chain-length distribution of the nanomaterials, can provide more information to produce nanomaterials with better emulsification properties. Finally, the use of green methods to modify starch nanomaterials and their effect on the stability of Pickering emulsions should also be analyzed.

7.5 Conclusions

Modified and unmodified starch nanocrystals and nanoparticles exhibit functional characteristics that stabilize Pickering emulsions. The small size of both nanomaterials improves functionality compared with starch granules, although it has been suggested that starches with a small granule size (quinoa, amaranth, barley, rice) can stabilize emulsions, a capability increased by esterification with octenyl succinic anhydride (OSA). Although acid treatment is preferred to produce nanocrystals and nanoparticles, other treatments, as ball milling, that reduce the granule size are also used. The treatments reduce the effect of the granule size because large, medium, and small starch granules are used without evident effects in the size of the nanomaterials. OSA esterification has been improved by ultrasound during starch modification. The use of starch nanomaterials to stabilize Pickering emulsions although barely studied, is an area of opportunity in the food industry.

Acknowledgments The authors gratefully acknowledge the Instituto Politécnico Nacional, CEPROBI for the support.

Conflicts of Interest The authors declare no conflicts of interest.

References

1. Rayner M, Marku D, Eriksson M, Sjöö M, Dejmek P, Wahlgren M. Biomass-based particles for the formulation of Pickering type emulsions in food and topical applications. Colloids Surf A Physicochem Eng Asp. 2014;458:48–62.
2. Tavernier I, Wijaya W, Van der Meeren P, Dewettinck K, Patel AR. Food-grade particles for emulsion stabilization. Trends Food Sci Technol. 2016;50:159–74.
3. Timgren A, Rayner M, Sjöö M, Dejmek P. Starch particles for food based Pickering emulsions. Procedia Food Sci. 2011;1:95–103.
4. Rayner M, Sjöö M, Timgren A, Dejmek P. Quinoa starch granules as stabilizing particles for production of Pickering emulsions. Faraday Discuss. 2012;158(0):139–55.
5. Timgren A, Rayner M, Dejmek P, Marku D, Sjöö M. Emulsion stabilizing capacity of intact starch granules modified by heat treatment or octenyl succinic anhydride. Food Sci Nutr. 2013;1 (2):157–71.
6. Marefati A, Rayner M. Starch granule stabilized Pickering emulsions: an 8-year stability study. J Sci Food Agric. 2020;100(6):2807–11.
7. Agama-Acevedo E, Bello-Perez LA. Starch as an emulsions stability: the case of octenyl succinic anhydride (OSA) starch. Curr Opin Food Sci. 2017;13:78–83.

8. Lopez-Silva M, Bello-Perez LA, Agama-Acevedo E, Alvarez-Ramirez J. Effect of amylose content in morphological, functional and emulsification properties of OSA modified corn starch. Food Hydrocoll. 2019;97:105212.
9. Sun Q. Chapter 18—Starch nanoparticles. In: Sjöö M, Nilsson L, editors. Starch in food. 2nd ed. Duxford: Woodhead Publishing; 2018. p. 691–745.
10. Angellier H, Molina-Boisseau S, Dole P, Dufresne A. Thermoplastic starch−waxy maize starch nanocrystals nanocomposites. Biomacromolecules. 2006;7(2):531–9.
11. Zhou J, Tong J, Su X, Ren L. Hydrophobic starch nanocrystals preparations through crosslinking modification using citric acid. Int J Biol Macromol. 2016;91:1186–93.
12. Xu Y, Ding W, Liu J, Li Y, Kennedy JF, Gu Q, et al. Preparation and characterization of organic-soluble acetylated starch nanocrystals. Carbohydr Polym. 2010;80(4):1078–84.
13. Thielemans W, Belgacem MN, Dufresne A. Starch nanocrystals with large chain surface modifications. Langmuir. 2006;22(10):4804–10.
14. Alila S, Aloulou F, Thielemans W, Boufi S. Sorption potential of modified nanocrystals for the removal of aromatic organic pollutant from aqueous solution. Ind Crop Prod. 2011;33(2): 350–7.
15. Habibi Y, Dufresne A. Highly filled bionanocomposites from functionalized polysaccharide nanocrystals. Biomacromolecules. 2008;9(7):1974–80.
16. Namazi H, Dadkhah A. Convenient method for preparation of hydrophobically modified starch nanocrystals with using fatty acids. Carbohydr Polym. 2010;79(3):731–7.
17. Sanchez de la Concha BB, Agama-Acevedo E, Nuñez-Santiago MC, Bello-Perez LA, Garcia HS, Alvarez-Ramirez J. Acid hydrolysis of waxy starches with different granule size for nanocrystal production. J Cereal Sci. 2018;79:193–200.
18. Sánchez de la Concha BB, Agama-Acevedo E, Agurirre-Cruz A, Bello-Pérez LA, Alvarez-Ramírez J. OSA esterification of amaranth and maize starch nanocrystals and their use in "Pickering" emulsions. Starch – Stärke. 2020;72(7–8):1900271.
19. Hao Y, Chen Y, Li Q, Gao Q. Synthesis, characterization and hydrophobicity of esterified waxy potato starch nanocrystals. Ind Crop Prod. 2019;130:111–7.
20. Qian X, Lu Y, Ge L, Yin S, Wu D. Starch nanocrystals as the particle emulsifier to stabilize caprylic/capric triglycerides-in-water emulsions. Carbohydr Polym. 2020;245:116561.
21. Qian X, Peng G, Ge L, Wu D. Water-in-water Pickering emulsions stabilized by the starch nanocrystals with various surface modifications. J Colloid Interface Sci. 2022;607:1613–24.
22. Ahmad A, Fazial FF, Khalil HPSA, Fazry S, Lazim A. Synthesis and characterization of sago starch nanocrystal laurate as a food grade particle emulsifier. Int J Biol Macromol. 2023;242: 124816.
23. Ren L, Dong Z, Jiang M, Tong J, Zhou J. Hydrophobization of starch nanocrystals through esterification in green media. Ind Crop Prod. 2014;59:115–8.
24. Saari H, Heravifar K, Rayner M, Wahlgren M, Sjöö M. Preparation and characterization of starch particles for use in Pickering emulsions. Cereal Chem. 2016;93(2):116–24.
25. Chen H, Jiang Y, Zhang B, Fang Y, Lin Q, Ding Y. Application of Pickering emulsions stabilized by corn, potato and pea starch nanoparticles: effect of environmental conditions and approach for curcumin release. Int J Biol Macromol. 2023;238:124115.
26. Remanan MK, Zhu F. Encapsulation of rutin in Pickering emulsions stabilized using octenyl succinic anhydride (OSA) modified quinoa, maize, and potato starch nanoparticles. Food Chem. 2023;405:134790.
27. Kou T, Faisal M, Song J, Blennow A. Stabilization of emulsions by high-amylose-based 3D nanosystem. Food Hydrocoll. 2023;135:108171.
28. Gautam G, Talukdar D, Mahanta CL. Sonochemical effect on the degree of substitution of octenyl-succinic anhydride into waxy rice starch nanoparticles and study of gastro-intestinal hydrolysis using INFOGEST in vitro digestion method. Food Res Int. 2023;165:112348.
29. Ye F, Miao M, Jiang B, Hamaker BR, Jin Z, Zhang T. Characterizations of oil-in-water emulsion stabilized by different hydrophobic maize starches. Carbohydr Polym. 2017;166: 195–201.

30. Yang T, Zheng J, Zheng B-S, Liu F, Wang S, Tang C-H. High internal phase emulsions stabilized by starch nanocrystals. Food Hydrocoll. 2018;82:230–8.
31. Azfaralariff A, Fazial FF, Sontanosamy RS, Nazar MF, Lazim AM. Food-grade particle stabilized Pickering emulsion using modified sago (Metroxylon sagu) starch nanocrystal. J Food Eng. 2020;280:109974.
32. Azfaralariff A, Farahfaiqah F, Joe LS, Fazry S, Mohamed M, Nazar MF, et al. Sago starch nanocrystal-stabilized Pickering emulsions: stability and rheological behavior. Int J Biol Macromol. 2021;182:197–206.
33. López-Hernández RE, García-Solís SE, Monroy-Rodríguez I, Cornejo-Mazón M, Calderón-Domínguez G, Alamilla-Beltrán L, et al. Preparation and characterization of canola oil-in-water Pickering emulsions stabilized by barley starch nanocrystals. J Food Eng. 2022;326:111037.
34. Ko EB, Kim J-Y. Application of starch nanoparticles as a stabilizer for Pickering emulsions: effect of environmental factors and approach for enhancing its storage stability. Food Hydrocoll. 2021;120:106984.
35. Choi H-D, Hong JS, Sm P, Ko E, Shin H-Y, Kim J-Y. Starch nanoparticles produced via acidic dry heat treatment as a stabilizer for a Pickering emulsion: influence of the physical properties of particles. Carbohydr Polym. 2020;239:116241.
36. Ge S, Xiong L, Li M, Liu J, Yang J, Chang R, et al. Characterizations of Pickering emulsions stabilized by starch nanoparticles: influence of starch variety and particle size. Food Chem. 2017;234:339–47.

Chapter 8
Starch-Based Ingredients and Functional Foods

Sarah L. Paz-Arteaga, Anna María Polanía, Natalia S. Vargas-Marín, Valeria Cedeño-Fierro, Juan F. Grisales-Mejia, Nathiely Ramírez-Gúzman, and Cristian Torres-León

8.1 Introduction

Foods have served as a valuable tool in preventing and treating diseases in ancient civilizations because of their distinct properties linked to specific compounds. They play a primary role in human nutrition and also possess various beneficial qualities [1].

Functional ingredients are part of the formulation of a food product and have the special characteristics of providing additional health benefits beyond their basic nutritional value. These ingredients are designed to improve the nutritional and technological quality of food, promote health, and participate in disease prevention [2]. Various compounds can be considered functional ingredients, and their structure plays a fundamental role in their potential application, e.g., vitamins such as vitamin A and their influence on vision, lipid metabolism, immune function, and modulation of the gut microbiota [3, 4]; soluble and insoluble fibers and their impact on digestive

S. L. Paz-Arteaga
Faculty of Sciences, Universidad Nacional de Colombia, Medellín, Antioquia, Colombia

A. M. Polanía
Faculty of Engineering, School of Food Engineering, Universidad del Valle, Tuluá, Valle del Cauca, Colombia

N. S. Vargas-Marín
Agroindustrial Engineering, Faculty of Engineering, Corporación Universitaria Minuto de Dios – UNIMINUTO, Buga, Valle del Cauca, Colombia

V. Cedeño-Fierro
Faculty of Engineering and Administration, Universidad Nacional de Colombia, Palmira, Valle del Cauca, Colombia

J. F. Grisales-Mejia
Centro de Desarrollo Tecnológico Agroindustrial, Universidad Tecnológica de Pereira, Pereira, Colombia

G. Ayala Valencia (ed.), *Starch Nanomaterials and Food Applications*,
https://doi.org/10.1007/978-3-031-60086-9_8

health [5–7]; essential fatty acids such as omega-3 and their impact on visual, cardiovascular, brain, and neurological health [8, 9]; probiotics such as certain lactobacilli and their immune system reinforcement, as well as their role in maintaining a healthy balance of gut microbiota [10, 11]; and phytochemicals present in a variety of fruits and vegetables that have antioxidant, anti-inflammatory, anticancer, and neuroprotective properties [12, 13].

Among the group of functional ingredients most commonly used in the food industry, those based on starches are prominent. These ingredients provide specific benefits, such as improving texture and even digestive health. They can be categorized based on their digestibility into three categories: rapidly digestible starches, digested within 20 min; slowly digestible starches, digested between 20 and 120 min [14]; and resistant starches, fermented in the large intestine [15].

Resistant starch (RS), which is not digested in the small intestine, can be fermented by intestinal flora in the colon, converting it into short-chain fatty acids. These fatty acids have a prebiotic effect, promoting the growth of beneficial intestinal microbiota, reducing the likelihood of colon cancer, and enhancing the absorption of vital minerals [16]. Furthermore, within this group, there are five subdivisions based on specific characteristics and resistance mechanisms [17].

Type 1 resistant starch (RS1) is a physically inaccessible starch, found in coarsely ground or whole grains, seeds, and legumes, with a barrier-like cell wall that prevents hydrolysis. Type 2 resistant starch (RS2) is a raw starch with a crystalline structure, making it indigestible, as found in raw potatoes or green bananas. Type 3 resistant starch (RS3) is known as retrograded starch, where the helical organization of its amylopectin chains resists enzymatic digestion. It is present in starchy foods that are cooked and subsequently cooled, such as potatoes, rice, pasta, and oats. Type 4 resistant starch (RS4), which is achieved through chemical modification of its original structure, is commonly derived from potatoes, corn, and yams. These starches serve as functional ingredients in food formulations and have shown positive effects on carbohydrate metabolism in humans. Finally, type 5 resistant starch (RS5) results from the combination of long and unbranched starch chains with free fatty acids, forming a complex structure that is difficult to digest. Foods

N. Ramírez-Gúzman
Facultad de Ciencias Biológicas, Universidad Autonoma de Coahuila, Unidad Torreón, Torreón, Coahuila, Mexico

Agri-food and Agro-industrial Bioeconomy Research Group, Universidad Autónoma de Coahuila, Torreón, México

C. Torres-León (✉)
Reaserch Center and Ethnobiological Garden (CIJE), Universidad Autonoma de Coahuila, Unidad Torreón, Viesca, Coahuila, Mexico

Agri-food and Agro-industrial Bioeconomy Research Group, Universidad Autónoma de Coahuila, Torreón, México
e-mail: ctorresleon@uadec.edu.mx

containing naturally occurring or artificially produced amylose-lipid complexes fall into this category [17].

8.2 Functional Properties of Starch

Starches can be derived from various sources and exhibit great versatility. They possess a range of properties known as techno-functional properties, which enable them to perform specific functions in food and related products. Furthermore, their origin, genetics, structural conformation, and the proportion of amylose and amylopectin are variables that can influence these properties in different ways and make them suitable for specific product requirements.

8.2.1 Swelling and Solubilization

The structural organization of starch granules naturally renders them firm and impermeable under ambient conditions. However, when exposed to water between 30 and 60 °C, a series of interactions between the granules and water molecules occur, leading to structural modifications. This process involves water penetration into the more amorphous regions of the granule, resulting in expansion in all directions and greatly influencing the viscosity of the system [18]. It has been observed that the presence of high amylose content decreases this property, whereas high amylopectin content increases it [19]. In addition, the swelling capacity is influenced by the size of the granule and can be affected by other accompanying molecules such as lipids and proteins. Consequently, starches derived from cereals tend to exhibit lower swelling compared than those derived from tubers [20].

8.2.2 Gelatinization

Swelling is a reversible stage that precedes starch gelatinization. Once the granule initiates swelling, it reaches a point where structural changes become irreversible. This is facilitated by an increase in temperature, which promotes water absorption, leading to granule disruption and loss of birefringence [21]. This stage, during which the granule loses its molecular organization, typically occurs between 60 and 80 °C and is directly influenced by its source. However, it is important to note that this temperature range can be influenced by interactions, particularly in the amorphous regions, with other compounds. An example of such interactions is the effect of sugars on increasing the gelatinization temperature [18].

8.2.3 Retrogradation

The phenomenon in which starch molecules undergo structural rearrangements following cooling, after gelatinization, is known as retrogradation. This process involves the recrystallization of amylose and the reassociation of amylopectin. During this reorganization, amylose molecules tend to form helical structures, whereas amylopectin chains align with each other, resulting in a series of modifications in starch that can be either desirable or undesirable depending on the context [18]. Retrogradation can contribute to increased firmness and even the production of resistant starch, but it is also responsible for the staling of baked goods. The retrogradation process is influenced by the amylose/amylopectin ratio, with the former undergoing retrogradation at a shorter time scale. Regular corn starch exhibits slower retrogradation compared to starches with a higher amylose content derived from the same source [22].

8.3 Conventional Sources of Starch-Based Functional Ingredients

In recent years, there has been a growing interest in the development of functional ingredients derived from starch. Starch, a widely available and low-cost polysaccharide, is abundant in various plant sources such as cereals, tubers, and legumes. Because of its physical and chemical properties, starch can be modified to obtain ingredients with specific functional properties that can be used in a wide range of industrial food applications.

8.3.1 Corn Starch

Corn starch is a common source of starch-based functional ingredients. It is obtained from the endosperm of corn and is used in a variety of food products, including sauces, dressings, baked goods, and baby foods. Modified corn starch is also used as a thickener and stabilizer in processed foods. According to a study by Hussain et al. [23] modified corn starch has improved rheological properties, making it ideal for use in bakery products. In addition, corn starch can also act as an encapsulating agent to protect sensitive ingredients in dietary supplements [24].

8.3.2 *Potato Starch*

Potato starch is another widely used starch in the food industry. It is extracted from the parenchyma cells of potato tubers and is used as a thickener in soups, sauces, and dairy products. Potato starch also exhibits gelation properties, making it suitable for producing foods with gel-like textures, such as puddings and fillings. A recent study by Kaul et al. [25] demonstrated that potato starch can be chemically modified to improve its water-holding capacity, making it a promising functional ingredient in the food industry. In addition, potato starch can also be used as a thickening agent in pharmaceutical and food formulations [26].

8.3.3 *Rice Starch*

Rice starch is obtained from the endosperm of rice grains and is used in various industrial applications. Because of its low viscosity and excellent thermal stability, rice starch is widely used as a thickener and stabilizer in food, cosmetic, and pharmaceutical products. Recent research by Oyom et al. [27] investigated the possibility of using modified rice starch as an edible coating agent to extend the shelf life of fresh fruits. The results showed that modified rice starch had improved barrier properties, thereby helping to maintain fruit quality during extended storage periods. Furthermore, rice starch has also been used as a drug carrier in pharmaceutical formulations [28].

8.3.4 *Cassava Starch*

Cassava starch is extracted from the roots of the cassava plant. It is widely used in the food industry as a thickener, stabilizer, and gelling agent [29]. Cassava starch is also used in the production of gluten-free foods as an alternative to wheat. The latest studies on starch have investigated the possibility of using modified cassava starch as an encapsulating agent for the controlled release of bioactive compounds in pharmaceutical and food products [30]. The results showed that modified cassava starch had good retention and release capacity for the bioactive compounds. In addition, cassava starch can become a new material with antioxidant properties and applicability in functional foods, biomedical materials, and cosmetics [31].

8.4 Novel Sources of Starch-Based Functional Ingredients

In the ongoing search for new functional ingredients, starch has been the subject of numerous research and development efforts in recent years. In this text, we will describe some novel sources of starch-based functional ingredients and recent research supporting their usefulness.

8.4.1 Pea Starch

Pea starch is a novel source of starch-based functional ingredients that has gained attention recently. It is derived from peas, a legume rich in starch. Pea starch has demonstrated thickening and gelling properties, making it suitable for use in food products such as sauces, dressings, and baked products [32]. Furthermore, recent studies have shown that pea starch may have prebiotic properties, promoting the growth of beneficial bacteria in the gut [16].

8.4.2 Green Banana Starch

Green banana starch is another novel source of starch-based functional ingredients. It is obtained from unripe bananas, which contain resistant starch, a type of starch that is not digested in the small intestine and has health benefits. Green banana starch has shown blood glucose-reducing properties and can be used in products targeted at people with diabetes or those seeking to control blood sugar levels. In addition, its potential as an ingredient for low-calorie food production has also been investigated [33].

8.4.3 Purple Sweet Potato Starch

Purple sweet potato starch is a promising source of functional ingredients based on starch. Purple sweet potato, known for its anthocyanin content, a type of antioxidant, also contains starch with functional properties. Purple sweet potato starch has demonstrated antioxidant properties and can be used in the formulation of food and cosmetic products that aim for skin health benefits and protection against oxidative stress [34].

8.5 Use of Starch Nanomaterials to Develop Functional Foods

8.5.1 Functional Foods

The term "functional food" originated in Japan back in 1988 when referring to a fiber-rich beverage known as "Fiber mini" [35]. Nevertheless, an officially established definition of "functional food" is still lacking. Several authors agree that this term defines food products that contribute to health benefits, being specific in some cases, in addition to providing nutrition [36, 37]. Therefore, the consumption of these foods has been associated with a reduced risk of cancer, neurodegenerative diseases, cardiovascular diseases, gastrointestinal disorders, and metabolic syndrome, among others [38].

From 2009 to 2013, the global market for functional products witnessed a remarkable expansion of approximately 26.7%, followed by an annual growth rate of approximately 8.5% [39]. This significant growth can be attributed to shifts in consumer behavior, driven by various factors such as changing social dynamics (social networks), greater access to information, the influence of medical professionals, and the alarming mortality rates associated with cardiovascular diseases among others [37]. Consequently, today's consumers are actively seeking ways to prevent the occurrence of diseases by adopting healthier lifestyles.

8.5.2 Starch Nanomaterials

Starch nanomaterials (SNMts) are defined as crystalline particles ranging in size from 1 to 1000 nm, resulting from the disruption of the semicrystalline structure of starch granules [40–42]. SNMts can be classified into various categories, including nanocrystals, nanofibers, nanomicelles, nanoparticles, and nanovesicles [40]. These materials exhibit distinct characteristics compared with starch, including enhanced solubility, absorption capacity, and penetration rates, among others, making them highly desirable for the development of products associated with emulsions or encapsulation systems [41].

These nanomaterials can be obtained through various "top-down" approaches, including acid and enzymatic hydrolysis, high-pressure homogenization (HPH), anti-solvent precipitation (ASP), and ultrasound. These methods involve breaking down starch granules to obtain the desired nanostructures. In contrast, a "bottom-up" approach can be used to synthesize nanomaterials by accumulating starch chain precursors, such as monomers and oligomers, through physicochemical interactions. Methods such as self-assembly, emulsification, and electrospinning are employed within this approach [40].

8.5.3 Physicochemical Properties of the Starch Nanomaterials

The development of SNMts leads to the modification of the previously mentioned physicochemical properties of native starch. Firstly, there is a notable decrease in particle size, ranging between 1 and 1000 nm [40]. These nanomaterials generally exhibit spherical, lamellar, rod-like, or irregular shapes, which depend on both the type (i.e. nanoparticle, nanocrystal, etc.) of the nanomaterial obtained and the botanical source [43]. The amylose content and chain length, as well as those of amylopectin decrease because the acid or enzymatic hydrolysis employed for their obtention. Consequently, pasting and rheological properties are also affected. The paste of starch nanomaterials obtained through acid hydrolysis exhibits reduced maximum, minimum, and final viscosity, which can be attributed to decreased granular swelling, leaching of amylose, and interactions among polysaccharide chains [44]. Therefore, a higher concentration of the nanomaterial is necessary to increase the viscosity of the paste. Likewise, various studies have reported that the rheological behavior of nanomaterial-based fluids can vary depending on their size, concentration, and distribution [45].

Starch nanomaterial solutions have attracted attention in the development of functional foods due to their large surface area, lower viscosity, higher concentration, biodegradability, minimal toxicity, and biocompatibility. One potential application is the nanoencapsulation of bioactive compounds, such as polyphenols and carotenoids, which allows the masking of undesirable flavors associated with these secondary metabolites and greatly enhances their stability [45]. The above, combined with simulated digestion assays, have demonstrated that they can greatly contribute to the delivery of this type of compound [41]. Another application focuses on food emulsions. Emulsions are known to be a mixture of immiscible liquids, where one is dispersed (dispersed phase) in the other (continuous phase), which is made possible by the reduction in the interfacial tension between them due to the presence of a surfactant. In this regard, the use of solid particles, such as starch nanomaterials, would be an environmentally friendly alternative for replacing these substances [46]. Films are another area within the food industry where these SNMts can be employed. It has been discovered that their integration can enhance the mechanical strength of these materials by increasing tensile strength and Young's modulus while reducing elongation percentage. In addition, it leads to increased hydrophobicity, resulting in improved water resistance [46]. Some studies with these applications have been referenced in Table 8.1.

Table 8.1 Physicochemical properties and main applications of starch nanomaterials

Starch source/ Nanomaterial type	Physicochemical properties	Application	Reference
Taro/Nanoparticle	Emulsifier and stabilizer	Encapsulation of polyphenols	[47]
Corn/Nanoparticle		Pickering sunflower oil in water emulsion	[48]
Corn, potato, and wheat/Nanocrystals	Reinforcing nanofillers	Polymer nanocomposites	[49]
	Thermal stability	Stability in dry processes at temperatures between 150 and 200 °C	
Pea/Nanoparticles	High tensile strength and low permeability	Films	[50]
Corn/Nanoparticles	Biocompatibility	Adsorbents of polyphenols	[51]

8.6 Starch Nanomaterials in Functional Foods

Starch nanoparticles (SNPs) possess several beneficial attributes when used as functional ingredients in food. These include excellent water dispensability, high biocompatibility, low immunogenicity, minimal toxicity, small size, and a large specific surface area [16]. Furthermore, resistant SNPs offer additional advantages as they resist digestible enzymatic hydrolysis in the upper gastrointestinal tract, particularly by amylase. The robustness of resistant starch nanoparticles can be attributed to their compact crystalline structure. Consequently, they exhibit prebiotic properties, aid in managing diabetes and obesity, promote colon health and nutrient absorption, and can serve as a delivery system for bioactive compounds [52].

Resistant SNPs have been investigated as a control for obesity-related diabetes because these nanomaterials allow controlled nutrient absorption and a decrease in blood glucose and insulin. Production of short-chain fatty acids from resistant starch provides upregulation of genes related to glucose and lipid metabolism [53].

On the other hand, the benefits of resistant SNPs in the colon consist of stimulating the growth of beneficial intestinal microbiota such as *Bifidobacteria* y *Lactobacillus.* Furthermore, the formation of fatty acids contributes to intestinal pH by preventing the proliferation of pH-sensitive bacteria such as *Escherichia coli* y *Salmonella.* Fermentation of this starch fraction can produce high amounts of butyrate, reducing the risk of colorectal disease and cellular malignancy [54].

The high surface-to-volume ratio of the nanoparticles allows digestive enzymes, especially α-amylase to bind to their surfaces via hydrogen bridges. This alteration prevents the interaction of enzymes with other substrates in the gastrointestinal tract. In addition, digestive enzymes can undergo substantial conformational changes after adsorption that alter their activities due to their encapsulation by the starch nanoparticles [55].

SNPs have also been used as part of drug delivery systems. In recent years, SNPs have been used for the production of improved pharmaceuticals [56]. used SNPs to make topical pharmaceuticals with lipophilic bioactive molecules. This drug proved to be a suitable nanocarrier for topical use due to its high stability. Xu et al. [7] realized a pH-sensitive nanocarrier with starch nanoparticles for an antitumor drug. In addition, these nanoparticles have been used to improve drug release [57].

8.7 Interaction of Starches with Foods

8.7.1 Types of Interaction

Starch-protein interactions are of great interest especially for the development of bakery products, desserts, snacks, etc., for their determinant role in the formation of gels and final rheology of the product [58]. Likewise, the interaction between starches and lipids also plays a determining role in the functional properties of starch such as solubility, swelling capacity, water absorption rate, retrogradation, and paste properties, among others, all these characteristics are influenced by the amylose-lipid complexes [59].

High-starch foods may eventually present retrogradation problems that affect their functional properties and reduce their shelf life. In this sense, alternatives have been investigated to control or avoid this process. The use of polysaccharides as an alternative hydrocolloid has been analyzed as well as exopolysaccharide-type dextrans, the latest used to improve viscoelastic properties in products (Table 8.2).

8.8 Conclusions

The growing interest in functional foods has developed a solid and innovative industry that focuses on meeting the needs of today's society that seeks to have specific properties for requirements. Starch in the food industry represents one of the largest sources of carbohydrates of vegetable origin and is found in everyday foods in the kitchen, such as rice and cereals. This molecule represents an energy reserve in plants, similar to that in humans.

This vegetable polysaccharide is the only one capable of being assimilated by the human digestive system, as long as it is cooked, which is why starches are essential ingredients in the culinary industry since the types of starches each one has interesting characteristics for a different type of audience: slow digestion that takes longer to digest and turn into glucose, for which research and products are being carried out for people with metabolic problems such as obesity and diabetes, fast-digesting starches that due to their rapid adsorption and converted into glucose is used to combat malnutrition and weight problems, and finally resistant starch that, when not digested, works as non-soluble fiber, helps processes of low-calorie diets, in addition, the starch that, when not digested, reaches the intestine, where the microbiota converts it into a short-chain fatty acid that is capable of reducing colon inflammation.

Table 8.2 Recent research on the role of starch-macromolecule interactions

Description	Compounds	Conditions	Major findings	Reference
Starch-protein interactions				
Study on starch-protein interactions and their effects on physicochemical and digestible properties of the blends	Native corn starch-WPI	The effect of the addition of WPI (0–10%) in a 10% (w/w) starch solution on physicochemical properties was evaluated	Significant differences were found in the thermal properties of the mixture due to the addition of protein, with higher gelatinization temperatures (76.95 °C) at 10% (w/w). The addition of protein acts as a barrier that encapsulates the starch granules and leads to slow digestibility compared to native starches	[22]
Interactions of soy protein isolate with common and waxy corn starches and their effects on acid-induced cold gelation properties of complexes	SPI-waxy corn starch	The interactions of SPI and waxy corn starch produced on the cold gelation properties of the formed complexes were analyzed	The SPI-starch complex gels improved gel strength, viscoelasticity, and thermal stability with increasing starch concentration. The complex with the best gelation properties was the one with 2% of starch. SPI-starch gels exhibited improved water-holding capacity and thermal stability, but gel strength was reduced	[60]
Starch-lipid interactions				
Amylose-lipid complex formation from extruded maize starch mixed with fatty acids	Native corn starch (30% amylose)—corn oil or oleic acid or stearic acid (0%, 5%, and 10%)	Extrusion (15 rpm–85 °C zone 1, 100 °C zone 2) and drying of mixtures (40 °C–24 h)	The unsaturation degree of the fatty acids was a determining factor for the formation of complexes, this being more viable	[61]

(continued)

Table 8.2 (continued)

Description	Compounds	Conditions	Major findings	Reference
			in fatty acids of higher saturation (stearic acid) with an effect on the viscosity profile being lower compared to the other complexes evaluated	
Mechanisms underlying the Formation of Complexes between maize starch and lipids	Native maize starch- palmitic acid, monopalmitate glycerol, dipalmitate glycerol, and tripalmitate glycerol	Starch-lipid mixtures were prepared to analyze the effect of the addition of these lipids on the structure and functionality of corn starch	The results showed that the solubility and structure of lipid components have significant effects on the formation and structural order of starch-lipid complexes. The normal starch-palmitic acid complex presented more ordered structures than the monopalmityl glyceride complex. However, monoglyceride formed a greater amount of starch-lipid complex than palmitic acid. Finally, no complexes were detected between corn starch and the di- and triglycerides	[62]
Starch-polysaccharides interactions				
Interactions between soluble soybean polysaccharide and starch during the gelatinization and retrogradation: Effects of selected starch varieties	SSPS, maize starch, wheat starch, rice starch, and tapioca starch	The effect of the pasting and retrograding characteristics was investigated	After 14 days under refrigeration, SSPS decreased the relative crystallinities of maize starch and wheat but increased that of tapioca starch.	[63]

(continued)

Table 8.2 (continued)

Description	Compounds	Conditions	Major findings	Reference
			V-type structures were formed which inhibited starch retrogradation (maize and wheat starches). The retrogradation process was accelerated due to the porous structure of tapioca starch and the electrostatic interaction with SSPS	
Determination of viscoelastic and physicochemical interactions of dextran-type Exopolysaccharides with different starch samples	Standard maize starch, wheat starch, and waxy maize starch, Dextran-type glucans from *Leuconostoc mesenteroides* (BI-20), *Weissella cibaria* (MED17), and *Weissella cibaria* (PDER21) strains	The rheological properties of three distinct dextrans with different levels of (1 → 6)-linked α-D-glucose/(1 → 3)-linked α-D-glucose units from three lactic acid bacteria strains were determined	Standard corn starches containing PDER21 dextran exhibited more elastic behavior. In the 4–70 °C range, the solid property of the starch and dextran mixtures was observed. Likewise, the polymeric structure of the interaction between starch and PDER21 dextran did not change concerning the control sample	[64]
Starch-protein-lipid interactions				
Prior interaction of protein and lipid affects the formation of ternary complexes with starch	Wheat starch, lauric acid, β-lactoglobulin (βLG)	The effects of prior interaction between βLG and LA on their formation of ternary complexes with wheat starch were studied	Heating induces interactions between βLG and LA, which were significant at 75, 85 and 95 °C. In addition, the prior interaction between βLG and LA hinders the formation of ternary complexes with starch	[65]

(continued)

Table 8.2 (continued)

Description	Compounds	Conditions	Major findings	Reference
New insight into the interactions among starch, lipid, and protein in model systems with different starches	Waxy maize starch, Maize starch, Potato starch, wheat starch, βLG and LA	The starch-lipid and starch-lipid-protein complexes were produced by Rapid Visco Analyser (RVA)	During the cooling and profile maintenance stages in the RVA test, the maize starch-LA and wheat starch-LA systems exhibited a viscosity peak, whereas the potato starch-LA system showed only an increase in final viscosity. Wheat and corn starch formed crystalline complexes with LA, but no complexes formed between waxy maize starch and LA When βLG was incorporated, the final viscosity increased in all starch-LA systems	[66]

LA lauric acid, *SSPS* soluble soybean polysaccharide, *SPI* soy protein isolate, *WPI* whey protein isolate

Starch is an important alloy in the food industry because it has important functions in this industry that are related to the types of interactions that are: starch-protein, starch-lipid, starch-polysaccharide, and starch-protein-lipid. Each one of them performs different important functions in food preparation, for example, thickening, covering, preserving, sticking, and energizing, among other more complex functions, such as the nanoencapsulation of bioactive compounds that help to preserve their stability and mask unwanted flavors, as well as use them as nanomaterials such as bioplastics that contribute to the physicochemical characteristics of substances of industrial interest. There is no doubt that the search for better materials and technologies in the generation of ingredients and functional foods will continue to evolve and starch and its derivatives will form an important part of this.

Acknowledgments S.L. Paz-Arteaga, V. Cedeño-Fierro, J. F. Grisales-Mejía and N. S. Vargas-Marín thank the National University of Colombia for the support provided to develop their postgraduate degrees, and S.L. Paz-Arteaga appreciates the financial support to carry out his stay at the Autonomous University of Coahuila (Resolution F.C 0349 of 2022).

Conflicts of Interest The authors declare no conflicts of interest.

References

1. Pund S, Joshi A, Banerjee R. Engineered nanomaterials in functional foods. In: Innovative food processing technologies: a comprehensive review. Amsterdam: Elsevier; 2021. p. 546–64.
2. Gouveia L, Marques AE, Bandarra NM. Microalgae—source of natural bioactive molecules as functional ingredients. Food Sci Technol. 2008. https://doi.org/10.1616/1476-2137.15884.
3. Huang Q, Wang X, Lin Z, Liu J, Wang H, Zhang C, et al. Regulatory role of vitamin A in lipid metabolism and health of Eriocheir sinensis fed different vegetable oil sources. Aquaculture. 2023. https://doi.org/10.1016/j.aquaculture.2023.739324.
4. Pham VT, Dold S, Rehman A, Bird JK, Steinert RE. Vitamins, the gut microbiome and gastrointestinal health in humans. Nutr Res. 2021. https://doi.org/10.1016/j.nutres.2021.09.001.
5. Pradyumna T, Quartarone G. Role of guar fi ber in improving digestive health and function. Nutrition. 2019;59:158–69. https://doi.org/10.1016/j.nut.2018.07.109.
6. Tariq A, Sahar A, Usman M, Sameen A, Azhar M, Tahir R, et al. Extraction of dietary fiber and polyphenols from mango peel and its therapeutic potential to improve gut health. Food Biosci. 2023;53:102669. https://doi.org/10.1016/j.fbio.2023.102669.
7. Xu Z, Yang D, Long T, Yuan L, Qiu S, Li D, et al. pH-sensitive nanoparticles based on amphiphilic imidazole/cholesterol modified hydroxyethyl starch for tumor chemotherapy. Carbohydr Polym. 2022;277(24):118827. https://doi.org/10.1016/j.carbpol.2021.118827.
8. Britten-jones AC, Craig JP, Downie LE. Omega-3 polyunsaturated fatty acids and corneal nerve health: current evidence and future directions. Ocul Surf. 2023;27:1–12.
9. Pisaniello AD, Psaltis PJ, King PM, Liu G, Gibson RA, Tm J, et al. Omega-3 fatty acids ameliorate vascular inflammation: a rationale for their atheroprotective effects. Atherosclerosis. 2021;324:27–37. https://doi.org/10.1016/j.atherosclerosis.2021.03.003.
10. Maske BL, De MGV, Vale S, Neto DC, Grace S, Viesser A, et al. A review on enzyme-producing lactobacilli associated with the human digestive process: from metabolism to application. Enzym Microb Technol. 2021;149 https://doi.org/10.1016/j.enzmictec.2021.109836.
11. Cotta KI, Addo RT, Souza MJD. Evaluation of the intestinal colonizing potential and immunomodulating capacity of lactobacilli microspheres. J Pharm Sci. 2016:2–13. https://doi.org/10.1016/j.xphs.2016.02.016.
12. Arruda E De, Lima L De, Jos C, Lúcia V, Galv A, Lima D, et al. In vitro anticancer properties of anthocyanins: a systematic review. Biochim Biophy Acta (BBA) Rev Cancer. 2022;1877. https://doi.org/10.1016/j.bbcan.2022.188748.
13. Shu C, Wu S, Li H, Tian J. Health benefits of anthocyanin-containing foods, beverages, and supplements have unpredictable relation to gastrointestinal microbiota: a systematic review and meta-analysis of random clinical trials. Nutr Res. 2023;116(120):48–59. https://doi.org/10.1016/j.nutres.2023.04.002.
14. Englyst N, Cummings H. Digestion of the carbohydrates of banana (Musa paradisiaca sapientum) human small intestine. Am J Clin Nutr. 1986:42–50. https://doi.org/10.1093/ajcn/44.1.42.
15. Bello-Perez LA, Flores-Silva PC, Agama-Acevedo E, Tovar J. Starch digestibility: past, present, and future. Sci Food Agric. 2018. https://doi.org/10.1002/jsfa.8955.
16. Wang C, Julian D, Jiao A, Wang J, Jin Z, Qiu C. Resistant starch and its nanoparticles: recent advances in their green synthesis and application as functional food ingredients and bioactive delivery systems. Trends Food Sci Technol. 2022;119:90–100. https://doi.org/10.1016/j.tifs.2021.11.025.
17. Bojarczuk A, Sk S, Mousavi A, Marsza K. Health benefits of resistant starch: a review of the literature. J Funct Foods. 2022;93. https://doi.org/10.1016/j.jff.2022.105094.

18. Hermiati E, Sondari D, Sunarti TC. Extraction and classification of starch from different sources: structure, properties, and characterization. In handbook of natural polymers. Handb Nat Polym. 2023;1:19–60. https://doi.org/10.1016/B978-0-323-99853-6.00012-7.
19. Vamadevan V, Bertoft E. Observations on the impact of amylopectin and amylose structure on the swelling of starch granules. Food Hydrocoll. 2020;103:105663. https://doi.org/10.1016/j.foodhyd.2020.105663.
20. Debet MR, Gidley MJ. Three classes of starch granule swelling: influence of surface proteins and lipids. Carbohydr Polym. 2006;64:452–65. https://doi.org/10.1016/j.carbpol.2005.12.011.
21. Li C. Recent progress in understanding starch gelatinization—an important property determining food quality. Carbohydr Polym. 2022;293:119735. https://doi.org/10.1016/j.carbpol.2022.119735.
22. Yang C, Zhong F, Goff HD, Li Y. Study on starch-protein interactions and their effects on physicochemical and digestible properties of the blends. Food Chem. 2018. https://doi.org/10.1016/j.foodchem.2018.12.028.
23. Hussain S, Mohamed A, Alamri M, Ibraheem M, Qasem A, Shahzad S, et al. Use of gum cordia (Cordia myxa) as a natural starch modifier; effect on pasting, thermal, textural, and rheological properties of corn starch. Foods. 2020. https://doi.org/10.3390/foods9070909.
24. Timilsena Y, Haque MA, Adhikari B. Encapsulation in the food industry: a brief historical overview to recent developments. Food Nutr Sci. 2020;11:481–508. https://doi.org/10.4236/fns.2020.116035.
25. Kaul S, Kaur K, Kaur J, Mehta N, Kennedy JF. Properties of potato starch as influenced by microwave, ultrasonication, alcoholic-alkali and pre-gelatinization treatments. Int J Biol Macromol. 2023;226:1341–51. https://doi.org/10.1016/j.ijbiomac.2022.11.246.
26. Cui Y, Li C, Guo Y, Liu X, Zhu F, Liu Z, et al. Rheological & 3D printing properties of potato starch composite gels. J Food Eng. 2022;313:110756. https://doi.org/10.1016/j.jfoodeng.2021.110756.
27. Oyom W, Zhang Z, Bi Y, Tahergorabi R. Application of starch-based coatings incorporated with antimicrobial agents for preservation of fruits and vegetables: a review. Prog Org Coatings. 2022;166:106800. https://doi.org/10.1016/j.porgcoat.2022.106800.
28. Tao X, Zhan L, Huang Y, Li P, Liu B, Chen P. Preparation, characterization and evaluation of capsaicin-loaded indica rice starch nanoparticles. Food Chem. 2022;386:132692. https://doi.org/10.1016/j.foodchem.2022.132692.
29. dos Santos Alves MJ, Rodrigues Monteiro A, Ayala VG. Antioxidant nanoparticles based on starch and the phenolic compounds from propolis extract: production and physicochemical properties. Starch/Stärke. 2022.;2100289;74:1–9. https://doi.org/10.1002/star.202100289.
30. Ordoñez M, Herrera A. Morphologic and stability cassava starch matrices for encapsulating limonene by spray drying. Powder Technol. 2014;253:89–97. https://doi.org/10.1016/j.powtec.2013.11.005.
31. Fang K, He W, Jiang Y, Li K, Li J. Preparation, characterization and physicochemical properties of cassava starch-ferulic acid complexes by mechanical activation. Int J Biol Macromol. 2020;160:482–8. https://doi.org/10.1016/j.ijbiomac.2020.05.213.
32. Rong L, Shen M, Wen H, Xiao W, Li J, Xie J. Effects of xanthan, guar and Mesona chinensis Benth gums on the pasting, rheological, texture properties and microstructure of pea starch gels. Food Hydrocoll. 2022;125:107391. https://doi.org/10.1016/j.foodhyd.2021.107391.
33. Das M, Rajan N, Biswas P, Banerjee R. A novel approach for resistant starch production from green banana flour using amylopullulanase. LWT. 2022;153:112391. https://doi.org/10.1016/j.lwt.2021.112391.
34. Wang H, Yang Q, Ferdinand U, Gong X, Qu Y, Gao W, et al. Isolation and characterization of starch from light yellow, orange, and purple sweet potatoes. Int J Biol Macromol. 2020;160:660–8. https://doi.org/10.1016/j.ijbiomac.2020.05.259.
35. Takahashi T, Singh RB, Chibisov S, Tomar RS, Charkrabarti T, Chauhan AK, et al. The Singh's concept of functional foods and functional farming (4 F) for world health. In: The role

of functional food security in global health; 2019. p. 43–56. https://doi.org/10.1016/B978-0-12-813148-0.00003-7.

36. Alongi M, Anese M. Re-thinking functional food development through a holistic approach. J Funct Foods. 2021;81:104466. https://doi.org/10.1016/j.jff.2021.104466.
37. Das A, Sen CK. Chapter 2. Nutritional supplements and functional foods—functional significance and global regulations. In: Nutraceutical and functional food regulations in the United States and around the world. 2nd ed. Elsevier Inc.; 2014. p. 13–40. https://doi.org/10.1016/B978-0-12-816467-9.00002-2.
38. Marinho L, Vilar M, Betim C, Maróstica M. Functional food consumption and its physiological effects. Bioactive compounds. Elsevier Inc.; 2019. p. 203–225.
39. Bogue J, Collins O, Troy AJ. Chapter 2—Market analysis and concept development of functional foods. In: Developing new functional food and nutraceutical products, vol. 2. Amsterdam: Elsevier Inc.; 2017. p. 29–45.
40. Alves MJ dos S, Chacon WDC, Gagliardi TR, Henao ACA, Monteiro AR, Valencia GA. Food applications of starch nanomaterials: a review. Starch/Stärke. 2021;73:2100046. https://doi.org/10.1002/star.202100046.
41. Campelo PH, Sant'Ana AS, Pedrosa Silva Clerici MT. Starch nanoparticles: production methods, structure, and properties for food applications. Curr Opin Food Sci. 2020;33:136–40. https://doi.org/10.1016/j.cofs.2020.04.007.
42. Lecorre D, Bras J, Dufresne A. Influence of native starch's properties on starch nanocrystals thermal properties. Carbohydr Polym. 2012;87:658–66.
43. . Estrada Gasca CA, Arancibia Bulnes CA. Las Energías Renovables: la Energía Solar y sus Aplicacion. 2010. https://doi.org/10.1016/j.carbpol.2011.08.042.
44. Chavan P, Sinhmar A, Nehra M, Thory R, Kumar A, Allwyn A, et al. Impact on various properties of native starch after synthesis of starch nanoparticles: a review. Food Chem. 2021;364:130416. https://doi.org/10.1016/j.foodchem.2021.130416.
45. Kringel DH, Dias ARG, Zavareze EDR, Gandra EA. Fruit wastes as promising sources of starch: Extraction, properties, and applications. Starch-Stärke. 2020;72(3–4). https://doi.org/10.1002/star.201900200.
46. Torres FG, De-la-torre GE. Synthesis, characteristics, and applications of modified starch nanoparticles : a review. Int J Biol Macromol. 2022;194:289–305. https://doi.org/10.1016/j.ijbiomac.2021.11.187.
47. Shao P, Zhang H, Niu B, Jin W. Physical stabilities of taro starch nanoparticles stabilized Pickering emulsions and the potential application of encapsulated tea polyphenols. Int J Biol Macromol. 2018;118:2032–9. https://doi.org/10.1016/j.ijbiomac.2018.07.076.
48. Su P, Nikfarjam N, Deng Y, Taheri-qazvini N. Pickering emulsion stabilized by amphiphilic pH-sensitive starch nanoparticles as therapeutic containers. Colloids Surf B Biointerfaces. 2019;181:244–51. https://doi.org/10.1016/j.colsurfb.2019.05.046.
49. Dufresne A. Crystalline starch based nanoparticles. Curr Opin Colloid Interface Sci. 2014;19(5):397–408. https://doi.org/10.1016/j.cocis.2014.06.001.
50. Jiang S, Liu C, Wang X, Xiong L, Sun. Physicochemical properties of starch nanocomposite films enhanced by self-assembled potato starch nanoparticles. LWT Food Sci Technol. 2016;69:251–7. https://doi.org/10.1016/j.lwt.2016.01.053.
51. Lei W, Liang J, Tan P, Yang S, Fan L, Han M, et al. Preparation of edible starch nanomaterials for the separation of polyphenols from fruit pomace extract and determination of their adsorption properties. Int J Biol Macromol. 2022:2054–64. https://doi.org/10.1016/j.ijbiomac.2022.10.004.
52. Qiu C, Wang J, Zhang H, Qin Y, Xu X, Jin Z. Functional structure/activity relationships a novel approach with controlled nucleation and growth for green synthesis of size-controlled cyclodextrin-based metal -organic frameworks based on short-chain starch nanoparticles. 2018.
53. Zhou Z, Wang F, Ren X, Wang Y, Blanchard C. Resistant starch manipulated hyperglycemia/hyperlipidemia and related genes expression in diabetic rats. Int J Biol Macromol. 2015;75:316–21. https://doi.org/10.1016/j.ijbiomac.2015.01.052.

54. Wen L, Afifah N, Zulkipli M, Idayu I, Mohd A, Khair Z, et al. Dietary management for healthier batter formulations. Trends Food Sci Technol. 2021;113:411–22. https://doi.org/10.1016/j.tifs.2021.03.054.
55. Jiang S, Li M, Chang R, Xiong L, Sun Q. In vitro inhibition of pancreatic α-amylase by spherical and polygonal starch nanoparticles. Food Funct. 2018;9(1):355–63.
56. Martoa J, Gouveiaa LF, Gonc LM, Gaspar DP, Pintoa P, et al. A Quality by design (QbD) approach on starch-based nanocapsules: a promising platform for topical drug delivery. Colloids Surf B Biointerfaces. 2016;143:177–85. https://doi.org/10.1016/j.colsurfb.2016.03.039.
57. Putro JN, Ismadji S, Gunarto C, Soetaredjo FE, Ju YH. A study of anionic, cationic, and nonionic surfactants modified starch nanoparticles for hydrophobic drug loading and release. J Mol Liq. 2020;298. https://doi.org/10.1016/j.molliq.2019.112034.
58. Fan M, Hu T, Zhao S, Xiong S, Xie J, Huang Q. Gel characteristics and microstructure of fish myofibrillar protein/cassava starch composites. Food Chem. 2016. https://doi.org/10.1016/j.foodchem.2016.09.068.
59. Kong L, Perez-santos DM, Ziegler GR. Effect of guest structure on amylose-guest inclusion complexation. Food Hydrocoll. 2019;97:105188. https://doi.org/10.1016/j.foodhyd.2019.105188.
60. Zuritz C, Muñoz E, Malthey H, Pérez E, Gascón A, Rubio L, et al. Density, viscosity and coefficient of thermal expansion of clear grape juice at different soluble solid concentrations and temperatures. J Food Eng. 2005;75. https://doi.org/10.1016/j.jfoodeng.2004.10.026.
61. Cervantes-Ramírez JE, Cabrera-Ramirez AH, Morales-Sánchez E, Rodriguez-García ME, De M, Reyes-Vega L, et al. Amylose-lipid complex formation from extruded maize starch mixed with fatty acids. Carbohydr Polym. 2020;246:116555. https://doi.org/10.1016/j.carbpol.2020.116555.
62. Chao C, Yu J, Wang S, Copeland L, Wang S. Mechanisms underlying the formation of complexes between maize starch and lipids. Agric Food Chem. 2017. https://doi.org/10.1021/acs.jafc.7b05025.
63. Zhao Q, Tian H, Chen L, Zeng M, Qin F, Wang Z, et al. Interactions between soluble soybean polysaccharide and starch during the gelatinization and retrogradation: effects of selected starch varieties. Food Hydrocoll. 2021;118. https://doi.org/10.1016/j.foodhyd.2021.106765.
64. Gokcan H, Ozmen D, Yalcin MY, Dertli E, Toker OS. Determination of viscoelastic and physicochemical interactions of dextran type exopolysaccharides (EPS) with different starch samples. MDPI. 2023. https://doi.org/10.3390/su15064934.
65. Li X, Wang C, Chao C, Yu J, Copeland L, Liu Y, et al. Prior interaction of protein and lipid affects the formation of ternary complexes with starch. Food Chem. 2023;426. https://doi.org/10.1016/j.foodchem.2023.136500.
66. Cai J, Chao C, Niu B, Copeland L, Yu J, Wang S, et al. New insight into the interactions among starch, lipid and protein in model systems with different starches. Food Hydrocoll. 2021;112(29):106323. https://doi.org/10.1016/j.foodhyd.2020.106323.

Chapter 9
Starch Nanomaterials as Functional Packaging Ingredients

Pedro Augusto Invernizzi Sponchiado, Samile Bezerra de Aguiar, Carla Arias Ivonne La Fuente, and Bianca Chieregato Maniglia

9.1 Introduction

In recent times, the field of nanotechnology has provided valuable information on the synthesis of materials, leading to the development of new substances with desired properties. Using nanoscience techniques, traditional functional properties have been improved, in addition to revealing new features and functionality. This breakthrough paved the way for manufacturing nanomaterials from natural and sustainable sources, thus expanding the potential applications of nanotechnology in the packaging sector.

Nanotechnology involves the production and characterization of materials with at least one dimension ranging from 1 to 100 nm in length [1]. In the realm of packaging and other sectors, nanomaterials integrated into the matrix phase must possess not only these nanoscale dimensions but also exhibit distinct physical and chemical properties that differ significantly from their macroscale counterparts [2]. This distinction has significant implications and offers various applications. Typically, organic additives offer several advantages, such as a positive environmental impact due to their low energy consumption during manufacturing, as well as

P. A. I. Sponchiado
Departamento de Química, Faculdade de Filosofia, Ciências e Letras de Ribeirão Preto – FFCLRP, Universidade de São Paulo – USP, Ribeirão Preto, SP, Brazil

São Carlos Institute of Chemistry, University of São Paulo – USP, São Carlos, SP, Brazil

S. B. de Aguiar · B. C. Maniglia (✉)
São Carlos Institute of Chemistry, University of São Paulo – USP, São Carlos, SP, Brazil
e-mail: biancamaniglia@usp.br

C. A. I. La Fuente
Research Institute of Food Engineering for Development (IIAD), Universitat Politècnica de València (UPV), Valencia, Spain

G. Ayala Valencia (ed.), *Starch Nanomaterials and Food Applications*,
https://doi.org/10.1007/978-3-031-60086-9_9

ease of combustion and recycling [3]. In addition, their relatively simple processability and non-abrasive nature contribute to notable cost savings [3].

In this context, starch nanomaterials have emerged as promising functional packaging ingredients in recent years, offering a range of benefits for the development of sustainable and efficient packaging solutions. Derived from renewable resources, such as corn, potatoes, and cassava, starch nanomaterials (SNMts) possess unique properties at the nanoscale, making them highly desirable for enhancing packaging performance [4]. With advancements in nanotechnology, these materials have demonstrated remarkable potential in improving the barrier properties, mechanical strength, and biodegradability of packaging materials. In addition, SNMts have shown promise in addressing food safety concerns through their antimicrobial and oxygen-scavenging capabilities.

There are two main types of nanostarches: starch nanocrystals (SNCs) and starch nanoparticles (SNPs). SNCs are crystalline portions formed by the breakdown of the amorphous regions within starch granules [5]. In contrast, SNPs are derived from gelatinized starch and may contain both amorphous and crystalline regions [5]. These SNMts can be obtained using methods such as acid and enzymatic hydrolysis, antisolvent precipitation, high-pressure homogenization, ultrasound, electrospinning, self-assembly, and emulsification [6, 7].

This chapter discusses the use of SNMts as functional packaging ingredients, highlighting their role in enabling eco-friendly and innovative packaging solutions that extend the shelf life of products while reducing environmental impact. In addition, this chapter explores how starch modification from the microscale to the nanoscale helps in the mechanical and biological properties (antimicrobial and release of bioactive compounds) of packaging.

9.2 Fundamentals of Starch

Starch is a granular homopolysaccharide mainly composed of two natural polymers known as amylose and amylopectin, as represented in Fig. 9.1. These polymers are both composed of D-glucose residues linked together through $\alpha(1 \rightarrow 4)$ glycosidic bonds, differing from each other in molecular size and structural disposition: while amylose has a size distribution of approximately 10^5–10^6 Da and generally exhibits a linear chain organized in a helical structure, amylopectin is approximately 4–5 times larger than amylose (10^7–10^8 Da) and has a branched chain with a branching level of 4–5% with $\alpha(1 \rightarrow 6)$ linkages [8]. This homopolysaccharide is normally composed of 70–80% amylopectin, which comprises the semicrystalline regions of starch granules, and approximately 20–30% of amylose in amorphous zones. However, the proportion of amylose can range from less than 1% in some waxy starches to more than 70% in the high-amylose type [9], with the amylose content varying according to the botanical origin [10].

In terms of its biological functions, there are two types of this polysaccharide: transitory and storage starch [11]. Transitory starch, also known as temporary or

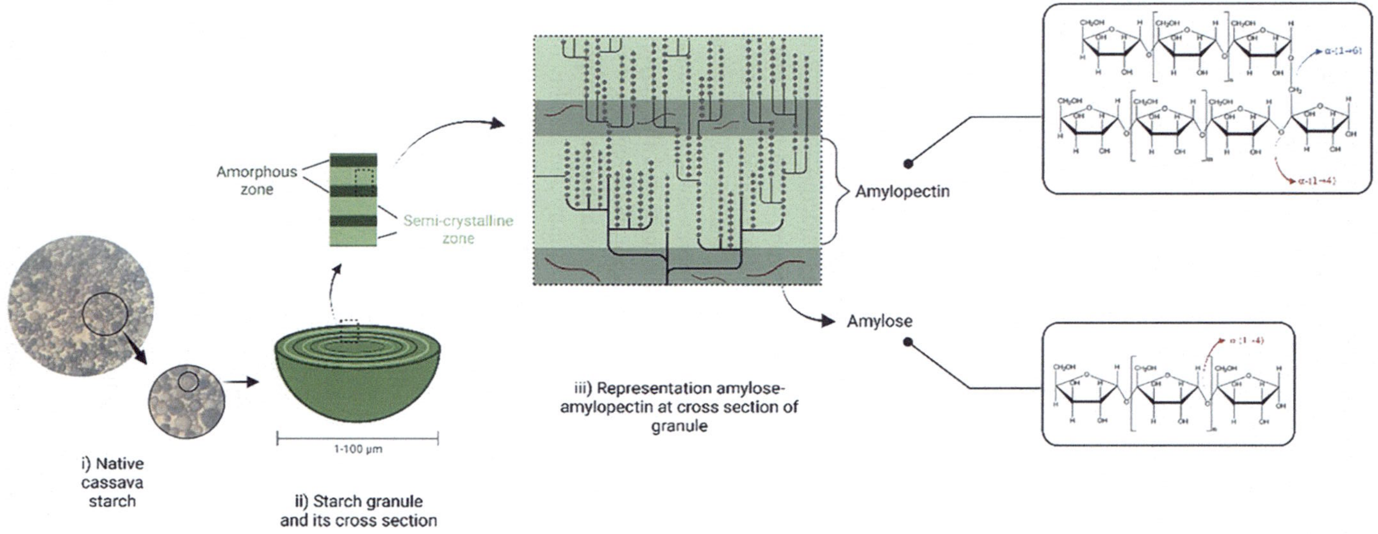

Fig. 9.1 Schematic representation of (i) Cassava starch granules; (ii) Cross section of starch granules with the alternate distribution of the crystalline and amorphous lamellae; (iii) Representation of the inner structure of the starch granules with the distribution of amylose and amylopectin

transient starch, is synthesized and accumulated during the day in photosynthetic tissues such as leaves, to be subsequently used during the night for respiration. It has a relatively simple structure with smaller granules [12, 13]. In plants, the transient starch serves two main functions: (a) as an alternative carbon source for the photosynthesis (faster than by sucrose synthesis); and (b) as a carbon supply at the night. On the other hand, storage starch is synthesized and found in non-photosynthetic tissues such as roots, tubers, and seeds, and is stored for long periods and has a complex structure with different proportions of amylose and amylopectin [14]. Owing to its characteristics, abundance, and multiple functionalities, this type of starch has a significant commercial value and finds application in various industrial sectors [14]. These applications range from being an additive in the food industry [15, 16] to serving as a raw material for biodegradable plastics production [4, 16].

As a storage carbohydrate in plants, starch is quantitatively one of the most abundant materials in nature and can be obtained from many resources such as cassava, potato, sweet potato, corn, and cereal grains [17]. The different sources of starch can directly influence their functional properties, including gelatinization temperature, paste viscosity, and retrogradation speed. This is because starches obtained from different botanical sources may undergo some structural variations, such as differences in the granule size, composition, and molecular architecture of amylose and amylopectin [18]. Once amylopectin comprises the crystalline region, a higher proportion of this polymer leads to starches with greater crystallinity [19]. Moreover, the gelatinization properties (represented by transition temperatures) depend on the molecular structure of amylopectin and the amylose-amylopectin ratio. Once that gelatinization initiates in the amorphous regions, it slowly spreads to the periphery, causing strain and tearing the crystalline zone. High transition temperatures correspond to a high degree of crystallinity and stability of the granule to gelatinization [20]. Naturally, starch granules can exhibit different shapes and sizes, ranging from spherical or polygonal to elongated, and they can have sizes ranging from submicron to approximately 100 μm, depending on their source [21]. For example, cassava starch has an amylose content (AC) of approximately 15% and flake-shaped, irregular granules with a mean size of 20 μm. Potato starch, on the other hand, has a mean AC of 21% and irregular, oval-shaped granules with a mean size of 35 μm. Corn starch contains 28% of amylose and angular granules with 16 μm [22–24].

9.3 Starch-Based Nanomaterials and Their Production

In addition to the traditional modifications applied to this polysaccharide at the microscale, such as physical [25–27] and chemical [28–30] methodologies, starch at the nanoscale has emerged as a new type of changed starch with the advantages associated with size reduction and increased surface area, along with the quantum effects relative to the nanoscale [31]. In the food industry, nanomaterials can act as a reinforcement for packaging materials, enhancing their mechanical and thermal properties. In this context, starch nanomaterials can be used to improve the thermal and mechanical resistance of food packaging, additionally with the potential to incorporate biocompatibility, biodegradability, and non-toxicity into materials [32–34].

SNMts can be categorized as follows:

(a) **nanocrystals (SNCs)**, which are crystalline sheets produced from the disruption and disorganization of the semicrystalline structure of starch granules [33, 34];
(b) **nanoparticles (SNPs)**, defined as nanosized amorphous solid particles with less than 10% of crystallinity degree [35, 36];
(c) **nanovesicles (SNVs)**, or nanoliposomes, are lipid vesicles coated with starch nanomaterials [37];
(d) **nanofibers (SNFs)**, which are cylindrical nanosized structures [38];
(e) **nanomicelles (SNMs)**, which are defined as colloidal dispersions consisting of a hydrophobic core and a hydrophilic shell based on starch nanomaterials [5].

However, some authors distingue the SNMts into only two major groups in terms of crystallinity, namely nanocrystals (SNCs), and nanoparticles (SNPs). This categorization is based on the distinct structural characteristics and degree of crystallinity exhibited by these two nanomaterials: while SNCs possess a crystalline platelet-like structure with defined and ordered crystalline regions, the SNPs are amorphous solid particles with less structural definition and unorganized regions [4, 34].

In general, starch nanomaterials can be obtained through two main approaches: "top-down" and "bottom-up". In the first approach, the starch-based nanomaterials are produced by breaking down starch granules using methods such as acid and enzymatic hydrolysis (AH and EH, respectively), and physical treatments [5]. Although these methods are usually simple and inexpensive, they are less effective for starch granules with irregular shapes and tiny sizes [39]. In contrast, in the "bottom-up" approach, the starch nanomaterials are generated through physical and/or chemical interactions and the accumulation of starch chain precursors (monomers and oligomers), such as nanoprecipitation, electrospinning, and self-assembly [40]. These methods are capable of producing nanostructured starch with a controllable shape and size [39]. The production of starch nanomaterials using some of the methods mentioned above is presented in Table 9.1.

As a "top-down" approach, acid hydrolysis (AH) is a typical methodology for the production of SNCs and is based on the use of acids such as sulfuric acid, hydrochloric acid, and citric acid, which penetrate the amorphous regions of starch granules to break the glycosidic bonds of the polymers while leaving the crystalline zones undamaged [4]. In general, the AH occurs in three steps: (i) surface granule erosion; (ii) radial diffusion of the acid, which causes a collapse of the granule wall; and (iii) fragmentation of the regions [31]. To ensure an effective procedure, it is important to consider factors such as temperature (below the gelatinization point of starch), acid and starch concentrations, time of reaction, and rotation [46]. Velásquez-Castillo et al. [47] showed that varying the temperature of acid hydrolysis of quinoa starch (30, 35, and 40 °C) affected the physical properties of the

Table 9.1 Starch nanomaterial production methods and their applications

Starch source	Material type	Method of production	Major findings	Reference
Cassava starch (17% amylose)	Cassava starch-based nanocomposite films with cassava SNCs	Cassava SNCs prepared by AH	Enhanced mechanical and barrier properties, as well as film solubility; reduced WVP, while increased traction resistance and elastic modulus of films	[33]
Waxy maize starch (99% amylopectin)	Nanocomposite films made of normal corn starch and debranched SNPs	Debranched SNPs prepared by ultrasonication	Addition of debranched SNPs in films enhance its mechanical properties; reduce the water vapor penetrability of the films	[36]
Normal corn starch (26.5% amylose)				
HI-MAIZE® 260 starch	Starch-based films with SNFs	SNFs prepared by electrospinning methodology with posterior cross-linking	The cross-linking improved the tensile strength and the water contact angle of nanocomposite films (with SNFs obtained by electrospinning)	[38]
Cassava starch (28% amylose)	Cassava starch-based nanocomposite films filled with waxy maize SNPs or cassava SNPs	Waxy SNPs obtained by AH; waxy SNPs and cassava SNPs produced by gamma irradiation	Waxy SNPs and cassava SNPs formed by gamma irradiation exhibited the same thermal degradation behavior; the addition of cassava SNPs as cassava matrix reinforcement did not significantly improve its mechanical properties, while the use of waxy SNPs let to an enhancement of 102% in the storage modulus of nanocomposite films	[41]
Waxy maize starch (99% amylopectin)				
Potato starch (31% amylose)	Nanocomposite films made of pea starch and potato starch nanoparticles	SNPs fabricated through recrystallization (self-assembly of short amylose)	Enhanced physicochemical properties of pea starch-based nanocomposite film with the addition of SNPs; increased tensile	[42]

Pea starch (40% amylose)			strength and thermal properties; the addition of SNPs decreases the WVP of films	
Cassava starch (30% amylose)	Starch-based nanocomposite films with SNPs and cellulose nanoparticles	SNPs obtained by ultrasound (SNPs-U) and AH (SNPs-AH)	The addition of SNPs and cellulose nanoparticles influenced the morphology, WVP, and mechanical properties of the nanocomposite films; SNPs-AH had higher crystallinity than SNPs-U	[43]
Quinoa starch (22.8% amylose)	Nanocrystals made of quinoa starch and waxy maize starch	SNCs obtained by AH	Influence of AH temperature over morphological characteristics, crystallinity, and yield of SNCs; better properties were obtained for SNCs at 35 and 40 °C	[44]
Waxy maize starch (99% amylopectin)				
Maize corn starch (29% amylose)	Polyurethane-based nanocomposite films, filled with maize corn SNPs	SNPs synthesized by ultrasound-assisted AH method	Nanocomposite films showed 99% resistance to bacterial transformation; a reduction of 60% WVP in films with addition of 20% SNPs; mechanical and thermal properties of the polyurethane-based films were improved with addition of SNPs	[45]

AH acid hydrolysis, *SNCs* starch nanocrystals, *SNFs* starch nanofibers, *SNPs* starch nanoparticles, *WVP* water vapor permeability nanocrystals produced. The SNCs developed at 30 °C exhibited micrometric dimensions and an irregular shape, unlike those developed at 35 and 40 °C, which presented better morphologies and properties suitable for use as reinforcement materials. In addition, acid hydrolysis also has been combined with other methods. Amini and Razavi [48] used simultaneous ultrasound and sulfuric acid hydrolysis to prepare corn SNCs. By adjusting the temperature, acid/starch concentration, and duration of the process, the researchers obtained the best nanocrystals within 45 min of reaction (using both methods, ultrasound, and hydrolysis), yielding up to a 22% and resulting in particle sizes smaller than 100 nm.

In addition to chemical hydrolysis, the use of enzymes such as α-amylase, β-amylase, and pullulanase to break and de-branch starch chains is also an interesting pathway. Similar to acid hydrolysis, the enzymatic method (EH) proceeds in two stages: initially attacking the starch chains in the amorphous regions, rapidly, followed by a slower action on the crystalline lamella [39]. Owing to this similarity in mechanisms, EH, and chemical hydrolysis, also tends to result in the development of starch nanomaterials with high crystallinity. However, according to Marta et al. [39], the use of enzymes to produce SNMts has the advantage of being carried out in a shorter production time compared with AH, along with generally yielding a higher amount of SNMts. Dukare et al. [49] developed nanostarches from three different sources (maize, potato, and cassava) using the EH technique with α-amylase, an enzyme that catalyzes the cleavage of α-1,4-glycosid bonds in starch. The researchers achieved an improved yield compared with the nanomaterials obtained by AH (18, 29, and 41% compared to 16, 25, 35%, respectively). However, the size of the nanostarches produced by EH was significantly larger than that obtained by AH, and this difference may be attributed to the larger molecular size of α-amylase compared to the acid molecules, which could difficult its penetration and the attack on starch granule regions. Moreover, a combination of acid and enzymatic technique can also be employed to improve the properties of starch nanomaterials developed. Hao et al. [50] prepared SNCs from waxy potato starch by glucoamylase hydrolysis followed by acid hydrolysis. The results demonstrated enhanced stability of SNCs with the combination of these two methods, along with improved crystallinity and reduced time required for acid hydrolysis. This improvement was attributed to the prior enzymatic attack, which possibly created pores facilitating acid penetration into the granules.

In addition to the mentioned "top-down" approaches, various physical methods can also be used to produce SNPs, including gamma irradiation (GI), high-pressure homogenization (HPH), and ultrasonication. These physical methods offer several advantages over other "top-down" techniques, as they are less expensive, time-efficient, and require a reduced use of chemical reagents [39]. In this context, gamma irradiation is widely employed since is a fast and convenient method that is independent of catalysts. Gamma radiation is a form of electromagnetic radiation with high energy and short wavelength, and it is used in GI methodology to generate free radicals that can cleave the glycosidic linkages of starch polymers, mostly in the amorphous region, leading to the formation of smaller fragments [41, 51]. Despite being a radical-based mechanism, it is important to emphasize that the free radicals involved in this process are easily recombined in water, resulting in final solutions with free radicals [52]. Lamanna et al. [41] employed gamma radiation to prepare SNPs from cassava and waxy maize starches, using a dose of 20 kGy. The authors obtained nanoparticles with an average size of approximately 20 and 30 nm. Additionally, they observed that incorporating waxy maize starches-gamma radiated as a reinforcement in the cassava matrix led to the formation of starch nanocomposites with improved mechanical properties (an increase of 102% in storage modulus).

As mentioned previously, two other advantageous physical methods used to produce SNPs are high-pressure homogenization (HPH) and ultrasonication. HPH is considered a "green" technique that enables the production of smaller particles with better homogeneity. It involves the flow of a fluid such as a starch suspension, through a small orifice (throttle valve) under high pressure [53]. The reduction in particle size to the nanoscale can be explained as a result of the disruption of linkages inside the granules, especially the hydrogen bonds, induced by the mechanical shear forces generated during the flow [51]. Ahmad et al. [54] developed SNPs from sago starch using the environmentally friendly HPH procedure, at a pressure of 250 MPa. Ultrasonication was used at the end to ensure the proper dispersion of the produced nanoparticles. The obtained SNPs exhibited a highly narrow size distribution, and when incorporated as reinforcement material in starch films, they effectively reduced the water vapor permeability and improved the transparency of the films.

In addition to HPH, ultrasonication is a rapid and environmentally friendly physical method for producing SNPs through the application of high-frequency waves. The process involves the formation of gas bubbles in the medium, at a process called as cavitation, which subsequently collapse, generating a high local pressure and shockwaves that will break down the starch granules into smaller nanoparticles [39]. To maintain the integrity of the modification during the process, it is important to control certain parameters such as starch dispersion, frequency of ultrasound, temperature, and total time [51]. SNPs from waxy maize and standard maize starches were produced using ultrasonication at low temperatures (8–10 °C). After 75 min of ultrasonication, the SNPs were obtained with an average size between 100 and 200 nm and exhibited a decrease in crystallinity [48]. Alternatively, this method can also be used combined with another "top-down" method, such as Bajer [55] who developed SNPs from corn and waxy corn starches using both AH and ultrasonication. Applying only one of these methods it was obtained nanoparticles with an average size of around 3–4 nm. However, by employing both methods, even smaller particles were formed, with average sizes of up to 3 nm, without the formation of aggregates.

A second approach to producing SNPs is known as the "bottom-up" approach and is based on the principle of building the nanoparticles from smaller components and gradually assembling them into the desired nanomaterials. Therefore, they are methods that have the advantage of developing particles with precise control over their size and composition [7]. Among the various "bottom-up" approaches, in recent years, the most used methods explored as alternatives for producing SNPs are nanoprecipitation, electrospinning, and self-assembly [5]. In this context, nanoprecipitation is a simple and scalable technique that involves the rapid addition of a non-solvent, such as ethanol, propanol, or acetone, into a starch solution under a vigorous agitation. Rapid addition leads to the formation of a supersaturated starch solution, resulting in nucleation and growth of particles, culminating in the precipitation of nanoscale particles [4, 56]. Factors such as polymer concentration, mixing speed, and temperature can affect the size and morphology of the SNPs. Furthermore, Wu et al. [57] reported that the proportion of solvent/non-solvent used in the nanoprecipitation process can affect the size of the nanoparticles, as an increase in the amount of non-solvent usually leads to a decrease in particle size. In addition, the source of starch also can influence the formation of SNPs, as shown by Qin et al. [58]. In their work, SNPs were prepared from seven native starches, including waxy corn, normal corn, high-amylose corn, potato, tapioca, sweet potato, and pea, by precipitating with ethanol. This process leads to the formation of nanoparticles with a mean particle size ranging from 30 to 75 nm, and a positive correlation was observed between the size of the native starch granule and the resulting SNP. In other words, the smaller the starch granules in the source material, the smaller the resulting nanoparticles were produced.

Recently, the electrospinning technique, also known as electrospraying, has been widely used to produce starch nanomaterials, particularly starch nanofibers with diameters of up to 100 nm, depending on the parameters used [5]. With the advantage of developing SNFs with precise control over fiber diameter and morphology, the nanofibers produced by this method show several advantageous properties, such as high surface-area-to-volume ratio, high porosity with small pore size, and superior mechanical properties depending on the source material [59]. In this "bottom-up" method, a polymer solution is extruded through a small opening called a spinneret, while a high voltage (usually between 100 and 500 kV/m) is applied between this solution and a collector. As the voltage increases, the droplet of the solution undergoes deformation, changing from a pointed shape to a conical shape, and is then ejected from the syringe pump toward the collector, where it is deposited as randomly oriented nanofibers [60, 61]. Among the various applications of starch nanofibers, including biomedical delivery systems and materials for tissue engineering [61, 62], SNFs have also been developed for the production of films in the food packaging industry. For example, Cai et al. [62] synthesized SNF-based films for food packaging, that were coated with a hydrophobic surface made of stearic acid. SNFs were obtained using 25% of starch with a mean diameter of 365 ± 97 nm and uniform distribution, exhibiting a smooth surface and random orientation. Moreover, the films prepared from these SNFs exhibited excellent flexibility, making them suitable for use in food packaging.

In addition to the mentioned "bottom-up" approaches, self-assembly is a versatile method that allows obtaining different types of SNMts, such as nanoparticles and nanovesicles [5]. In general, self-assembly is a process in which molecules, such as starch, organize themselves into stable nanostructures spontaneously, driven by intermolecular interactions such as hydrophobic and Van der Waals interactions, hydrogen bonding, electrostatic forces, and π- π aromatic stacking [63]. In order to enhance the properties of starch chains and facilitate self-assembly, in some cases, a pretreatment of starch is applied. This can involve the use of combined methods, such as chemical or enzymatic hydrolysis. From this perspective, a starch-based film of pea starch composed of SNPs derived from potato starch was prepared by Jiang et al. [42]. The process involved the self-assembly of short amylose combined with prior enzymatic treatment of the potato starch. The SNPs obtained through the self-assembly displayed a globular shape with particle sizes ranging from 15 to 30 nm under optimized conditions. In addition, the incorporation of SNPs into the pea-starch films decreased water vapor permeability and an improvement in the mechanical properties of the materials.

9.4 Applications in Food Packaging and Coating

9.4.1 Thermal Stability Agents

To develop improved food packaging, several film parameters can influence the final quality of the packaging, such as optical, mechanical, barrier, and thermal properties [64]. In this case, the thermal characteristics of materials are crucial for their use as commercial products, considering diverse aspects ranging from the transportation process to their storage and final usage of these manufactured items.

In general, during film characterization, the thermal stability can be determined by a simple method known as thermogravimetric analysis (TGA), which is based on the measurement of the mass variation of a sample upon heating over time. In other words, TGA allows for obtaining information about the thermal degradation behavior of materials [65]. In addition to TGA, differential scanning calorimetry (DSC) can also be used to investigate how the materials respond to heating, providing information about the melting point (T_m) of the films and their enthalpy change (ΔH) [45].

As an alternative to improving the thermal properties of films to be used in food packaging, different types of nanomaterials can be used in their matrices. According to Basavegowda and Baek [64], the presence of nanomaterials in polymeric films, for example, can act as an obstacle to heat and mass transfer, thus reducing the molecular mobility of polymers in the matrix and preventing heat-induced polymer degradation. In addition to the applications of starch nanomaterials in the food packaging systems, such as reinforcement materials and antimicrobial agents, these nanoscale starches can also be used to optimize the thermal properties of nanocomposite films. Hakke et al. [45] produced SNPs from maize corn starch and used them as nanofillers to improve several of properties of polyurethane (PU) films. TGA and DSC analysis of the SNP-PU nanocomposite films revealed an increase in ΔH values (ranging from 3.34 to 17.33 J/g for the pure PU film and the 30% SNP-PU material, respectively), and a decrease in thermal degradation as the SNP concentration increased (for the first state of weight loss of 10%: in a temperature below of 100 °C for PU film, compared to 200–300 °C for the SNP-PU films). Moreover, the difference between the melting (T_m) and glass transition (T_g)

temperatures of the nanocomposite film with the incorporation of 30 wt% of SNPs was higher than that of the PU film (T_m = 64.7 °C and 93.3 °C, T_g = 33.6 °C and 51.2 °C, respectively for pure PU and PU with 30 wt% of SNPs). All these observations were attributed to the presence of SNPs in the PU films. According to the authors, the overall increase in the operating range of temperature with the addition of SNPs will help to improve the thermal resistance of the nanocomposite films in coating applications.

In the research of Jiang et al. [42], the gradual addition of potato SNPs synthesized by enzymatic hydrolysis to the pea starch-based films was also correlated with an improvement in the thermal properties of the nanocomposite film. When compared to the control material, the melting temperature of all films with SNPs was increased (ranging from T_m of 226.91 to 235.81 °C for materials with 3 to 12 wt% content of SNPs; and T_m of 225.81 °C for the control). According to the authors, this observed improvement in the thermostability of starch films can be attributed to the presence of strong interactions between nanoparticles and the film matrix.

Nevertheless, it is important to consider that not only the presence of starch nanomaterials that can improve the thermal properties of the materials, but also the amount of starch nanomaterials used as nanofillers can influence them. As reported in the study of Li et al. [66], a gradual improvement in the thermal properties of a film with an increased amount of starch nanomaterials is not always observed. In this study, pea starch-based films (PS) with glycerol were produced, and SNCs developed by acid hydrolysis of waxy maize starch were used as nanofillers. Results obtained from TGA and DSC measurements revealed that the nanocomposite films with 1, 3, and 5 wt% of SNCs showed higher values of ΔH (ranging from 24.34 to 26.76 J/g for 1 and 5 wt% SNCs in the films, respectively; and 23.51 J/g for the native PS material) and higher thermal decomposition temperatures compared to the pure PS film. However, with the addition of 7 and 9 wt% of SNCs, the SNC-PS films showed poorer thermal performance and lower ΔH values (around 22 J/g for both materials), which was attributed to the catalyzed depolymerization of the materials due to the presence of a large number of sulfate groups on the surface of the SNCs. In this case, it was evident that the addition of SNCs could improve the thermal stability of the pea starch-based films when suitable amounts of SNCs were used, i.e., less than 5 wt%.

The main advantages of using SNPs as thermal stability agents are their biocompatibility, biodegradability, and low toxicity compared with synthetic alternatives. However, it is essential to note that the effectiveness of SNPs as thermal stability agents depends on the specific application and the processing techniques used to incorporate them into the desired materials or products. Research in this field is ongoing, and further advancements may lead to even more diverse and efficient applications of SNPs in thermal stability enhancement, such as heat-resistant coatings, thermal stabilizers in food applications, and flame-retardant materials.

9.4.2 Reinforcing Material and Mechanical Properties

SNMts have been described as an excellent option as one reinforcing material; when well dispersed on the matrix, they increase the interfacial area between matrix/fillers, and thus influences the material properties [67]. Mainly, SNPs and SNCs have attracted the attention of food packaging researchers [5].

The selection of the matrix depends on some parameters, such as the compatibility between components, the operation process, the cost, and the final application [68]. Common functional groups of nanomaterials could form links with the surface of the matrix material, holding the percolating network together [69]. In fact, nanoparticles are likely to bond with hydroxyl groups, strengthening the molecular forces between nanoparticles and starch [70].

From an industrial perspective, certain industries have already begun manufacturing nanoparticles. For instance, *Novamont* (Novara—Italy), collaborated with Goodyear Tire and Rubber to develop materials using nanoparticles derived from cornstarch. These nanoparticles have been used to partially substitute conventional carbon black and silica particles in their products [71]. This development showcases the growing trend of incorporating sustainable and renewable resources in industrial applications, leading to more eco-friendly and innovative materials.

The literature has reported that these materials improve the mechanical, thermal, and WVP properties, as a result of the formation of a filler-matrix interface [5]. The hierarchical organization of starch with its semicrystalline structure has been a key factor in the preparation of nanocrystals through different processes, i.e., controlled acid hydrolysis, as in the case of potato and waxy maize starch granules [71].

Several studies conducted have compared SNCs and SNPs. In one particular study [72], researchers examined the morphological, structural, and thermal characteristics of SNCs obtained via acid hydrolysis and compared them with SNPs derived from the same waxy maize starch, using ultrasound treatment. The findings revealed distinct differences between the two materials. The SNPs were observed to be completely amorphous, slightly smaller, and lacked any surface charge. On the other hand, the SNCs exhibited the expected platelet-like morphology and possessed a negative surface charge. Despite these variations in morphology and structure, both materials demonstrated excellent potential as reinforcing agents.

Table 9.2 summarizes some of the recent research on SNPs and SNCs from different sources and their effect on the mechanical properties of films produced from different matrices.

Although most of the studies reported increased mechanical performance due to the addition of these nanomaterials, the effect of SNPs on the mechanical properties has a strong concentration dependance. Indeed, Santana et al. [43] pointed out that low concentrations of SNPs (1 wt%) are probably unable to significantly increase the tensile strength. Moreover, Limin Dai et al. [81] reported that the aggregation of SNCs at higher concentrations, could weaken the interface adhesion between starch and nanocrystals. Therefore, and optimum concentrations must be determined to achieve the desired results.

Table 9.2 Representative studies of starch nanoparticles and starch nanocrystals reinforcing polymeric matrices and its effect on mechanical properties

Polymer matrix	Nanomaterial	Nanocomposite production	Major findings	Reference
Cassava starch	Quinoa SNCs obtained by acid hydrolysis at concentrations of (0, 2.5, 5.0, 7.5 and 10%w/w)	Casting	5 wt% of quinoa SNCs was the best concentration regarding the tensile strength and Young's modulus. These properties were lower for the quinoa SNCs concentrations above or below 5 wt%	[44]
Sago powder	Sago SNPs from sago (*Metroxylon Sagu*) starch at concentration of 0–8% (w/w)	Casting	Sago SNPs enhanced the tensile strength of films up to only 4% (w/w) loading of nanoparticles	[70]
Corn starch	Taro SNPs content of 0.5–15 wt%	Casting	Taro SNPs enhanced the tensile strength of films from 1.11 to 2.87 MPa	[73]
Maize starch	Corn SNPs content of 10–15 wt%	Casting	With 10% of corn SNPs, the tensile strength of films increased about 45% and the Young's modulus about 72%.	[74]
Green banana starch	Acetylated banana SNPs at concentrations of 0.01%, 0.05%, and 0.1% (w/w)	Casting	Nanoparticle concentration increased the tensile strength and reached the highest values at higher concentration of nanoparticles.	[75]
Native water chestnut starch	Water chestnut SNPs prepared by acid hydrolysis in concentrations of 0.5, 1, 2, 5 and 10 wt%	Casting	In nano starch composite films, more force was required to burst the film samples	[76]
PPC	Core–shell SNPs concentrations between 10 and 40 wt%	Melt-blending	With 20 wt% of SNPs, the tensile strength of the composite reached 21.5 MPa, which was 3.2 times that of neat PPC	[77]
PLA	Corn SNPs synthesized via gelatinization of the esterified starch in concentration 10–30 wt%	Melt-blending	PLA nanocomposites reinforced with corn SNPs resulted in a favorable balance between high tensile strength and elongation at break	[78]

(continued)

Table 9.2 (continued)

Polymer matrix	Nanomaterial	Nanocomposite production	Major findings	Reference
Mango seeds starch	SNCs from 0 to 10 wt%	Casting	The optimized conditions (8.5 wt% of SNCs) resulted in a film with enhanced strength, modulus, although the elongation has been impaired	[79]
Commercial cross-linked cassava starch	SNCs concentrations of 2–8 wt%	Casting	Film with 6% SNCs showed the best mechanical performance	[80]
Waxy maize starch	SNCs prepared with heat-moisture treatment and acid hydrolysis at concentrations of 0–12 wt%	Casting	When the SNCs content reached 10%, the nanocomposite film had the best overall performance	[81]
Corn starch: chitosan biopolymers mixes	Corn SNCs in concentrations of 2, 4 and 6 wt%	Casting	Tensile strength was directly related to the corn SNCs ratio	[82]

PLA poly (lactic acid), *PPC* poly(propylene carbonate), *SNCs* starch nanocrystals, *SNPs* starch nanoparticles

Nanomaterials have also been employed to improve the properties of coating materials [83], resulting in an interesting tool for addressing changes in packaging materials and improving the quality, shelf life, safety, and security of food products [84]. Indeed, Bizymis and Tzia [85] reported that starch nanomaterials can increase the barrier and the mechanical properties of edible coatings, in which natural ingredients with antimicrobial and antioxidant activity can be incorporated to increase their beneficial effects on the fresh produce quality [86]. Wang et al. [87] cited that SNPs added to polydimethylsiloxane were used to fabricate coatings; their results indicate the promising application of SNPs in super wettability systems and show the potential of the coating. In another study, cross-linked starch–BTCAD–NHS (starch–butane tetracarboxylic acid dianhydride–N-hydroxysuccinimide) was used to modify gelatin film to be used as a coating; the nanoparticles increased the tensile strength and the elongation at break (50–300%, respectively) [88].

Limin Dai, Zhang, and Cheng [80] produced a starch-based nanocomposite film/coating made from cross-linked cassava starch reinforced by SNCs from 0 to 8 wt% of starch. The coatings based on 6 wt% showed the best mechanical and best coating to preserve pears, and maintain the color, texture, cell membrane permeability, total phenolic, soluble solids, and titratable acid contents of the fruit.

In general, SNPs, when used as reinforcing materials, can significantly affect the mechanical properties of the materials in which they are incorporated. It is also essential to optimize the concentration and dispersion of SNPs within the material to achieve the desired mechanical enhancements effectively. In addition, factors such as particle size, shape, and surface modification can also influence the overall mechanical performance of the composite material. SNPs show promising potential as sustainable and biodegradable reinforcements for various applications, including bioplastics, coatings, and composites, where improved mechanical properties are sought.

9.4.3 Biodegradability

It is relevant to mention new approaches that have been explored, such as the addition of nanosized materials as a tool to improve the biodegradation rate, in which the structure and composition of the biomaterials significantly influence this rate in different stages. The common claim is that these nanomaterials are biodegradable and make materials more biodegradable, which makes them even more attractive. However, data to confirm this assumption are missing [84]. Synthetic nanoparticles have already been discovered in the aquatic environment. Therefore, knowledge of their biodegradability is of maximum importance for risk estimation [84].

Few studies in the literature have reported the effect on biodegradation behavior of different fillers in different matrices. For example, in the PLA matrix was observed that fillers enhanced the hydrophilicity of the composite material, promoting the polymer hydration capacity and the degradative action of microorganisms [89]. Likewise, these additives might inhibit the crystallization of the polymer, promoting the degradation rate of the PHBV polymer [89]. According to Mehmood et al. [90], low-density polyethylene (LDPE) was reinforced with titanium nanoparticles in concentrations of containing 5% (w/w), and the addition of titanium nanoparticles increased the degree of biodegradation. However, although the major studies related increased in the biodegradability due to the action of these nanomaterials, the contrary effect is also valid. Babaee et al. [91] related that a dense structure was obtained in films prepared with plasticized starch (PS)/Chitosan nanoparticles (CNPs) loaded at 1, 2, 3, and 4 wt.% and that their complete biodegradation occurred in more days than neat PS.

Regarding nanomaterials produced from starch, the literature reports that nanoparticles and nanocrystals are found to degrade faster than their macroscopic counterparts due to their higher surface area [92]; however, few studies have reported the effect of these fillers on different polymeric matrices, some of which are shown in Table 9.3.

As shown in Table 9.3, all the studies related to biodegradation in compost or soil media. To our knowledge, no biodegradation studies in aquatic environments have been reported in the literature, which highlights the importance of these type of studies.

Table 9.3 Representative studies of starch nanoparticles influencing biodegradability

Polymer matrix	Nanomaterial	Media	Biodegradability measurement	Major findings	Reference
PLA	Grafted SNCs at 5 wt%	Soil	Weight loss measurement, XPS, FTIR, and DSC	Degradation of PLA/g-SNCs nanocomposites is considerably higher than neat PLA	[93]
HDPE	SNPs at 5 wt%	Air	Thermal degradation, and photodegradation	SNPs decreased the thermal degradation temperature of PE. The photodegradation tests under UV radiation for 28 days showed polymer degradation	[94].
PBAT + starch thermoplastic	SNPs at 0.6 wt%	Compost	Photo Evaluation	Starch nanoparticles promoted faster deterioration in comparison to neat matrix	[95]
Potato starch	Potato SNPs at 0.5, 1, 2, 5, and 10 wt%	Soil	Weight losses	The results showed that the addition of nanoparticles in starch films increases the biodegradability of starch films	[96]
PVA	Corn SNCs synthesized using acid hydrolysis at 0.2–2 wt%	Soil	Weight losses	SNCs improved the biodegradability of the with PVA/S films and was higher than PVA/S film	[97]
PLA	Waxy maize SNCs at 1, 3 and 5 wt%	Dilute alkali-catalyzed hydrolysis	Weight losses	The abiotic degradation studies indicated that WSNC can accelerate the degradation process of PLA	[98]

DSC differential scanning calorimetry, *FTIR* Fourier-transform infrared spectroscopy, *HDPE* high-density polyethylene, *PBAT* poly(butylene adipate-co-terephthalate), *PLA* poly (lactic acid), *PVA* polyvinyl alcohol, *SNPs* starch nanoparticles, *SNCs* starch nanocrystals, *XPS* X-ray photo electron spectroscopy

9.4.4 Encapsulating Agents

The SNPs can be applied in encapsulation systems, which are an attractive alternative for bioactive compounds [31, 99]. The use of nanoamide as a superior encapsulation material is due to its biocompatibility, low viscosity at high concentrations, large surface area, non-toxicity, low cost, and optimal entrapment of bioactive materials [5, 100]. Thus, we have examples of various food ingredients and pharmaceutical application materials that have been encapsulated using SNPs by various encapsulation methods with different efficiencies in the encapsulation process.

According to Ahmad and Gani [101], encapsulation efficiency (EE) determines the amount of core material trapped in the carrier material, and the percentage depends on the number of compounds initially loaded during the encapsulation process. Some studies showed that the highest percentage of EE in starch without modification treatment was shown in vitamin E, with soluble SNPs reaching 91.63%; this finding also indicated that most vitamin E can be retained in SNPs [102]. The highest proportion of EE, above 97%, was produced in conjugated linoleic acid encapsulation using waxy corn starch nanoparticle encapsulation agents; conjugated linoleic acid was effectively trapped in nanostructured particles and can be absorbed with the initial modification treatment, which will effectively increase the absorption efficiency [103]. Similarly, the acetylation of banana starch nanoparticles shows better curcumin encapsulation ability than the nanoparticles without acetylation [104].

Ultrasonication assisted with acid hydrolysis of starch has currently been used to encapsulate antioxidant compounds such as L-ascorbic acid and oxalic acid with high encapsulation efficiency [105]. Horse chestnut, water chestnut and lotus, and lotus starch nanoparticles treated by hydrolysis with sodium hydroxide increased the stability and antioxidant activity of catechin, with the encapsulation efficiency reaching up to 59% and the degradation of catechin encapsulated in SNPs being lower than of unencapsulated catechin, both under digestion conditions [106]. These same starches can also efficiently release resveratrol to specific sites and exhibit higher anti-obesity and anti-diabetic activity than free resveratrol after digestion.

Quinoa SNPs loaded with piroxicam can improve anti-inflammatory activity in vitro, showing around 78% drug release in 0.1 M HCl after 2 h [107]. Unramified starch successfully incorporated epigallocatechin gallate (EGCG) into the nanoparticles with the highest encapsulation efficiency of 84.4%. Furthermore, the resulting nanoparticles showed a controlled release of EGCG in simulated gastric and intestinal fluids. The authors speculated that the developed SNPs could have good prospects as oral release nanocarriers for active compounds or drugs [108]. In recent studies, Escobar-Puentes et al. [109], used succinylated SNPs for the encapsulation of anthocyanins and observed that these systems could be used for the controlled release of bioactive compounds targeting applications such as biodegradable food packaging or as antioxidant additives in the food industry.

In general, results from the literature suggest that SNPs hold significant promise in enhancing the stability, solubility, and availability of bioactive compounds.

Moreover, they exhibit the potential to protect these valuable compounds during food processing and storage, thereby aiding in the development of functional foods.

9.4.5 Antimicrobial Agents

Starch has the remarkable ability to produce a diverse array of SNMts that can effectively stabilize bioactive compounds with antimicrobial properties against foodborne pathogens [110]. These starch-based nanomaterials, which act as stabilizers for antimicrobial compounds, exhibit superior characteristics compared with other natural polymers used as wall materials. They demonstrate faster diffusion rates, higher water solubility, enhanced absorption capacity, and superior penetration rates through biological barriers [110]. Incorporating these nanoparticles as antimicrobial agents in food products can significantly enhance food safety by inhibiting the growth of pathogenic microorganisms. Because of their large surface area, nanoparticles can efficiently attract and immobilize a higher number of microorganisms, leading to increased antimicrobial effectiveness [111].

Furthermore, the development of SNPs allows for the compartmentalization of active substances, particularly stabilizing antimicrobial compounds known for their efficacy [110, 112]. This opens new possibilities for the creation of safer and more stable functional foods with enhanced antimicrobial properties.

Potassium sorbate is considered a GRAS additive (generally recognized as safe) and is known for its antimicrobial properties against yeasts, molds, and various bacteria. However, it is vulnerable to oxidation when exposed to environmental conditions. On the other hand, carvacrol serves as another food preservative and is the primary component found in oregano or thyme oils. It exhibits strong antimicrobial activity against both pathogenic microorganisms and foodborne diseases [110]. Alzate et al. [110], have found that starch-based micro- and nanoparticles, created through antisolvent precipitation (AP), offer a viable method for incorporating potassium sorbate. In another study by the same authors [113], potassium sorbate was successfully stabilized using esterification-modified SNPs. The results showed that the retention capacity of the stabilized potassium sorbate ranged from 41.5 to 90 mg/g, indicating that it is a highly promising antimicrobial agent for food systems. Notably, this approach effectively inhibited potassium sorbate oxidation even under ambient conditions. Fonseca et al. [114] used SNFs to stabilize carvacrol through electrowinning. They prepared solutions containing different concentrations of carvacrol (0%, 20%, 30%, and 40% v/v) with SNFs. The results showed that starch nanofibers loaded with 30% carvacrol led to significant reductions in various pathogens: 89% for *Listeria monocytogenes,* 68.0% for *Salmonella Typhimurium*, 62% for *Escherichia coli*, and 49.0% for *Staphylococcus aureus.* Based on their findings, the authors concluded that starch nanofibers (SNFs) hold immense promise as a vehicle for the controlled release of carvacrol in antimicrobial and antioxidant food packaging applications. Nieto-Suaza et al. [75] employed acetylated SNPs loaded with curcumin, which were then incorporated into films based on banana starch and *aloe vera*. The resulting films demonstrated a controlled release of curcumin when applied to various food products. The researchers suggested that these films have the potential to effectively manage microbial growth in food items, making them promising candidates for food preservation and safety applications.

9.5 Future Trends

Biomass is one of the most promising renewable energy sources, demonstrating its potential as a valuable resource in producing competitive nanomaterials [115]. These nanomaterials are derived from diverse sources such as agricultural waste, forest residues, industrial byproducts, and sewage sludge [115]. Indeed, rice and wheat straws, cotton and corn stalks, sugarcane bagasse, bamboo biomass, shaddock, banana, and rice husk [116], stems, pulp, stem, shell, straw, seeds, husk, peel, stubble, roots, livestock slurry, waste generated during the maintenance of woodlands, and waste from vegetables and fruits hold potential as valuable raw materials to be used in various applications, making them a potential solution for economic, social, and environmental problems [117, 118].

One example pertains to the significant environmental burden posed by potato peel waste, with accumulations reaching several tons per year, resulting in substantial negative impacts [119], minor quantities are sold for supplementary animal feed at very low cost and the rest is discarded as waste [50]. SNPs produced from potato peels have unique properties that are suitable for large-scale applications [119]. Hence, the development of cost-effective techniques for transforming waste potato peel into valuable products is crucial [120]. Another example is cassava starch nanoparticles that could be produced from peels and stems and used in the pharmaceutical industry as a novel drug vehicle [121], which resulted in major advantages such as biocompatibility, biodegradability, easy modification, efficient carrier, and potential targeting [115].

The primary emphasis of incorporating innovative nanomaterials in food packaging has been to enhance the overall quality and safety of products, promote healthier lifestyles, and improve well-being [122]. The advantages of these nanomaterials have been widely studied, but their health implications with it are still in the process of feasibility and verification [122], mainly because of the toxicological effect [123] of nanomaterials due to the risk of migration from packaging to food matrix [122, 123]. Thus, the employment of these materials in the field of food packaging has yet to be allowed due to security concerns.

Consequently, to assess the potential risks, researchers should thoroughly investigate the interactions and stability of nanomaterials in food and feed, as well as their behavior within gastrointestinal systems and biological tissues. In addition, comprehensive studies on the toxicity evaluation of nanomaterials, encompassing chronic exposure and carcinogenicity, are imperative [124].

Another point is related to the use of unconventional starches in the production of nanoparticles, which is a promising and innovative strategy that aims to avoid competition with the food sector while taking advantage of the unique properties of these starches to create materials with diverse applications. To overcome this competition, researchers have explored non-conventional starches, which are obtained from alternative sources, such as agro-industrial residues, non-food plants, and even microorganisms [125].

These unconventional starches can be an abundant and affordable source for the production of nanoparticles with various industrial and medical applications [126]. Different unconventional starches have been explored for nanoparticle production such, as pine seed [127], lotus seed [128], quinoa [47], and pine [129]. By using unconventional starches to produce nanoparticles, several advantages can be achieved: sustainability, low cost, abundance, versatility, and non-food applications.

The production of nanoparticles from unconventional starches is still in the research and development phase. Challenges include optimizing starch extraction and modification processes to obtain the desired characteristics in nanoparticles and ensuring the safety and stability of the final products. In short, the use of unconventional starches in the production of nanoparticles is a promising and sustainable approach that allows exploration of new technological opportunities without compromising the food supply. This approach can contribute to the development of advanced materials with innovative applications in different sectors, thus boosting the economy and scientific research.

9.6 Conclusions

In conclusion, starch nanomaterials (SNMts) offer a promising and innovative solution for functional packaging ingredients in various industries. Starch is one of the most abundant biopolymers on earth, and its nanoscale presents exciting opportunities to enhance packaging material's performance and sustainability.

The unique properties of SNMts, including their biodegradability, renewable nature, and biocompatibility, make them an attractive alternative to conventional packaging additives derived from fossil fuels. These nanomaterials can improve packaging properties, such as barrier performance, mechanical strength, and moisture resistance, while also enabling the controlled release of active compounds for food preservation and other applications.

Moreover, the facility of modification and functionalization of SNMts allows for tailoring their characteristics to meet specific packaging requirements, thereby addressing the diverse needs of different products and industries. By incorporating SNMts into packaging formulations, companies can move toward greener and more sustainable practices, thereby reducing their environmental impact and promoting circular economy principles.

However, challenges related to large-scale production, cost-effectiveness, and compatibility with existing packaging processes must be addressed to fully realize the potential of SNMts in functional packaging applications. Continued research and development efforts are crucial to optimize processing techniques and ensure the safety and regulatory compliance of these novel ingredients.

In conclusion, SNMts represent a promising frontier in the field of functional packaging ingredients. Their eco-friendly nature, versatile properties, and potential for widespread applications make them an essential component in driving the future of sustainable packaging solutions, benefiting both industries and the environment alike. As advancements continue and adoption increases, starch nanomaterials are set to revolutionize the packaging landscape, ushering in a new era of greener, more efficient, and safer packaging materials.

Acknowledgments The authors thank the São Paulo State Research Foundation (FAPESP, Brazil, grant 2020/08727-0 and 2021/05947-2) and the National Council for Scientific and Technological Development (CNPq, Brazil, grant #162334/2021-4 and #88887.682741/2022-00). This study was financed in part by the Coordenação de Aperfeiçoamento de Pessoal de Nível Superior—Brasil (CAPES)—Finance Code 001 and COFECUB (Ph 1006/23), and the postdoctoral grant from the Valencian Government APOST/2021/19.

Conflict of interest The authors declare no conflicts of interest.

References

1. Sarfraz J, Gulin-Sarfraz T, Nilsen-Nygaard J, Pettersen MK. Nanocomposites for food packaging applications: an overview. Nanomaterials. 2020;11(1):10.
2. Kuswandi B. Environmental friendly food nano-packaging. Environ Chem Lett. 2017;15(2):205–21.
3. Moustafa H, Youssef AM, Darwish NA, Abou-Kandil AI. Eco-friendly polymer composites for green packaging: Future vision and challenges. Compos Part B Eng. 2019;172:16–25.
4. Maniglia BC, La Fuente CIA, Siqueira L, do V, Tadini CC. Carbohydrate nanomaterials addition to starch-based packaging: a review about fundamentals and application. Starch - Stärke. 2021;73(11–12):2100057.
5. Alves MJ dos S, Chacon WDC, Gagliardi TR, Henao ACA, Monteiro AR, Valencia GA. Food applications of starch nanomaterials: a review. Starch/Stärke. 2021;73:2100046.
6. Valencia GA, Zare EN, Makvandi P, Gutiérrez TJ. Self-assembled carbohydrate polymers for food applications: a review. Compr Rev Food Sci Food Saf. 2019;18(6):2009–24.
7. Marta H, Rizki DI, Mardawati E, Djali M, Mohammad M, Cahyana Y. Starch nanoparticles: preparation, properties and applications. Polymers (Basel) [Internet]. 2023;15(5). Available from: https://www.mdpi.com/2073-4360/15/5/1167
8. Bertoft E. Understanding starch structure: recent progress. Agronomy. 2017;7(3):56.
9. Martens BMJ, Gerrits WJJ, Bruininx EMAM, Schols HA. Amylopectin structure and crystallinity explains variation in digestion kinetics of starches across botanic sources in an in vitro pig model. J Anim Sci Biotechnol. 2018;9(1):91.
10. Chi K, Wang H, Catchmark JM. Sustainable starch-based barrier coatings for packaging applications. Food Hydrocoll. 2020;103:105696.
11. Pfister B, Zeeman SC. Formation of starch in plant cells. Cell Mol Life Sci. 2016;73:2781–807.
12. Sonnewald U, Kossmann J. Starches-from current models to genetic engineering. Plant Biotechnol J. 2013;11(2):223–32.

13. Chu KL, Koley S, Jenkins LM, Bailey SR, Kambhampati S, Foley K, et al. Metabolic flux analysis of the non-transitory starch tradeoff for lipid production in mature tobacco leaves. Metab Eng. 2022;69:231–48.
14. Lloyd JR, Kossmann J. Transitory and storage starch metabolism: two sides of the same coin? Curr Opin Biotechnol. 2015;32:143–8.
15. Onyeaka H, Obileke K, Makaka G, Nwokolo N. Current research and applications of starch-based biodegradable films for food packaging. Polymers (Basel). 2022;14(6):1126.
16. Cheng H, Chen L, McClements DJ, Yang T, Zhang Z, Ren F, et al. Starch-based biodegradable packaging materials: a review of their preparation, characterization and diverse applications in the food industry. Trends Food Sci Technol. 2021;114:70–82.
17. Gul K, Mir NA, Yousuf B, Allai FM, Sharma S. Starch: an overview. In: Food biopolymers: structural, functional and nutraceutical properties. Cham: Springer International Publishing; 2021. p. 3–17.
18. Cornejo-Ramírez YI, Martínez-Cruz O, Del Toro-Sánchez CL, Wong-Corral FJ, Borboa-Flores J, Cinco-Moroyoqui FJ. The structural characteristics of starches and their functional properties. CyTA J Food. 2018;16(1):1003–17.
19. Alcázar-Alay SC, Meireles MAA. Physicochemical properties, modifications and applications of starches from different botanical sources. Food Sci Technol. 2015;35(2):215–36.
20. Chakraborty I, Pooja N, Mal SS, Paul UC, Rahman MH, Mazumder N. An insight into the gelatinization properties influencing the modified starches used in food industry: a review. Food Bioprocess Technol. 2022;15(6):1195–223.
21. Gómez-Aldapa CA, Velazquez G, Gutierrez MC, Castro-Rosas J, Jiménez-Regalado EJ, Aguirre-Loredo RY. Characterization of functional properties of biodegradable films based on starches from different botanical sources. Starch-Stärke. 2020;72(11–12):1900282.
22. Saari H, Rayner M, Wahlgren M. Effects of starch granules differing in size and morphology from different botanical sources and their mixtures on the characteristics of Pickering emulsions. Food Hydrocoll. 2019;89:844–55.
23. Dai W, Kawazoe N, Lin X, Dong J, Chen G. The influence of structural design of PLGA/collagen hybrid scaffolds in cartilage tissue engineering. Biomaterials. 2010;31(8):2141–52.
24. Singh N, Singh J, Kaur L, Sodhi NS, Gill BS. Morphological, thermal and rheological properties of starches from different botanical sources. Food Chemistry. 2003;81:219–31. Elsevier.
25. Fonseca LM, Halal SLM El, Dias ARG, Zavareze E da R. Physical modification of starch by heat-moisture treatment and annealing and their applications: a review. Carbohydr Polym. 2021;274:118665.
26. Maniglia BC, Lima DC, Matta Junior MD, Le-Bail P, Le-Bail A, Augusto PED. Preparation of cassava starch hydrogels for application in 3D printing using dry heating treatment (DHT): a prospective study on the effects of DHT and gelatinization conditions. Food Res Int. 2020;128:108803.
27. Maniglia BC, Castanha N, Le-Bail P, Le-Bail A, Augusto PED. Starch modification through environmentally friendly alternatives: a review. Crit Rev Food Sci Nutr. 2021;61(15):2482–505.
28. Haq F, Yu H, Wang L, Teng L, Haroon M, Khan RU, et al. Advances in chemical modifications of starches and their applications. Carbohydr Res. 2019;476:12–35.
29. Nakkala K, Godiyal S, Ettaboina SK, Laddha KS. Chemical modifications of turmeric starch by oxidation, phosphorylation, and succinylation. Starch/Staerke. 2022;74:1–9.
30. Thakur Y, Thory R, Sandhu KS, Kaur M, Sinhmar A, Pathera AK. Effect of selected physical and chemical modifications on physicochemical, pasting, and morphological properties of underutilized starch from rice bean (Vigna umbellata). J Food Sci Technol. 2021;58(12):4785–94.
31. Campelo PH, Sant'Ana AS, Pedrosa Silva Clerici MT. Starch nanoparticles: production methods, structure, and properties for food applications. Curr Opin Food Sci [Internet]. 2020;33(i):136–40. Available from: https://doi.org/10.1016/j.cofs.2020.04.007

32. Wang Y, Zhang G. The preparation of modified nano-starch and its application in food industry. Food Res Int. 2021;140:110009.
33. Costa KDC, de Souza CO, da Silva JB, Druzian JI. Hydrolysis of part of cassava starch into nanocrystals leads to increased reinforcement of nanocomposite films. J Appl Polym Sci. 2017;134(41):1–9.
34. Adeyeye SAO, Babu AS, Guruprasath NGN, Ganesh PS. Starch nanocrystal and its food packaging applications. Curr Res Nutr Food Sci J. 2023;11(1):01–21.
35. Jiang T, Duan Q, Zhu J, Liu H, Yu L. Starch-based biodegradable materials: challenges and opportunities. Adv Ind Eng Polym Res. 2020;3(1):8–18.
36. Lin Q, Ji N, Li M, Dai L, Xu X, Xiong L, et al. Fabrication of debranched starch nanoparticles via reverse emulsification for improvement of functional properties of corn starch films. Food Hydrocoll. 2020;104(February):105760.
37. Sintov AC. AmyloLipid Nanovesicles: a self-assembled lipid-modified starch hybrid system constructed for direct nose-to-brain delivery of curcumin. Int J Pharm [Internet]. 2020;588 (July):119725. Available from: https://doi.org/10.1016/j.ijpharm.2020.119725
38. Zhu W, Zhang D, Liu X, Ma T, He J, Dong Q, et al. Improving the hydrophobicity and mechanical properties of starch nanofibrous films by electrospinning and cross-linking for food packaging applications. LWT. 2022;169:114005.
39. Marta H, Wijaya C, Sukri N, Cahyana Y, Mohammad M. A comprehensive study on starch nanoparticle potential as a reinforcing material in bioplastic. Polymers (Basel). 2022;14(22): 4875.
40. Chang R, Ji N, Li M, Qiu L, Sun C, Bian X, et al. Green preparation and characterization of starch nanoparticles using a vacuum cold plasma process combined with ultrasonication treatment. Ultrason Sonochem. 2019;58(February):104660.
41. Lamanna M, Morales NJ, García NL, Goyanes S. Development and characterization of starch nanoparticles by gamma radiation: potential application as starch matrix filler. Carbohydr Polym. 2013;97(1):90–7.
42. Jiang S, Liu C, Wang X, Xiong L, Sun Q. Physicochemical properties of starch nanocomposite films enhanced by self-assembled potato starch nanoparticles. LWT Food Sci Technol [Internet]. 2016;69:251–7. Available from: https://doi.org/10.1016/j.lwt.2016.01.053
43. Santana JS, de Carvalho Costa ÉK, Rodrigues PR, Correia PRC, Cruz RS, Druzian JI. Morphological, barrier, and mechanical properties of cassava starch films reinforced with cellulose and starch nanoparticles. J Appl Polym Sci. 2019;136(4):47001.
44. Velásquez-Castillo LE, Leite MA, Tisnado VJA, Ditchfield C, Sobral PJ, do A, Moraes ICF. Cassava starch films containing quinoa starch nanocrystals: physical and surface properties. Foods. 2023;12(3):576.
45. Hakke VS, Landge VK, Sonawane SH, Uday Bhaskar Babu G, Ashokkumar M, Flores EMM. The physical, mechanical, thermal and barrier properties of starch nanoparticle (SNP)/polyurethane (PU) nanocomposite films synthesised by an ultrasound-assisted process. Ultrason Sonochem. 2022;88:106069.
46. Hassan NA, Darwesh OM, Smuda SS, Altemimi AB, Hu A, Cacciola F, et al. Recent trends in the preparation of nano-starch particles. Molecules. 2022;27(17):5497.
47. Velásquez-Castillo LE, Leite MA, Ditchfield C, Sobral PJ do A, Moraes ICF. Quinoa starch nanocrystals production by acid hydrolysis: kinetics and properties. Int J Biol Macromol. 2020;143:93–101.
48. Mohammad Amini A, Razavi SMA. A fast and efficient approach to prepare starch nanocrystals from normal corn starch. Food Hydrocoll [Internet]. 2016;57:132–8. Available from: https://doi.org/10.1016/j.foodhyd.2016.01.022
49. Dukare AS, Arputharaj A, Bharimalla AK, Saxena S, Vigneshwaran N. Nanostarch production by enzymatic hydrolysis of cereal and tuber starches. Carbohydr Polym Technol Appl. 2021;2:100121.

50. Hao Y, Chen Y, Li Q, Gao Q. Preparation of starch nanocrystals through enzymatic pretreatment from waxy potato starch. Carbohydr Polym [Internet]. 2018;184(December 2017):171–7. Available from: https://doi.org/10.1016/j.carbpol.2017.12.042
51. Kaur J, Kaur G, Sharma S, Jeet K. Cereal starch nanoparticles—a prospective food additive: a review. Crit Rev Food Sci Nutr. 2018;58(7):1097–107.
52. Verma R, Jan S, Rani S, Jan K, Swer TL, Prakash KS, et al. Physicochemical and functional properties of gamma irradiated buckwheat and potato starch. Radiat Phys Chem. 2018;144: 37–42.
53. Ding Y, Kan J. Optimization and characterization of high pressure homogenization produced chemically modified starch nanoparticles. J Food Sci Technol. 2017;54(13):4501–9.
54. Ahmad AN, Lim SA, Navaranjan N, Hsu YI, Uyama H. Green sago starch nanoparticles as reinforcing material for green composites. Polymer (Guildf). 2020;202:122646.
55. Bajer D. Nano-starch for food applications obtained by hydrolysis and ultrasonication methods. Food Chem. 2023;402:134489.
56. Martínez Rivas CJ, Tarhini M, Badri W, Miladi K, Greige-Gerges H, Nazari QA, et al. Nanoprecipitation process: from encapsulation to drug delivery. International Journal of Pharmaceutics. 2017;532:66–81. Elsevier B.V.
57. Wu X, Chang Y, Fu Y, Ren L, Tong J, Zhou J. Effects of non-solvent and starch solution on formation of starch nanoparticles by nanoprecipitation. Starch - Stärke. 2016;68(3–4):258–63.
58. Qin Y, Liu C, Jiang S, Xiong L, Sun Q. Characterization of starch nanoparticles prepared by nanoprecipitation: influence of amylose content and starch type. Ind Crop Prod [Internet]. 2016;87:182–90. Available from: https://doi.org/10.1016/j.indcrop.2016.04.038
59. Kong L, Ziegler GR. Fabrication of pure starch fibers by electrospinning. Food Hydrocoll. 2014;36:20–5.
60. Gomez-Caturla J, Ivorra-Martinez J, Lascano D, Balart R, García-García D, Dominici F, et al. Development and evaluation of novel nanofibers based on mango kernel starch obtained by electrospinning. Polym Test. 2022;106:107462.
61. Sill TJ, von Recum HA. Electrospinning: Applications in drug delivery and tissue engineering. Biomaterials. 2008;29(13):1989–2006.
62. Cai J, Zhang D, Zhou R, Zhu R, Fei P, Zhu Z-Z, et al. Hydrophobic interface starch nanofibrous film for food packaging: from bioinspired design to self-cleaning action. J Agric Food Chem. 2021;69(17):5067–75.
63. Fan Y, Liu Y, Wu Y, Dai F, Yuan M, Wang F, et al. Natural polysaccharides based self-assembled nanoparticles for biomedical applications—a review. Int J Biol Macromol. 2021;192:1240–55.
64. Basavegowda N, Baek K-H. Advances in functional biopolymer-based nanocomposites for active food packaging applications. Polymers (Basel). 2021;13(23):4198.
65. Akhila V, Badwaik LS. Recent advancement in improvement of properties of polysaccharides and proteins based packaging film with added nanoparticles: a review. Int J Biol Macromol. 2022;203:515–25.
66. Li X, Qiu C, Ji N, Sun C, Xiong L, Sun Q. Mechanical, barrier and morphological properties of starch nanocrystals-reinforced pea starch films. Carbohydr Polym. 2015;121:155–62.
67. Qin Y, Wang J, Qiu C, Hu Y, Xu X, Jin Z. Effects of degree of polymerization on size, crystal structure, and digestibility of debranched starch nanoparticles and their enhanced antioxidant and antibacterial activities of curcumin. ACS Sustain Chem Eng. 2019;7(9):8499–511.
68. Sapper M, Bonet M, Chiralt A. Wettability of starch-gellan coatings on fruits, as affected by the incorporation of essential oil and/or surfactants. LWT. 2019;116:108574.
69. Nazrin A, Sapuan SM, Zuhri MYM. Mechanical, physical and thermal propertiesof sugar palm nanocellulose reinforced thermoplastic starch (TPS)/poly (lactic acid) (PLA) blend bionanocomposites. Polymers (Basel). 2020;12(2216):18.
70. Shapi'i RA, Othman SH, Nordin N, Kadir Basha R, Nazli Naim M. Antimicrobial properties of starch films incorporated with chitosan nanoparticles: in vitro and in vivo evaluation. Carbohydr Polym. 2020;230:115602.

71. Mary SK, Pothan LA, Thomas S. Applications of starch nanoparticles and starch-based bionanocomposites. In: Biopolymer nanocomposites. Hoboken: Wiley; 2013. p. 293–308.
72. Bel Haaj S, Thielemans W, Magnin A, Boufi S. Starch nanocrystals and starch nanoparticles from waxy maize as nanoreinforcement: a comparative study. Carbohydr Polym. 2016;143: 310–7.
73. Dai L, Qiu C, Xiong L, Sun Q. Characterisation of corn starch-based films reinforced with taro starch nanoparticles. Food Chem [Internet]. 2015;174:82–8. Available from: https://www.sciencedirect.com/science/article/pii/S0308814614017300
74. Lu H, Ji N, Li M, Wang Y, Xiong L, Zhou L, et al. Preparation of borax cross-linked starch nanoparticles for improvement of mechanical properties of maize starch films. J Agric Food Chem. 2019;67(10):2916–25.
75. Nieto-Suaza L, Acevedo-Guevara L, Sánchez LT, Pinzón MI, Villa CC. Characterization of Aloe vera-banana starch composite films reinforced with curcumin-loaded starch nanoparticles. Food Struct. 2019;22:100131.
76. Dularia C, Sinhmar A, Thory R, Pathera AK, Nain V. Development of starch nanoparticles based composite films from non-conventional source—water chestnut (Trapa bispinosa). Int J Biol Macromol. 2019;136:1161–8.
77. Liu L, Wang Y, Hu Q, Li T, Ma P, Zhang H, et al. Core–shell starch nanoparticles improve the mechanical and thermal properties of poly(propylene carbonate). ACS Sustain Chem Eng. 2019;7(15):13081–8.
78. Wang Y, Hu Q, Li T, Ma P, Zhang S, Du M, et al. Core–shell starch nanoparticles and their toughening of polylactide. Ind Eng Chem Res. 2018;57(39):13048–54.
79. Silva APM, Oliveira AV, Pontes SMA, Pereira ALS, Souza Filho M de sá M, Rosa MF, et al. Mango kernel starch films as affected by starch nanocrystals and cellulose nanocrystals. Carbohydr Polym. 2019;211:209–16.
80. Dai L, Zhang J, Cheng F. Cross-linked starch-based edible coating reinforced by starch nanocrystals and its preservation effect on graded Huangguan pears. Food Chem. 2020;311 (November 2019):125891.
81. Dai L, Yu H, Zhang J, Cheng F. Preparation and characterization of cross-linked starch nanocrystals and self-reinforced starch-based nanocomposite films. Int J Biol Macromol. 2021;181:868–76.
82. Kalateh-Seifari F, Yousefi S, Ahari H, Hosseini SH. Corn starch-chitosan nanocomposite film containing nettle essential oil nanoemulsions and starch nanocrystals: optimization and characterization. Polymers (Basel). 2021;13(13):2113.
83. Taha IM, Zaghlool A, Nasr A, Nagib A, El Azab IH, Mersal GAM, et al. Impact of starch coating embedded with silver nanoparticles on strawberry storage time. Polymers (Basel). 2022;14(7):1439.
84. Chavan P, Sinhmar A, Sharma S, Dufresne A, Thory R, Kaur M, et al. Nanocomposite starch films: a new approach for biodegradable packaging materials. Starch - Stärke. 2022;74(5–6): 2100302.
85. Bizymis A-P, Tzia C. Edible films and coatings: properties for the selection of the components, evolution through composites and nanomaterials, and safety issues. Crit Rev Food Sci Nutr. 2022;62(31):8777–92.
86. Primožič M, Knez Ž, Leitgeb M. (Bio)Nanotechnology in Food science—food packaging. Nanomaterials. 2021;11(2):292.
87. Wang F, Chang R, Ma R, Tian Y. Eco-friendly and superhydrophobic nano-starch based coatings for self-cleaning application and oil-water separation. Carbohydr Polym. 2021;271 (March):118410.
88. Tao F, Shi C, Cui Y. Preparation and physicochemistry properties of smart edible films based on gelatin-starch nanoparticles. J Sci Food Agric. 2018;98(14):5470–8.
89. Hernández-García E, Vargas M, González-Martínez C, Chiralt A. Biodegradable antimicrobial films for food packaging: effect of antimicrobials on degradation. Foods. 2021;10(6):1256.

90. Mehmood CT, Qazi IA, Hashmi I, Bhargava S, Deepa S. Biodegradation of low density polyethylene (LDPE) modified with dye sensitized titania and starch blend using Stenotrophomonas pavanii. Int Biodeterior Biodegradation. 2016;113:276–86.
91. Babaee M, Garavand F, Rehman A, Jafarazadeh S, Amini E, Cacciotti I. Biodegradability, physical, mechanical and antimicrobial attributes of starch nanocomposites containing chitosan nanoparticles. Int J Biol Macromol. 2022;195:49–58.
92. Kümmerer K, Menz J, Schubert T, Thielemans W. Biodegradability of organic nanoparticles in the aqueous environment. Chemosphere. 2011;82(10):1387–92.
93. Sharafi Zamir S, Fathi B, Ajji A, Robert M, Elkoun S. Biodegradation of modified starch/poly lactic acid nanocomposite in soil. Polym Degrad Stab. 2022;199:109902.
94. Amigo N, Palza H, Canales D, Sepúlveda F, Vasco DA, Sepúlveda F, et al. Effect of starch nanoparticles on the crystallization kinetics and photodegradation of high density polyethylene. Compos Part B Eng. 2019;174:106979.
95. González Seligra P, Eloy Moura L, Famá L, Druzian JI, Goyanes S. Influence of incorporation of starch nanoparticles in PBAT/TPS composite films. Polym Int. 2016;65(8):938–45.
96. Gujral H, Sinhmar A, Nehra M, Nain V, Thory R, Pathera AK, et al. Synthesis, characterization, and utilization of potato starch nanoparticles as a filler in nanocomposite films. Int J Biol Macromol. 2021;186:155–62.
97. Mittal A, Garg S, Bajpai S. Assessing biodegradability of PVA/starch based composite films reinforced with long chain fatty acid grafted barley husk. Res Sq. 2021;(1):1–22.
98. Takkalkar P, Ganapathi M, Al-Ali M, Kao N, Griffin G. Morphological, structural, thermal and degradation properties of polylactic acid-waxy maize starch nanocrystals based nanocomposites prepared by melt processing. Adv Mater Lett. 2019;10(3):170–7.
99. Morán D, Gutiérrez G, Blanco-López MC, Marefati A, Rayner M, Matos M. Synthesis of starch nanoparticles and their applications for bioactive compound encapsulation. Appl Sci. 2021;11(10):4547.
100. Kumari S, Yadav BS, Yadav RB. Synthesis and modification approaches for starch nanoparticles for their emerging food industrial applications: A review. Food Res Int [Internet]. 2020;128(May 2019):108765. Available from: https://doi.org/10.1016/j.foodres.2019.108765
101. Ahmad M, Gani A. Ultrasonicated resveratrol loaded starch nanocapsules: characterization, bioactivity and release behaviour under in-vitro digestion. Carbohydr Polym. 2021;251(June 2020):117111.
102. Kheradvar SA, Nourmohammadi J, Tabesh H, Bagheri B. Starch nanoparticle as a vitamin E-TPGS carrier loaded in silk fibroin-poly(vinyl alcohol)-Aloe vera nanofibrous dressing. Colloids Surf B Biointerfaces. 2018;166:9–16.
103. Yang J, He H, Gu Z, Cheng L, Li C, Li Z, et al. Conjugated linoleic acid loaded starch-based emulsion nanoparticles: in vivo gastrointestinal controlled release. Food Hydrocoll. 2020;101:105477.
104. Acevedo-Guevara L, Nieto-Suaza L, Sanchez LT, Pinzon MI, Villa CC. Development of native and modified banana starch nanoparticles as vehicles for curcumin. Int J Biol Macromol. 2018;111:498–504.
105. Shabana S, Prasansha R, Kalinina I, Potoroko I, Bagale U, Shirish SH. Ultrasound assisted acid hydrolyzed structure modification and loading of antioxidants on potato starch nanoparticles. Ultrason Sonochem. 2019;51(March 2018):444–50.
106. Ahmad M, Mudgil P, Gani A, Hamed F, Masoodi FA, Maqsood S. Nano-encapsulation of catechin in starch nanoparticles: characterization, release behavior and bioactivity retention during simulated in-vitro digestion. Food Chem [Internet]. 2019;270(April 2018):95–104. Available from: https://doi.org/10.1016/j.foodchem.2018.07.024
107. Bhatia M, Rohilla S. Formulation and optimization of quinoa starch nanoparticles: quality by design approach for solubility enhancement of piroxicam. Saudi Pharm J. 2020;28(8):927–35.
108. Liu Q, Li M, Xiong L, Qiu L, Bian X, Sun C, Sun Q. Oxidation modification of debranched starch for the preparation of starch nanoparticles with calcium ions. Food Hydrocoll. 2018;85:86–92.

109. Escobar-Puentes AA, García-Gurrola A, Rincón S, Zepeda A, Martínez-Bustos F. Effect of amylose/amylopectin content and succinylation on properties of corn starch nanoparticles as encapsulants of anthocyanins. Carbohydr Polym. 2020;250(May):116972.
110. Alzate P, Zalduendo MM, Gerschenson L, Flores SK. Micro and nanoparticles of native and modified cassava starches as carriers of the antimicrobial potassium sorbate. Starch/Staerke. 2016;68(11–12):1038–47.
111. Raigond P, Sood A, Kalia A, Joshi A, Kaundal B, Raigond B, et al. Antimicrobial activity of potato starch-based active biodegradable nanocomposite films. Potato Res. 2019;62(1):69–83.
112. Jeong O, Shin M. Preparation and stability of resistant starch nanoparticles, using acid hydrolysis and cross-linking of waxy rice starch. Food Chem. 2018;256:77–84.
113. Alzate P, Gerschenson L, Flores S. Ultrasound application for production of nano-structured particles from esterified starches to retain potassium sorbate. Carbohydr Polym [Internet]. 2020;247(June):116759. Available from: https://doi.org/10.1016/j.carbpol.2020.116759
114. Fonseca LM, Cruxen CE dos S, Bruni GP, Fiorentini ÂM, Zavareze E da R, Lim LT, et al. Development of antimicrobial and antioxidant electrospun soluble potato starch nanofibers loaded with carvacrol. Int J Biol Macromol. 2019;139:1182–90.
115. Sangeetha J, Shettar AK, Thangadurai D. Nanomaterials from biomass: an update. In: Handbook of nanomaterials and nanocomposites for energy and environmental applications. Cham: Springer International Publishing; 2021. p. 1005–22.
116. Jampílek J, Kráľová K. Preparation of nanocomposites from agricultural waste and their versatile applications. In: Multifunctional hybrid nanomaterials for sustainable agri-food and ecosystems. Elsevier; 2020. p. 51–98.
117. Capanoglu E, Nemli E, Tomas-Barberan F. Novel approaches in the valorization of agricultural wastes and their applications. J Agric Food Chem. 2022;70(23):6787–804.
118. Dey A, Dhumal CV, Sengupta P, et al. Challenges and possible solutions to mitigate the problems of single-use plastics used for packaging food items: a review. J Food Sci Technol. 2021;58:3251–69.
119. Hasanin MS. Simple, economic, ecofriendly method to extract starch nanoparticles from potato peel waste for biological applications. Starch - Stärke. 2021;73(9–10):2100055.
120. Raigond P, Raigond B, Kochhar T, Sood A, Singh B. Conversion of potato starch and peel waste to high value nanocrystals. Potato Res. 2018;61(4):341–51.
121. Nizzy AM, Kannan S. A review on the conversion of cassava wastes into value-added products towards a sustainable environment. Environ Sci Pollut Res. 2022;29(46):69223–40.
122. Chaudhary P, Fatima F, Kumar A. Relevance of nanomaterials in food packaging and its advanced future prospects. J Inorg Organomet Polym Mater. 2020;30(12):5180–92.
123. Duan B, Huang Y, Lu A, Zhang L. Recent advances in chitin based materials constructed via physical methods. Prog Polym Sci. 2018;82:1–33.
124. Barage S, Lakkakula J, Sharma A, Roy A, Alghamdi S, Almehmadi M, et al. Nanomaterial in food packaging: a comprehensive review. R L, editor. J Nanomater. 2022;2022:1–12.
125. Troncoso OP, Torres FG. Non-conventional starch nanoparticles for drug delivery applications. Med Dev Sensors 2020;3(6).
126. Zhu F. Underutilized and unconventional starches: why should we care? Trends Food Sci Technol. 2020;100:363–73.
127. Leandro GC, Cesca K, Monteiro AR, Valencia GA. Morphological and physicochemical properties of pine seed starch nanoparticles. Starch - Stärke [Internet]. 2023;2200225:1–7. Available from: https://onlinelibrary.wiley.com/doi/10.1002/star.202200225
128. Chen C, Li G, Zhu F. A novel starch from lotus (Nelumbo nucifera) seeds: composition, structure, properties and modifications. Food Hydrocoll. 2021;120:106899.
129. Pinto VZ, Moomand K, Deon VG, Biduski B, Zavareze E da R, Lenhani GC, et al. Effect of physical pretreatments on the hydrolysis kinetic, structural, and thermal properties of pinhão starch nanocrystals. Starch/Staerke. 2021;73(7–8):1–9.

Chapter 10
Starch Nanomaterials and Digestibility

Wilson Daniel Caicedo Chacon, Milena Dutra Pierezan, Germán Ayala Valencia, and Silvani Verruck

10.1 Introduction

Nanotechnology has revolutionized the field of material development and manufacturing with the advent of nanomaterials, which are materials with sizes ranging from 1 to 100 nm [1]. Nanomaterials have potential applications in various fields, including medicine, electronics, energy, and food science [2–4]. In recent years, there has been a growing trend in developing new food products based on nanopolymers, such as maltodextrins, cellulose, starch, proteins, among others [5]. Among the various types of polymeric nanomaterials, starch-based nanomaterials have attracted significant attention due to their potential applications in food and biodegradable packaging [6, 7].

Once these consumable nanomaterials are developed, it is of utmost importance to evaluate their quality aspects, toxicity, and digestibility. This evaluation ensures food safety, compliance with regulations, development of functional foods, and optimization of formulations. In particular, the study of digestibility allows for the assessment of various important aspects such as the speed and extent of food decomposition in the gastrointestinal tract, nutrient release and absorption in the small intestine, as well as resistance or fermentation capacity in the large intestine [8].

W. D. C. Chacon · G. Ayala Valencia
Department of Chemical and Food Engineering, Federal University of Santa Catarina, Florianópolis, SC, Brazil
e-mail: g.ayala.valencia@ufsc.br

M. D. Pierezan · S. Verruck (✉)
Department of Food Science and Technology, Federal University of Santa Catarina, Florianópolis, SC, Brazil
e-mail: silvani.verruck@ufsc.br

G. Ayala Valencia (ed.), *Starch Nanomaterials and Food Applications*,
https://doi.org/10.1007/978-3-031-60086-9_10

Specifically, the digestibility of starch-based nanomaterials (SNMts) is an important parameter that needs to be studied. Starch is an important source of carbohydrates in the diet and is found in foods such as cereals, tubers, legumes, and fruits [9]. However, its digestibility can vary widely among the various types of nanomaterials related to different plant sources, varying different parameters such as starch structure and the presence of other nutrients (Table 10.1). Starch conformation consists of two carbohydrates, amylose and amylopectin, which differ in structure and directly affect digestibility (Table 10.1) [10–13]. Amylose, a linear glucose polymer, is relatively resistant to digestion and therefore has a lower glycemic index. On the other hand, amylopectin, a branched glucose polymer, is more easily digested and has a higher glycemic index [14].

Within the process of evaluating the *in vitro* digestibility of SNMts, a series of enzymes and compounds are commonly used. These include pepsin, pancreatin, glucoamylase, amylase, lipases, trypsin, and compounds such as bile salts [15–18]. These enzymes have been frequently used following different methods, such as *in vitro* dialysis, *in vitro* enzymatic hydrolysis, and *in vitro* static models, which simulate the oral, gastric, and intestinal digestion processes [19–21]. These methods have allowed for the study of the digestibility of SNMts, evaluating geometric parameters such as size, shape, and surface area [20, 22], structural parameters such as crystallinity, amylose, and amylopectin content, and retrogradation [20, 23, 24], and starch content based on its digestibility characteristics, such as rapidly digestible starch (RDS), slowly digestible starch (SDS), or resistant starch (RS) in the small intestine [25–27].

Understanding the digestibility of SNMts for food purposes can help improve the efficiency of food use in humans and animals, as well as contribute to the promotion of gastrointestinal health. Therefore, this chapter aims to gather and present the available information on the digestibility of SNMts by evaluating the different methods and parameters involved in this process.

10.2 Digestibility of Starch Nanomaterials

10.2.1 Enzymes and Methods Used for the Digestion of Starch-Based Nanomaterials

SNMts have gained significant attention for their potential applications in the food and pharmaceutical industries due to their unique physicochemical properties [3, 6]. However, understanding the digestibility of these materials is a crucial factor that needs to be considered before their widespread application in consumer products. Starch is a complex carbohydrate that undergoes various digestion processes in the body, which can be studied through laboratory-scale analysis using *in vitro* digestibility tests. Within the *in vitro* digestibility processes, various enzymes and methods have been employed to simulate the digestion process of starch

Table 10.1 Summary of studies on starch nanomaterials for food application and their physicochemical characteristics

Starch architecture type	Starch source	Amylose content (%)		Particle size (diameter, nm)		Relative crystallinity (%)		Reference
		Native	Treated	Native	Treated	Native	Treated	
Nanoparticles	Waxy maize	0	Ni	Ni	5–100	Ni	Ni	[15]
	Waxy rice	0.16	Ni	3492.14	SGC: 46.50 ASGC1: 215.53 ASGC2:168.90 ASGC3 116.38 ASGC4: 127.50 ASGC5: 74.43	Ni	Ni	[55]
	Corn resistant starch	Ni	Ni	Ni	40 ± 33.2	Ni	Ni	[17]
	Waxy maize	0.85	Ni	Ni	46.3–172.6 (25 °C) 96.7–348.9 (60 °C) 143–223.1 (90 ° C)[a]	Ni	Ni	[46]
	Maize	26.9	Ni	Ni				
	Jackfruit (*Artocarpus heterophyllus* L.)	24.35–28.29	Ni	7.09–15.98	4.28–8.11 (small) 7.01–15.01 (medium) 8.05–16.94 (large)	Ni	Ni	[56]
Nanocapsules	Corn	Ni	Ni	Ni	30–40	Ni	Ni	[19]
	Maize	2	Ni	5000–25,000	SNP: 100–300 SNCs: 80	36.02	SNPs: 46.32 SNCs: 62.21	[20]
	Horse chestnut	Ni	Ni	Ni	322.7 ± 78.3	Ni	13	
	Water chestnut	Ni	Ni	Ni	559.2 ± 130.4	Ni	26	[57]
	Lotus stem	Ni	Ni	Ni	615.6 ± 95.7	Ni	22	

(continued)

Table 10.1 (continued)

Starch architecture type	Starch source	Amylose content (%)		Particle size (diameter, nm)		Relative crystallinity (%)		Reference
		Native	Treated	Native	Treated	Native	Treated	
	Water chestnut	Ni	Ni	3400	271	Ni	Ni	[39]
	Guayabo green bananas (*Musa paradisiaca*)	Ni	Ni	33,400 ± 180	BSN: 120 ± 0.05 CABSN:220 ± 0.03	Ni	Ni	[58]
	Horse chestnut	Ni	Ni	Ni	419.59	Ni	Ni	[59]
	Water chestnut	Ni	Ni	Ni	691	Ni	Ni	
	Lotus stem	Ni	Ni	Ni	797.32	Ni	Ni	
	Sorghum	33.54 ± 0.58	45.05 ± 0.25	2076.7	271.12	30.23	16.08	[11]
	Foxtail millet	30.58 ± 0.59	42.53 ± 0.50	1784.89	467.98	33.48	18.56	
	Quinoa	6.3	Ni	Ni	99	Ni	Ni	[60]
	Maize	25.8	Ni	Ni	214	Ni	Ni	
Nanoemulsions	Waxy maize[b]	Ni	Ni	Ni	120	Ni	Ni	[13]
	Sago	26.11	Ni	34,471 ± 11,000	59.4 ± 0.7	18.85 ± 1.27	28.54 ± 2.14	[61]
	Corn	13.35	Ni	14,190 ± 8100	68.1 ± 1.2	25.80 ± 1.33	35.95 ± 1.63	
	Potato	11.67	Ni	44,652 ± 9000	55.5 ± 0.6	15.25 ± 0.97	25.70 ± 2.28	
Nanocarrier	Corn	Ni	Ni	Ni	220.8 ± 418.9	47.5	17.0–18.9	[26]

SGC short glucan, obtained after native starch hydrolysis mediated by pullulanase, *ASG* amphiphilic short glucans, obtained after native starch modification with octenyl-succinic anhydride, *Ni* not informed, *SNPs* starch nanoparticles, *SNCs* starch nanocrystals, *CABSN* citric-acid banana starch nanoparticles, *BSN* banana starch nanoparticle

[a]Results are shown as a range of particle size obtained with different degrees of starch polymerization formed at different reaction temperatures

[b]Particle size of the whole nanoemulsion (starch and oil phase containing b-carotene)

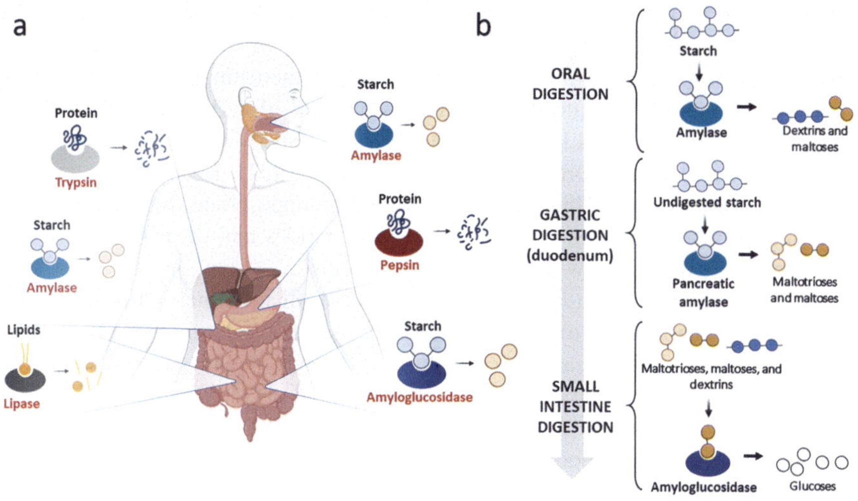

Fig. 10.1 Enzymes frequently used in *in vitro* digestibility for starch nanoparticle digestion for each of the human digestion phases

nanoparticles. Digestive enzymes such as pepsin, pancreatin, amylolytic glucoamylase, amylase, lipases, and trypsin, as well as compounds such as bile salts, have been frequently used in *in vitro* digestibility tests to evaluate the ability of starch-based nanomaterials to be digested and absorbed by the digestive system [16–18]. Each digestive enzyme has a specific function in the digestion process of starch-based nanomaterials, which are typically conjugated with proteins, lipids, or other carbohydrates (Fig. 10.1).

Firstly, amylase has been used to evaluate its ability to break down complex carbohydrates into simpler sugars. This is the first enzyme that acts on food, being produced and acting initially in the mouth. *In vitro* digestion usually involves the addition of amylase to a solution containing the food or compound being evaluated [25]. The solution is adjusted to an appropriate pH for amylase action, which is typically slightly alkaline, with an optimum pH range of 5–7, and they are usually used at 37 °C [25, 28]. In the case of starch-based nanomaterials, amylase could initiate the digestion process in them due to their carbohydrate-based composition (Fig. 10.1). In this case, amylase breaks down starch molecules into shorter and branched chains, known as dextrins and maltoses, which are easier to digest in the small intestine [28].

In simulated *in vitro* digestion processes, along with amylase, the use of pepsin is suggested. Pepsin is produced in the stomach and breaks down proteins into smaller peptides. It acts in an optimal pH range of 2–4 and this enzyme is typically used at 37 °C, simulating body conditions [29]. Its appropriate use can be given in SNMts that are conjugated with proteins.

Pancreatin is a mixture of enzymes produced by the pancreas, including amylase, lipase, and trypsin. Pancreatic amylase in this case is responsible for digesting starch that could not be digested by the amylase produced in the mouth. Lipase breaks down fats into fatty acids and glycerol, and trypsin breaks down proteins into smaller peptides, acting at optimal pH ranges of 7–11 and 8–9, respectively, and at an appropriate temperature of 37 °C for both [30]. Pancreatin is used in *in vitro* digestibility tests to simulate the pancreatic digestion of carbohydrates, fats, and proteins [31]. Amylase and lipase act in the small intestine, whereas trypsin acts in the duodenum, the first portion of the small intestine [31]. Because this mixture degrades various types of biopolymers, it would be appropriate for evaluating the digestibility of pure SNMts (producing oligosaccharides and dextrins of different sizes) or those conjugated with lipids and/or proteins.

Amyloglucosidase is an enzyme found in the small intestine and produced by cells in the intestinal villi, making it commonly used for *in vitro* digestion simulation. It is responsible for breaking down glucosidic bonds in glucose polysaccharides (such as starch and maltodextrin), acting at an optimal pH close to 4.5 [32]. As for the optimal temperature, its maximum performance occurs at 55–60 °C, but this does not mean it is inhibited at body temperature, because interaction with other enzymes allows its good functioning [32]. These enzymes may be suitable for the *in vitro* digestion evaluation of starch-based nanoparticles because this polysaccharide has already been broken down by amylases into dextrins, maltoses, and maltotrioses, and ultimately converted into glucose monomers.

Finally, bile salts are substances produced by the liver that help emulsify fats in the intestine [33]. They are used in *in vitro* digestibility tests to simulate fat digestion. They could aid in the digestion process of lipid-crosslinked starch nanoparticles (SNPs).

Collectively, these enzymes and compounds are essential for simulating the digestion and absorption of nutrients in the body, and their proper combined use of these can simulate the digestibility of SNMts, providing an opportunity to expand the understanding of the digestion process of these emerging nanomaterials. In recent years, different authors have employed some of these enzymes and compounds in different sequences to simulate the digestion process of starch-based nanomaterials. Cai et al. [15] in their research evaluated the digestibility process of waxy corn nano-starch at different degrees of debranching by amylase, sequentially using the enzymes pepsin, pancreatin, and amylolytic glucoamylase to emulate the breakdown of this polysaccharide from the stomach to the small intestine. It was observed that starch debranching favored digestion, suggesting that linear chains with lower molecular weights are more susceptible to enzymatic attack. Sivapragasam et al. [17] used pancreatic amylase and amylolytic glucoamylase to study the *in vitro* digestibility of soy flour nanoparticles, employing enzymes that specifically act on carbohydrates to simulate gastric and intestinal conditions. In this research, it was observed that, within 5 h of the simulated gastrointestinal process, the dissolution of the nanoparticles was less than 30%. Materials with less than 35% dissolution under simulated human physiological conditions can be used as enteric coatings for specific delivery to the colon [34]. Therefore, this study demonstrated the potential use of resistant SNPs from soy flour for colon-targeted delivery. Gautam et al. [16], in turn, used porcine amylase, porcine pepsin, porcine pancreatin, and bile salts to study the digestion process of octenyl succinylated modified waxy rice nano-starch, simulating the process from the oral phase to the intestinal phase. This gastrokinetic study revealed a reduction in the digestibility constant in the oral-gastric phase, with significantly improved kinetic constants in the intestinal phase, demonstrating sustained gastrointestinal stability. Octenyl succinic-modified starches with a higher degree of octenyl succinate group substitution have a more rigid and compact surface structure, providing superior protection against biochemical conditions in the gastric fluid. The results obtained by different authors have demonstrated that starch-based nanoparticles are promising compounds for use as vehicles for functional compounds, as they largely retain their integrity until the small intestine, allowing for the release of active compounds such as drugs.

The combination of some of these enzymes with special techniques to simulate digestibility describes methods that approach human conditions. Among the methods that have been most frequently employed for the *in vitro* simulation of starch nanoparticle (SNP) digestion are the dialysis method, enzymatic hydrolysis method, and static model method (Fig. 10.2).

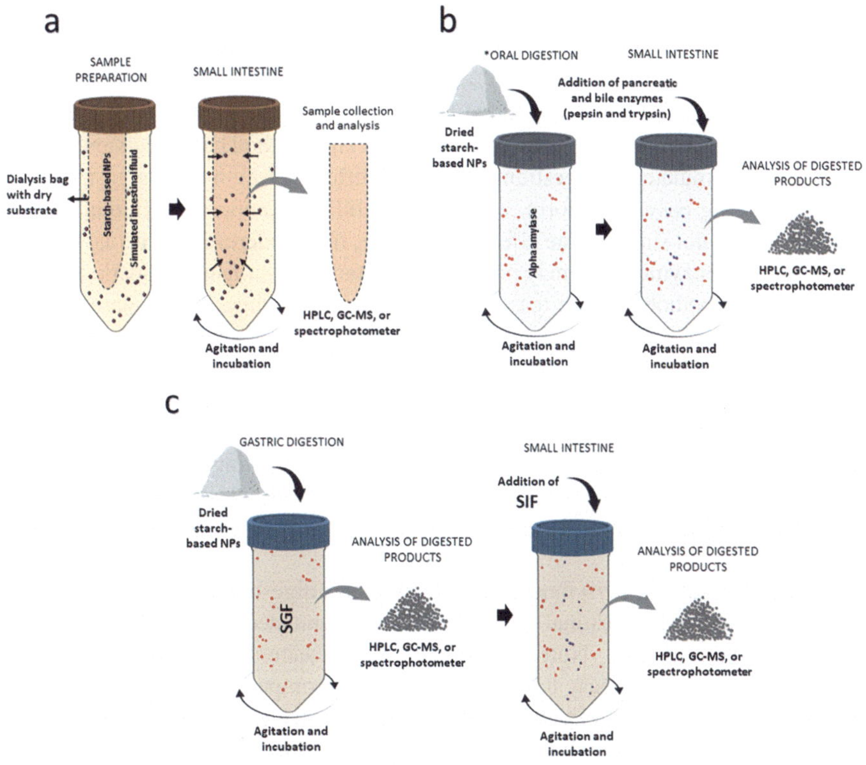

Fig. 10.2 Frequently used methods to studying the *in vitro* digestibility of starch nanoparticles. * The *in vitro* simulation of the oral phase has been optionally used for the three described methods

The *in vitro* dialysis method simulates the human digestion process by separating the soluble components from the insoluble components. A semipermeable membrane was used to perform this separation (Fig. 10.2a). Digestion occurs in different stages, with the addition of gastric and pancreatic juice, and nutrient release is monitored. The dialysis method is mainly used to simulate intestinal digestion and nutrient absorption in the small intestine [20]. The advantages of this method include its ease of use and its usefulness in comparing different foods, however, its disadvantages include the possibility of underestimating real digestibility, inability to simulate the effect of chewing, and inadequate enzyme secretion [20]. In the case of SNMts, it is feasible to use the dialysis method because pancreatic enzymes (amylase, proteases, and lipases) can be added, which allows for comprehensive breakdown of these types of nanomaterials. Hasanvand et al. [19] simulated the digestion process of starch nanoparticles as carriers for vitamin D fortification of milk, using the dialysis method. The main purpose of this study was to adjust the release kinetics of vitamin D3 using mathematical models such as Rigter-Peppas and Weibull. This study demonstrated that starch nanoparticles are suitable for improving the bioavailability of hydrophobic compounds in the gastrointestinal tract.

In the case of the *in vitro* enzymatic hydrolysis method, digestion is simulated by adding specific digestive enzymes to the food in a controlled environment [35]. Alpha-amylase was added to simulate digestion in the mouth, and then pancreatic and bile enzymes are added to simulate digestion in the small intestine (Fig. 10.2b). Finally, nutrient release is measured. This method has the advantages of being able to simulate the digestion of foods in different regions of the gastrointestinal tract, being easy to perform, and being useful for comparing different foods. However, it cannot simulate the effect of chewing, may underestimate actual digestibility, and the enzymes used may not be representative of human enzymatic activity [35, 36]. This method may be feasible for simulating the digestion of starch-based nanoparticles because it includes both oral and intestinal digestion and allows for the degradation of carbohydrates, proteins, and lipids. Liu et al. [20] employed this method to observe the digestion process of SNPs, starch nanocrystals (SNCs), cooked starch, and native corn starch. It was observed that native starch showed the lowest degradation in the gastrointestinal phase, exhibiting a digestion rate of 64% at 180 min, while cooked starch and SNPs, SNCs showed digestion rates of 99%, 96%, and 56%, respectively, indicating their degree of digestion based on crystallinity.

Finally, the static *in vitro* model method simulates digestion by incubating foods in a simulated gastrointestinal fluid (SGF) and simulated intestinal fluid (SIF) solutions (Fig. 10.2c). Nutrient release is measured at different stages [37, 38]. This method is easy to perform, useful for comparing different foods and optimizing processes, but it has the disadvantages of potentially underestimating real digestibility, not simulating the effect of chewing, and enzyme secretion may not be adequate. Unlike the dialysis and hydrolysis *in vitro* digestion methods, the static model method may better represent the digestibility of starch-based nanoparticles, as it includes the gastric stage [37, 38]. Ahmad et al. [21], used this method to perform *in vitro* digestion of horse chestnut, water chestnut, and lotus stem nano-starches, with the main objective of observing the release of encapsulated catechin. In this research, it was observed that a higher amount of catechin was released in the SGF phase compared with the SIF phase, indicating that in the first phase, the surface catechin is possibly released, followed by the release of the encapsulated compound. Furthermore, a controlled release of 0.5–1 h (SGF) and up to 4 h (SIF) was observed, suggesting that SNPs could serve as an effective protection for the slow release of catechin at the absorption site in the gastrointestinal tract when incorporated into potential functional foods. Ahmad et al. [39] also studied the release of probiotics from camel milk using the same method and observed a similar controlled release effect through the SGF and SIF phases. Similarly, Jhan et al. [11] studied the *in vitro* digestion of SNPs with folic acid, where they observed that the release of this compound was significantly lower under simulated oral conditions compared to gastrointestinal conditions, demonstrating that starch nanoparticles have a high potential for drug release at appropriate sites.

Through the various *in vitro* digestion methods employed by different researchers, it has been demonstrated that it is possible to conduct more detailed studies on the degradation of starch-based nanoparticles (which can be conjugated with proteins, lipids, or other polysaccharides) and the release of functional compounds, thus approaching closer to the real conditions of human digestion.

10.2.2 Particle Size, Shape and Surface Area

Once having knowledge about the enzymes and methods that have been used to simulate the digestibility of SNPs, we can move on to evaluating the different factors that influence the degradation of these types of materials. Geometric factors, such as particle size, shape, and surface area, can be parameters that mark significant differences in the digestibility of SNMts.

Particle size can influence the digestibility of SNMts. Decreasing the particle size increases the surface area available for the action of digestive enzymes, resulting in a higher rate of starch hydrolysis [40]. Larger starch granules can form a physical barrier that prevents the access of enzymes to the inner regions of the granule, which can reduce the rate of hydrolysis. Therefore, it has been demonstrated that reducing the particle size of starch can improve its digestibility and digestion rate [41]. However, it should be noted that particle size can also be influenced by other factors such as the presence of other compounds (other carbohydrates, proteins, and lipids), additives (surfactants), or processing form, which can affect the digestibility of starch differently. Such statements have already been confirmed, as shown in the research by Liu et al. [20], who observed that starch particle size impacts the digestion rate, with starch nanoparticles and nanocrystals being digested at nearly twice the speed of native starch granules. Li et al. [42] reported that starch nanoparticles were easily digested by amylases and intestinal enzymes, and the digestion rate increased as the nanoparticle size decreased. Luo et al. [43] found that smaller-sized starch nanoparticles exhibited higher solubility and greater digestibility than larger nanoparticles. However, Liu et al. [22] discovered that starch nanoparticles had lower digestibility than native starch due to their high surface area and resistance to enzymatic hydrolysis. Nanotechnology in the food area has highlighted this aspect as it has overseen the production of materials at the nanoscale, which significantly and positively impacts digestibility, as in the case of SNMts.

The particle shape of starch can also influence its digestibility. Specifically, starch-based nanoparticle particles can have different shapes, such as spherical, oval, cylindrical, among others. Although there are no scientific references on the influence of starch shape on digestion, the effect of this parameter on other types of materials can be used as a theoretical basis [44]. Spherical particles have a greater surface area than disk-shaped particles, which increases their accessibility to digestive enzymes and improves digestibility. Disk-shaped particles tend to form aggregates that can decrease the accessibility of digestive enzymes, thus reducing digestibility [44]. In the case of starch particles with irregular or elongated shapes, they may have greater mechanical resistance and require more time for complete digestion.

The surface area and chemical composition of SNMts also directly influence their digestibility. The greater the surface area, the more accessible and reactive the materials will be to digestive enzymes. Therefore, a larger surface area can increase the rate of digestion [45]. This is because a greater number of enzymes can bind to the surface. In addition, a larger surface area can also promote the formation of smaller, soluble degradation products, which can facilitate their absorption into the body [42, 45]. In the case of starch-based nanoparticles, they have an advantage over macroparticles because they have a greater exposed area due to their nanoscale size, which favors digestibility. However, it should be noted that the surface of starch-based nanomaterials may be composed of proteins and lipids, or it may have undergone chemical modifications, which can alter the interaction with the surrounding environment due to the functional groups on the starch. Wang et al. [46] demonstrated that the acetylation of SNPs decreased their digestibility by inhibiting the interaction between enzymes and starch molecules. Conversely, carboxylation of SNPs increased their digestibility by modifying their surface charge and enhancing enzyme adsorption. Gautam et al. [16] studied the digestion process of octenyl succinylated modified waxy rice nano-starch, simulating the process from the oral phase to the intestinal phase. This gastrokinetic study revealed that chemical modification allowed the creation of starch that is more resistant to biochemical conditions in the gastric fluid, enabling significant digestion to occur in the small intestine.

10.2.3 Crystallinity, Amylose–Amylopectin Ratio and Retrogradation Conditions

In addition to the geometric characteristics of SNMts, the distribution and organization of their molecules also affect the digestibility process, including factors such as crystallinity, the amount of amylose and amylopectin (Table 10.1), and retrogradation.

Crystallinity specifically refers to the molecular organization of starch in a highly ordered structure, where the chains of amylose and amylopectin are tightly bound together to form a crystalline structure [47]. The digestibility of starch is affected by its degree of crystallinity. In general, the more crystalline the starch, the more difficult it is to digest because the crystalline structure can protect the carbohydrate bonds from digestive enzymes, making the hydrolysis process slower [48]. On the other hand, less crystalline starches are easier to digest because their molecular structure is less organized and more accessible to digestive enzymes [48]. Starch digestibility can be affected by the different types of crystallinities it presents, including types A, B, C, and V [49, 50]. Type A crystallinity, which is the most common in starches, is characterized by a highly organized and compact structure, due to the parallel arrangement of the chains of amylose and amylopectin. This type of starch is the most common in plants and is easier to digest due to its less compact structure [48, 49]. Type A starch is further divided into two subtypes: type A and type C starch. Type C starch is less common and is mainly found in certain types of cereals. It is more resistant to digestion due to its more compact structure [49]. On the other hand, type B crystallinity is less organized and less compact, making it more susceptible to digestion. It is characterized by a perpendicular arrangement of the chains of amylose and amylopectin [48, 49]. This is less common than type A starch because its origin is animal. Finally, type V crystallinity is intermediate between type A and type B in terms of organization and compactness, and therefore, its digestibility can also vary [50, 51]. In general, starches with higher proportions of type B and V crystalline forms are considered to have greater digestibility than starches with higher proportions of type A crystalline form. Due to the modifications made to starch to reduce them to nanoscale particles, physical or chemical modifications are usually required, which can result in type V crystallinity. It should be noted that SNMts, due to the presence of other compounds such as proteins and lipids, may reduce their crystallinity and promote digestibility. Liu et al. [20] observed that the hydrolysis rate of SNPs with type B crystallinity was lower than that of SNCs with type A crystallinity. SNCs exhibited higher resistance to hydrolysis due to recrystallization. This recrystallization process may be attributed to the short linear chains of SNPs that are more prone to reorganization [52]. Despite SNCs having relatively higher crystallinity than starch nanoparticles, the hydrolysis rate of starch nanocrystals is higher than that of SNPs, possibly because SNCs show an increase in laminar structure and are of type A, exposing more interior regions susceptible to enzymes compared to SNPs, which are of type B.

Likewise, the amount of amylose and amylopectin can influence the digestibility of starch because amylose is more easily digested than amylopectin. Amylose is a linear chain of glucose, whereas amylopectin is a branched molecule with multiple binding sites [47]. Due to its structure, amylopectin is less accessible to digestive enzymes, which can limit its digestibility. Therefore, a higher amount of amylopectin in starch-based nanomaterials may result in reduced digestibility compared to a higher amount of amylose [24, 53]. The ratio of amylose to amylopectin varies depending on the source of starch and the type of process used to reduce starch granules to a nanometer size. Syahariza et al. [53], in their research on rice grain starch, concluded that the *in vitro* digestion rate tends to increase with longer amylose branches and smaller ratios between long amylopectin and amylose branches and short amylopectin branches.

Retrogradation refers to the process by which the amylose and amylopectin molecules in starch reassociate after being cooked and cooled. This can occur during the storage of starch-containing foods or during cooling after cooking [23]. Retrogradation can influence the digestibility of starch as the crystalline structure formed during this process can be more resistant to the action of digestive enzymes, thus reducing digestibility [27]. However, some studies suggest that retrogradation may also have health benefits, such as reducing carbohydrate digestibility and improving glycemic control [27, 54]. The level of retrogradation in SNPs depends on the mode of production, as bottom-up methods generally show more retrogradation than top-down methods. Therefore, when producing SNPs, it is important to define the type of processing to be used for their structural modification.

10.2.4 RDS, SDS, and RS in Different Types of Starch Nanomaterials

Starch is a complex polysaccharide found in many foods, and its digestibility can vary depending on its structure and properties. Three types of starch are of great interest in human nutrition: rapidly digestible starch (RDS), slowly digestible starch (SDS), and resistant starch (RS), which vary between different plant sources and undergo subsequent physicochemical modifications (Table 10.2).

RDS is quickly digested in the body due to its chemical and physical structure, meaning that it is rapidly converted into glucose in the small intestine. Foods containing a large amount of RDS tend to raise blood glucose levels faster than other types of carbohydrates. This is because they have a less complex structure and lower crystallinity, which facilitates their hydrolysis by digestive enzymes [62, 63]. Examples of foods high in RDS include French fries, white bread, and white rice.

On the other hand, SDS is more difficult to digest than RDS and is digested more slowly in the body. This is due to its physical and chemical structure, which has

Table 10.2 Summary of studies on starch nanomaterials for food application—hydrolysis or digestion evaluation

Starch architecture type	Starch source	Digestion condition (37 °C)	Studied effect	Digestion behavior		Reference
				Native	Treated	
Nanoparticles	Waxy maize	Pepsin (5.0 mg), pH 5.2 Pancreatin and amyloglucosidase (5 mL)	RDS, SDS, and RS content (%)	RDS: 4.3 ± 0.8 SDS: 66.7 ± 0.4 RS: 29 ± 0.4	RDS: 22.5 ± 0.3 (24 h), 4.9 ± 0 (ICM) SDS: 6.1 ± 0.1 (24 h), 8.2 ± 0.8 (ICM) RS: 71.4 ± 0.4 (24 h), 86.9 ± 0.8 (ICM)	[15]
	Waxy rice	OP: porcine amylase (75 U/mL), pH 7 GP: porcine pepsin (2000 U/mL), pH 3 GP: porcine pancreatin (8 x USP) and bile salts (8 mg/mL), pH 7	mg of reducing sugar per g of starch	GP: 55.39 IP: 66.10	GP: SGC 72.83, ASGC1 35, ASGC2 30, ASGC3 30 ASGC4 15, ASGC5 15 (2 h) IP: SGC 133.92, ASGC1 58.92, ASGC2 51.31, ASGC3 50.05, ASGC4 40.12, ASGC5 39.27 (3 h)	[55]
	Corn resistant starch	Pancreatic a-amylase (30 U/mL) and amyloglucosidase (300 U/mL)	Starch digestion rate after 16 h (%)	Ni	20.3	[17]
	Waxy maize	Ni	RDS, SDS, and RS content (%)	RDS: 28.1 ± 2.0 SDS: 64.4 ± 1.2 RS: 75 ± 1.2	RDS: 16.6–49.4 (C), 16.8–41.2 (25 °C), 14.5–35.1 (60 °C), 9.2–28.4 (90 °C) SDS: 27.5–32.7 (C), 39.0–50.4 (25 °C), 40.3–48.3 (60 °C), 35.3–43.9 (90 °C) RS: 12.5–53.3 (C), 19.8–32.8 (25 °C), 23.6–45.2 (60 °C), 27.7–55.5 (90 °C)[a]	[46]
	Maize			RDS: 76.8 ± 1.1 SDS: 13.0 ± 1.9 RS: 10.2 ± 0.8		

(continued)

Table 10.2 (continued)

Starch architecture type	Starch source	Digestion condition (37 °C)	Studied effect	Digestion behavior		Reference
				Native	Treated	
	Jackfruit (*Artocarpus heterophyllus* L.)	Porcine pancreatin (0.15 g/mL) amyloglucosidase (260 U/mL), pH 5.2	RDS, SDS, and RS content (%)	RDS: 9.54 ± 0.17 SDS: 13.68 ± 1.92 RS: 76.38 ± 2.1	RDS: 5.39 ± 0.17 (S), 8.83 ± 1.17 (M), 12.97 ± 0.17 (L) SDS: 27.18 ± 0.42 (S), 11.97 ± 0.50 (M), 4.21 ± 1.59 (L) RS: 67.02 ± 0.25 (S), 78.81 ± 0.67 (M), 82.42 ± 1.42 (L)	[56]
Nanocapsules	Corn	Dialysis bag method: GP: gastric buffer, pH 1 IP: pancreatic amylases, pH 7.4	Vitamin D3 release from starch nanocapsules (%)	Ni	GP: < 3.5 (8 h) IP: ~20–80	[19]
	Maize	Enzymatic hydrolysis: Porcine pancreatic α-amylase (290 U/mL) and amyloglucosidase (500 U/mL), pH 5.2	Starch digestion rate (%)	48.87 (30 min), 64.41 (180 min)	SNP: 39.09 (30 min), 56.32 (180 min) SNC: 75.89 (30 min), 96.35 (180 min)	[20]
	Horse chestnut	Static model: GP: pepsin (3 g/L), pH 2 IP: bile salts (10 g/L) and pancreatin, pH 8	Catechin released from starch nanocapsules (%)	Ni	GP: 5.94 ± 0.06 (0.5 h), 5.82 ± 0.03(1 h) IP: 3.87 ± 0.61(2 h), 4.8 ± 0.35(4 h)	[57]
	Water chestnut			Ni	GP: 5.53 ± 0.11 (0.5 h), 5.45 ± 0.42 (1 h) IP: 3.17 ± 0.48 (2 h), 3.39 ± 0.09 (4 h)	
	Lotus stem			Ni	GP: 5.57 ± 0.33 (0.5 h), 6.04 ± 0.11 (1 h) IP: 4.06 ± 0.96 (2 h), 5.35 ± 0.84 (4 h)	
	Water chestnut	Static model: GP: pepsin (3 g/L), pH 2 IP: bile salts (10 g/L) and pancreatin, pH 8	Probiotic cell viability in starch capsules (Log CFU/g)	8.40 ± 0.49	7.84 ± 0.18	[39]
	Guayabo green bananas (*Musa paradisiaca*)	GP: pepsin (3 g/L), pH 2 IP: bile salts (3 g/L) and pancreatin (1 g/L), pH 8	β-carotene release from starch nanoparticles (%)	Ni	GP: BSN: ~3 (20 min–2 h), CABS: ~3 (20 min-2 h) IP: BSN: ~10 (1 h 20 min), ~17 (1 h	[58]

					40 min–4 h), CABS: ~30 (1 h 20 min), ~35 (1 h 40 min), ~40 (2–4 h)	
	Horse chestnut	Static model GP: pepsin (3 g/L), pH 2.0 IP: bile salts (10 g/L) and pancreatin, pH 8	Resveratrol release from starch nanocapsules (%)	Ni	GP: 16.13 ± 0.7 (0.5 h), 16.30 ± 0.48 (1 h) IP: 11.62 ± 0.19 (2 h), 64.51 ± 1.64 (2.5 h)	[59]
	Water chestnut			Ni	GP: 19.15 ± 0.15 (0.5 h), 19.23 ± 0.23 (1 h) IP: 12.84 ± 0.07 (2 h), 54.54 ± 0.98 (2.5 h)	
	Lotus stem			Ni	GP: 16.38 ± 0.98 (0.5 h) 19.24 ± 0.17 (1 h) IP: 56.38 ± 0.64 (2 h), 75.83 ± 0.86 (2.5 h)	
	Sorghum	Static model: OP: α-amylase (0.2%), pH 7.2 GP: pepsin (3 g/L), pH 3 IP: bile salts (3 g/L) and pancreatin (10 g/L)	Folic acid released from total encapsulated product (%)	OP: ~5 (5 min) GP: 45 (0.5 h), ~40 (1 h) IP: 42 (0.5 h), 48 (1 h), 48 (2 h)	OP: ~5 (5 min) GP: 15 (0.5 h), ~20 (1 h) IP: ~56 (0.5 h), 60 (1 h), 80 (2 h)	[11]
	Foxtail millet			OP: ~5 (5 min) GP: 34 (0.5 h), ~36 (1 h) IP: ~54 (0.5 h), 50	OP: ~5 (5 min) GP: 17 (0.5 h), ~22 (1 h) IP: ~48 (0.5 h), 50 (1 h), 70 (2 h)	

(continued)

Table 10.2 (continued)

Starch architecture type	Starch source	Digestion condition (37 °C)	Studied effect	Digestion behavior		Reference
				Native	Treated	
				(1 h), 56 (2 h)		
	Quinoa	OP: salivary amylase (0.2%), pH 6.8 GP: pepsine:HCl ratio (1:3, g/mL), pH 2 IP: bile salts (10 g/L) and pancreatin (3 g/L)	Rutin released from starch nanoparticles (%)	Ni	OP: 0.57 (5 min) GP: 5.22 (0.5 h), 5.23 (1 h) IP: 3.50 (2 h), 3.50 (4 h), 3.59 (6 h)	[60]
	Maize			Ni	OP: 0.58 (5 min) GP: 5.22 (0.5 h), 5.23 (1 h) IP: 4.19 (2 h), 4.14 (5 h), 4.24 (6 h)	
Nanoemulsions	Waxy maize	Amano lipase (0.5 g), amyloglucosidase (5.3 μL/mL) and pancreatin (2.4 mg/mL). pH 7	Content of digested starch (%)	Ni	30–45 (10 min), 45–60 (20 min), 70–80 (120 min)	[13]
	Sago	GP: pH 1.2 IP: trypsin (20 mg/L), pH 7.5	Curcumin release from Pickering nanostarch emulsion (%)	GP: 76.44 (2 h)	GP: 32.15 (2 h) IP: ~15 (1 h), ~25% (2 h), ~50% (3 h)	[61]
	Corn			GP: 87.67 (2 h)	GP: 35.32 (2 h) IP: ~20% (1 h), ~30% (2 h), ~ 45% (3 h)	
	Potato			GP: 76.37 (2 h)	GP: 30.67 (2 h) IP: ~10% (1 h), ~35% (2 h), ~ 45% (3 h)	
Nanocarrier	Corn	Amyloglucosidase (500 U/mL) and α-amylase (290 U/mL), pH 5.2	Polyphenols release from starch nanocarrier (%)	Ni	GP: 78.06 (15 min), 87.15 (0.5 h), 93.82 (1 h); IP: 417.45 (2 h), 431.35 (1.5 h), 451.39 (4 h)	[26]

OP oral phase, *GP* gastric phase, *IP* intestinal phase, *Ni* not informed, *RDS* rapidly digestible starch, *SDS* slowly digestible starch, *RS* resistant starch, *ICM* isolated crystalline material, obtained after filtering debranched waxy maize starch product after 24 h of crystallization and drying at 40 °C in an oven overnight, *C* control, *S* small, *M* medium, *L* large

[a]Results are shown as a range of particle size obtained with different degrees of starch polymerization formed at different reaction temperatures

higher crystallinity and more complex structures. Slowly digestible starch can help maintain stable blood glucose levels and can provide energy over a longer period [62, 63]. Examples of foods containing SDS are brown rice, oatmeal, and legumes.

Finally, resistant starch (RS) is not digested in the small intestine due to its more complex structure, which prevents digestive enzymes from adhering to it, allowing it to reach the colon intact, where it is fermented by gut bacteria. Resistant starch has been associated with health benefits, such as reducing the risk of chronic diseases and promoting intestinal health [62, 64]. Examples of foods that contain RS include green bananas, cooked and cooled potatoes, and legumes. Therefore, the efficiency with which starch-based nanoparticles are digested depends on their origin and the physical and chemical treatment they have undergone. According to previous studies, jackfruit starch nanoparticles showed a high content of resistant starch, approximately 67%, and low content of RDS and SDS, with 5% and 27%, respectively [18]. In the same research, it was observed that with an increase in particle size, there was an increase in SDS content, whereas the content of RDS and RS decreased. This result indicates that the small granules of these starches exhibit the highest enzyme sensitivity. This is because smaller-sized starch granules have a larger surface area ratio, leading to a greater contact area between the enzyme and substrate. As a result, small granular starches are easier to digest. Overall, the content of RDS, SDS, and RS in starch nanomaterials varies depending on the source and the modification process. Cai et al. [15] observed that the RS content varied with the starch modification process. The debranching process of native maize starch for nanoparticle production increased the RS content from <5% to 87% after 24 h of treatment. Therefore, RS formation could be characterized as the aggregation and arrangement of double helices from short linear chains into a crystalline structure. The same effect was observed in the research by Qin et al. [46], where debranching and precipitation of starch to produce nanoparticles increased the RS content from 7% to 55%. The differences in enzymatic digestion behaviors between debranched waxy maize starches suggest that the debranching technique combined with controlled crystallization could be used to design starch structures with different digestibility levels. Therefore, understanding and optimizing these components in SNMts can have implications for the development of functional foods and other products. In summary, the digestibility of SNMts is a complex process that needs to be carefully considered before their application in food products.

10.3 Conclusions

The digestibility of starch nanomaterials (SNMts) is a crucial aspect to consider before their widespread application in the food and pharmaceutical industries, as it can provide the final endorsements to approve the consumption of a newly developed product. To evaluate the quality of SNMts, it is of great importance to understand the enzymes and their functionality in order to simulate the real conditions of human and animal digestibility. The review revealed that the most commonly used enzymes were amylases, pepsins, pancreatin, amylolytic glucoamylases, lipases, and trypsin, along with methods such as dialysis, enzymatic hydrolysis, and the static model method. These methods have helped to understand the process of digestive degradation of starch nanoparticles, considering that each method has its advantages and disadvantages in terms of simulating human digestion conditions.

These methods have helped to understand the digestibility of SNMts based on various factors such as the botanical origin of the starch, size, shape, and surface area, crystallinity, amylose and amylopectin content, and the type of compounds with which these materials are conjugated, such as lipids, proteins, or other carbohydrates. It has also been observed that the modification of starch by reducing it to nanoscale promotes the improvement of its digestibility, as resistant starch (RS) increases in some processes, helping regulate the glycemic index in the blood and providing better control over the metabolism of these types of nanomaterials.

In the future, more research is needed to explore the digestibility of SNMts and optimize their use in food and pharmaceutical applications. Understanding the interactions between enzymes and SNMts will help design nanocomposites with improved digestibility and controlled release properties. In addition, the development of advanced *in vitro* digestion models that better mimic human conditions will contribute to more accurate assessments of the digestibility of these nanomaterials. Overall, the study of enzymes and methods used for the digestion of SNMts provides valuable information about their potential applications and paves the way for the development of innovative and digestible nanomaterials in the future.

Acknowledgments W.D.C. Chacon gratefully acknowledges the Fundação de Amparo à Pesquisa e Inovação do Estado de Santa Catarina (FAPESC) for the PhD fellowship (3003/2021), as well as the Hermes 57983 project of the Universidad Nacional de Colombia sede Palmira, the opportunity to learn skills for academic writing. G. A. Valencia would like to thank the National Council for Scientific and Technological Development (CNPq) (grant 302434/2022-4) for financial support. The authors gratefully acknowledge the Federal University of Santa Catarina for the support.

Conflicts of Interest The authors declare no conflicts of interest.

References

1. Lizarazo-Salcedo CG, González-Jiménez EE, Arias-Portela CY, Guarguati-Ariza J. Nanomaterials: being closer to basics. Nanomater: Artículo especial Med Segur Trab (Internet). 2018;64(251):109–27.
2. Hossain A, Skalicky M, Brestic M, Mahari S, Kerry RG, Maitra S, et al. Application of nanomaterials to ensure quality and nutritional safety of food. J Nanomater. 2021;2021: 9336082
3. Damodharan J. Nanomaterials in medicine—an overview. Mater Today Proc. 2020;37(Part 2): 383–5.
4. Baig N, Kammakakam I, Falath W, Kammakakam I. Nanomaterials: a review of synthesis methods, properties, recent progress, and challenges. Mater Adv. 2021;2(6):1821–71.

5. Caicedo Chacon WD, Verruck S, Monteiro AR, Valencia GA. The mechanism, biopolymers and active compounds for the production of nanoparticles by anti-solvent precipitation: A review. Food Res Int. 2023;168:112728. Elsevier Ltd.
6. Nile SH, Baskar V, Selvaraj D, Nile A, Xiao J, Kai G. Nanotechnologies in food science: applications, recent trends, and future perspectives. Nano-Micro Lett. 2020;12:45. Springer.
7. Emamhadi MA, Sarafraz M, Akbari M, Thai VN, Fakhri Y, Linh NTT, et al. Nanomaterials for food packaging applications: a systematic review. Food Chem Toxicol. 2020;146:111825.
8. Chernukha IM, Meliashchenia AV, Kaltovich IV, Vasilevskaya ER, Aryzina MA, Smaliak TM, et al. Evolution of in vitro digestibility techniques: a systematic review. Theory Pract Meat Process. 2022;6(4):300–10.
9. Villarroel Heise P, Gómez C, Vera C, Torres J. Resistant starch: technological characteristics and physiological interests. Revista Chilena de Nutricion Sociedad Chilena de Nutricion Bromatologia y Toxilogica. 2018;45:271–8.
10. Ahmad M, Gani A. Ultrasonicated resveratrol loaded starch nanocapsules: characterization, bioactivity and release behaviour under in-vitro digestion. Carbohydr Polym. 2021;251: 117111.
11. Jhan F, Gani A, Noor N, ul Ashraf Z, Gani A, Shah A. Characterisation and utilisation of nano-reduced starch from underutilised cereals for delivery of folic acid through human GI tract. Sci Rep. 2021;11(1):1–15.
12. Remanan MK, Zhu F. Encapsulation of rutin using quinoa and maize starch nanoparticles. Food Chem. 2021;353:128534.
13. Lin Q, Liang R, Zhong F, Ye A, Singh H. Effect of degree of octenyl succinic anhydride (OSA) substitution on the digestion of emulsions and the bioaccessibility of ß-carotene in OSA-modified-starch-stabilized-emulsions. Food Hydrocoll. 2018;84:303–12.
14. Jenkins DJA, Wolever TMS, Jenkins AL. Starchy foods and glycemic index. Diabetes Care [Internet]. 1988;11(2):149–59. Available from: http://diabetesjournals.org/care/article-pdf/11/2/149/526237/11-2-149.pdf
15. Cai L, Shi YC, Rong L, Hsiao BS. Debranching and crystallization of waxy maize starch in relation to enzyme digestibility. Carbohydr Polym. 2010;81(2):385–93.
16. Gautam G, Talukdar D, Mahanta CL. Sonochemical effect on the degree of substitution of octenyl-succinic anhydride into waxy rice starch nanoparticles and study of gastro-intestinal hydrolysis using INFOGEST in vitro digestion method. Food Res Int. 2023;165:112348.
17. Sivapragasam N, Thavarajah P, Ohm JB, Margaret K, Thavarajah D. Novel starch based nano scale enteric coatings from soybean meal for colon-specific delivery. Carbohydr Polym. 2014;111:273–9.
18. Dong S, Fang G, Luo Z, Gao Q. Effect of granule size on the structure and digestibility of jackfruit seed starch. Food Hydrocoll. 2021;120:106964.
19. Hasanvand E, Fathi M, Bassiri A, Javanmard M, Abbaszadeh R. Novel starch based nanocarrier for Vitamin D fortification of milk: production and characterization. Food Bioprod Process. 2015;96:264–77.
20. Liu C, Jiang S, Han Z, Xiong L, Sun Q. In vitro digestion of nanoscale starch particles and evolution of thermal, morphological, and structural characteristics. Food Hydrocoll. 2016;61: 344–50.
21. Ahmad M, Mudgil P, Gani A, Hamed F, Masoodi FA, Maqsood S. Nano-encapsulation of catechin in starch nanoparticles: characterization, release behavior and bioactivity retention during simulated in-vitro digestion. Food Chem. 2019;270:95–104.
22. Liu W, Pan H, Zhang C, Zhao L, Zhao R, Zhu Y, et al. Developments in methods for measuring the intestinal absorption of nanoparticle-bound drugs. Int J Mol Sci. 2016;17:1171. MDPI AG.
23. Martínez P, Peña F, Gómez Y, Vargas G, Velezmoro C. Propiedades fisicoquímicas, funcionales y estructurales de almidones nativos y acetilados obtenidos a partir de la papa (Solanum tuberosum) var. "ÚNICA". Rev Soc Quím Perú. 2019;85(3):338–51.
24. Vetrani C, Sestili F, Vitale M, Botticella E, Giacco R, Griffo E, et al. Metabolic response to amylose-rich wheat-based rusks in overweight individuals. Eur J Clin Nutr. 2018;72(6):904–12.

25. Zheng M, Ye A, Singh H, Zhang Y. The in vitro digestion of differently structured starch gels with different amylose contents. Food Hydrocoll. 2021;116:106647.
26. Lei W, Liang J, Tan P, Yang S, Fan L, Han M, et al. Preparation of edible starch nanomaterials for the separation of polyphenols from fruit pomace extract and determination of their adsorption properties. Int J Biol Macromol. 2022;222:2054–64.
27. Wang S, Li C, Copeland L, Niu Q, Wang S. Starch retrogradation: a comprehensive review. Compr Rev Food Sci Food Saf. 2015;14(5):568–85.
28. Gómez-Villegas P, Vigara J, Romero L, Gotor C, Raposo S, Gonçalves B, et al. Biochemical characterization of the amylase activity from the new haloarchaeal strain haloarcula sp. Hs isolated in the odiel marshlands. Biology (Basel). 2021;10(4):337.
29. Salelles L, Floury J, Le Feunteun S. Pepsin activity as a function of pH and digestion time on caseins and egg white proteins under static: in vitro conditions. Food Funct. 2021;12(24): 12468–78.
30. Murthy SNS, Kostman J, Dinoso VP. Effect of pH, substrate, and temperature on tryptic activity of duodenal samples. Dig Dis Sci. 1980;25(4):289–94.
31. Ito K, Matsuura K, Mihara Y, Sakamoto Y, Hasegawa K, Kokudo N, et al. Delivery of pancreatic digestive enzymes into the gastrointestinal tract by pancreatic exocrine tissue transplant. Sci Rep. 2019;9(1):5922.
32. Sánchez HC, Mejía GC, Figueroa MC, Esquivia MM, Agudelo EL, Zapata GN, et al. ESTUDIO DE CEPAS NATIVAS AMILOLÍTICAS. Revista de la Facultad de Química Farmacéutica. 2005;12(2):21–8.
33. Xu F, Hu XJ, Singh W, Geng W, Tikhonova IG, Lin J. The complex structure of bile salt hydrolase from Lactobacillus salivarius reveals the structural basis of substrate specificity. Sci Rep. 2019;9(1):12438.
34. Dimantov A, Greenberg M, Kesselman E, Shimoni E. Study of high amylose corn starch as food grade enteric coating in a microcapsule model system. Innov Food Sci Emerg Technol. 2004;5(1):93–100.
35. Cruz-Casas DE, Aguilar CN, Ascacio-Valdés JA, Rodríguez-Herrera R, Chávez-González ML, Flores-Gallegos AC. Enzymatic hydrolysis and microbial fermentation: the most favorable biotechnological methods for the release of bioactive peptides. Food Chem Mol Scis. 2021;3: 100047.
36. Bernal Bustos CR, Morales D, Cuellar L, Jaramillo S. Enzymatic hydrolysis of starch. Fundacién Universitaria de América. 2017;10(1):129–40.
37. Xavier AA, Mariutti LR. Static and semi-dynamic in vitro digestion methods: state of the art and recent achievements towards standardization. Curr Opin Food Sci. 2021;41:260–73. Elsevier Ltd.
38. Brodkorb A, Egger L, Alminger M, Alvito P, Assunção R, Ballance S, et al. INFOGEST static in vitro simulation of gastrointestinal food digestion. Nat Protoc. 2019;14(4):991–1014.
39. Ahmad M, Gani A, Hamed F, Maqsood S. Comparative study on utilization of micro and nano sized starch particles for encapsulation of camel milk derived probiotics (Pediococcus acidolactici). LWT. 2019;110:231–8.
40. Al-Rabadi GJS, Gilbert RG, Gidley MJ. Effect of particle size on kinetics of starch digestion in milled barley and sorghum grains by porcine alpha-amylase. J Cereal Sci. 2009;50(2):198–204.
41. Amaral NO, Amaral LGM, Cantarelli VS, Fialho ET, Zangeronimo MG, Rodrigues PB. Influence of maize particle size on the kinetics of starch digestion in the small intestine of growing pigs. Anim Prod Sci. 2015;55(10):1250–4.
42. Li C, Hu Y, Li S, Yi X, Shao S, Yu W, et al. Biological factors controlling starch digestibility in human digestive system. Food Sci Human Wellness. 2023;12(2):351–8.
43. Luo K, Park KH, Lee DH, Hong CE, Song YW, Yoo SH, et al. Self-assembly kinetics of debranched short-chain glucans from waxy maize starch to form spherical microparticles and its applications. Colloids Surf B Biointerfaces. 2019 Apr 1;176:352–9.
44. Wang J, Zhang X, Jiang J. Healthy-angular, unhealthy-circular: effects of the fit between shapes and healthiness on consumer food preferences. J Bus Res. 2022;139:740–50.

45. Yang Z, Zhang Y, Wu Y, Ouyang J. Factors influencing the starch digestibility of starchy foods: a review. Food Chem. 2023;406:135009.
46. Qin Y, Wang J, Qiu C, Hu Y, Xu X, Jin Z. Effects of degree of polymerization on size, crystal structure, and digestibility of debranched starch nanoparticles and their enhanced antioxidant and antibacterial activities of curcumin. ACS Sustain Chem Eng. 2019;7(9):8499–511.
47. Jacobs H, Delcour JA. Hydrothermal modifications of granular starch, with retention of the granular structure: a review. J Agric Food Chem. 1998;6(8):2896–2905.
48. Gaenssle ALO, Satyawan CA, Xiang G, van der Maarel MJEC, Jurak E. Long chains and crystallinity govern the enzymatic degradability of gelatinized starches from conventional and new sources. Carbohydr Polym. 2021;260:117801.
49. Martens BMJ, Gerrits WJJ, Bruininx EMAM, Schols HA. Amylopectin structure and crystallinity explains variation in digestion kinetics of starches across botanic sources in an in vitro pig model. J Anim Sci Biotechnol. 2018;9(1):91.
50. Gutiérrez TJ, Bello-Pérez LA. Self-assembled and assembled starch V-type complexes for the development of functional foodstuffs: a review. Food Hydrocoll. 2022;125:107453. Elsevier B.V.
51. Dries DM, Gomand SV, Delcour JA, Goderis B. V-type crystal formation in starch by aqueous ethanol treatment: the effect of amylose degree of polymerization. Food Hydrocoll. 2016;61: 649–61.
52. Shi YC, Cui XC, Birkett AM, Thatcher MG. Resistant starch prepared by isoamlase debranching of low amylose starch. European Patent Office. 2003:EP1362869A1.
53. Syahariza ZA, Sar S, Hasjim J, Tizzotti MJ, Gilbert RG. The importance of amylose and amylopectin fine structures for starch digestibility in cooked rice grains. Food Chem. 2013;136 (2):742–9.
54. Parada JAS, Rozowski JN. Relationship between glycemic response of starch and its microstructural state. Revista chilena de nutrición. 2008;35(2):1–13
55. Gautam G, Talukdar D, Mahanta CL. Sonochemical effect on the degree of substitution of octenyl-succinic anhydride into waxy rice starch nanoparticles and study of gastro-intestinal hydrolysis using INFOGEST in vitro digestion method. Food Res Int. 2023;165(May 2022): 112348.
56. Dong S, Fang G, Luo Z, Gao Q. Effect of granule size on the structure and digestibility of jackfruit seed starch. Food Hydrocoll. 2021;120(March):106964.
57. Ahmad M, Mudgil P, Gani A, Hamed F, Masoodi FA, Maqsood S. Nano-encapsulation of catechin in starch nanoparticles: characterization, release behavior and bioactivity retention during simulated in-vitro digestion. Food Chem. July 2018;2019(270):95–104.
58. Santoyo-Aleman D, Sanchez LT, Villa CC. Citric-acid modified banana starch nanoparticles as a novel vehicle for ß-carotene delivery. J Sci Food Agric. 2019;99(14):6392–9.
59. Ahmad M, Gani A. Ultrasonicated resveratrol loaded starch nanocapsules: characterization, bioactivity and release behaviour under in-vitro digestion. Carbohydr Polym. 2021;251 (September 2020):117111.
60. Remanan MK, Zhu F. Encapsulation of rutin using quinoa and maize starch nanoparticles. Food Chem. 2021;353(November 2019):128534.
61. Du C, Jiang F, Hu W, Ge W, Yu X, Du S, kui. Comparison of properties and application of starch nanoparticles optimized prepared from different crystalline starches. Int J Biol Macromol. 2023;235(February):123735.
62. Copeland L. Starch structure, functionality and application in foods. In: Starch structure, functionality and application in foods. Springer Singapore; 2020. p. 169–79.
63. Copeland L, Blazek J, Salman H, Tang MC. Form and functionality of starch. Food Hydrocoll. 2009 Aug;23(6):1527–34.
64. Lehmann U, Robin F. Slowly digestible starch—its structure and health implications: a review. Trends Food Sci Technol. 2007;18(7):346–55.

Chapter 11
Degradation and Environmental Impacts of Starch Nanomaterials

Shirley Duarte, Omayra Ferreiro, and José Vega-Baudrit

11.1 Introduction

Starch is a biodegradable polysaccharide with several food applications and can produce biodegradable materials [1, 2]. Starch nanomaterials (SNMts) have received increasing attention in recent years because of their unique properties and because they offer the opportunity for distinctive use in various fields, such as food engineering, medicine, the textile industry, cosmetics, and others. SNMts have been used to develop nanocomposite films because these nanomaterials can improve mechanical and barrier properties [2–6].

Carbon dioxide, methane, water, inorganic compounds, or biomass could be produced during biodegradation. In addition, there is modification of the physicochemical and surface properties of the material, reduction of mechanical properties, and weight loss. It is a desirable process favored by microorganisms and environmental factors [1]. However, their environmental effects still need to be studied, as biodegradable materials may not be compostable, i.e., they could be disintegrated to reduced particle sizes, but these particles could still cause ecotoxic damage [7].

S. Duarte (✉)
Faculty of Chemistry, National University of Asunción, San Lorenzo, Paraguay
e-mail: sduarte@qui.una.py

O. Ferreiro
Faculty of Chemistry, National University of Asunción, San Lorenzo, Paraguay

Bio and Materials Laboratory, Polytechnic School, National University of Asuncion, San Lorenzo, Central, Paraguay
e-mail: oferreiro@qui.una.py

J. Vega-Baudrit (✉)
Centro Nacional de Alta Tecnología "Dr. Franklin Chang Díaz", Laboratorio Nacional de Nanotecnología (LANOTEC), San José, Costa Rica

POLIUNA, Escuela de Química, Universidad Nacional, Heredia, Costa Rica

G. Ayala Valencia (ed.), *Starch Nanomaterials and Food Applications*,
https://doi.org/10.1007/978-3-031-60086-9_11

Despite their advantages, SNMts can also harm the environment, which raises concerns about their proper use and disposal [8–10]. In addition, the potential toxicity of these materials must also be considered.

The present chapter aims to review the biodegradability standards and critically analyze the effect of the incorporation of nanomaterials on the biodegradability process of starch-based materials and their potential nanotoxicity and environmental impact.

11.2 Worldwide Degradation and Biodegradation Standards for Starch-Based Materials

A polymer may be degradable but not biodegradable. Various processes, such as physical disintegration, chemical degradation, and biodegradation by biological mechanisms, can disintegrate or degrade a polymer. However, biodegradable polymers degrade under the action of microorganisms such as molds, fungi, and bacteria within a specific period and environment [2].

Based on the American Society for Testing and Materials (ASTM), biodegradable is defined as "capable of undergoing decomposition into carbon dioxide, methane, water, inorganic compounds, or biomass," in which the predominant mechanism is the enzymatic action of microorganisms that can be measured by standard tests, in a specified period, reflecting available disposal conditions (ASTM standard D-5488-94d).

There are two categories of biodegradation standards for plastic materials: testing methods and performance specifications [2]. These provide information on starch-based materials' degradation standards and biodegradability, including SNMts. These standards define the environment and time of biodegradation. Thus, plastic materials can be considered biodegradable in a compost environment if they biodegrade in one growing season or 6 months. Alternatively, materials can be properly defined as biodegradable in marine environments if they partially biodegrade within 6 months. Two essential components of biodegradation are that the material must be a food source for the bacteria in the disposal environment. That biodegradation must occur within a short period [2].

The main normalization institutes active in biodegradable materials, with respect to setting standards for biodegradable and compostable plastics, are the ASTM, operating in the USA and Canada, and the International Organization for Standardization (ISO), operating worldwide.

The biodegradation standards established for the United States and Europe for the biodegradation of plastic materials under industrial compost, marine, anaerobic digestion, active landfill, and soil biodegradation environments are presented in Table 11.1.

Biodegradation standards require performance specifications and test method standards. For industrial compost, the biodegradation standards are ISO 14855 and

Table 11.1 Disintegration (DIS) and biodegradation (BIO) standards applicable to starch-based materials

	United States		Europe	
Biodegradation environments	Performance specification	Test method	Performance specification	Test method
Industrial compost	ASTM D6400 (plastic), ASTM D 6868 (Paper coating)	ASTM D 5338 (BIO)	ISO 17088 (plastic), ISO 18606 (packaging), EN*13432 (packaging), EN 14995 (plastic)	ISO 14855 (BIO), ISO 20200 (DIS), ISO 16929 (DIS), EN 14045 (DIS)
Marine	ASTM D7081-05	ASTM D6691 (BIO), ASTM D7474/ D7473M (BIO)	Ni	ISO 14851, ISO 148521999, ISO 16221 (BIO)
Anaerobic digestion	ASTM D7081-05	ASTM D5511-02 (BIO)	Ni	ISO 148532005 (BIO)
Active landfill	Ni	ASTM D5526 (BIO)	Ni	ISO 15985 (BIO), ISO 11734 (BIO)
Soil	Ni	ASTM 5988 (BIO)	Ni	EN 17033 (BIO), ISO 17556 (BIO)

ASTM American Society for Testing and Materials, *EN* European Standards, *ISO* International Organization for Standardization, *Ni* not informed

ASTM D5338, whereas the evaluation of disintegration during composting follows three main test standards: EN 14045, ISO 20200, and ISO 16929. For marine biodegradation, the standards are ASTM D7474/D7473M, ASTM D6691, ISO 14852-1999, ISO 14851, or ISO 16221. However, the first specific standards for marine biodegradation of plastic were published by the Organization for Economic Cooperation and Development (OECD 306 standard) [3]. For anaerobic digestion, the standards are ASTM D5511-02 and ISO 14853-2005. For active landfills, the standards are ASTM D5526, ISO 15985, or ISO 11734. Marine biodegradation, anaerobic digestion, active landfill, and European soil standards exist only for test methods and not performance specifications.

The standard covers four main aspects: Disintegration during composting and assessing how the material breaks down into smaller pieces during the composting process. Ultimate aerobic biodegradation: evaluating the complete biodegradation of the material in the presence of oxygen. There are no adverse effects of compost on terrestrial organisms to ensure that the resulting compost does not harm land-based organisms. Finally, to control constituents, monitor the components of the plastic material.

The use of these standards for biodegradability analysis of starch-based materials has already been reported in the literature. Hubackova et al. [4] performed biodegradability tests in an aqueous anaerobic environment using the ISO 11734 standard to determine the amounts of carbon dioxide and methane in biogas produced because of the anaerobic degradation in mesophilic sludge of blends of polycaprolactone/starch particles.

In another study, Ghasemlou et al. [5] measured the amount of carbon dioxide produced because of soil microorganisms' decomposition of starch films to determine biodegradation using an adapted method from ASTM D-5988-18. Nayak [6] determined the biodegradability of the PBAT/TPS/nano clay blend nanocomposite using a method from ASTM D 5338, in which the total biological oxygen demand (BOD) in an aerobic compost medium was measured. Cano et al. [11] analyzed the ultimate aerobic biodegradation level of films of starch/PVA under controlled composting conditions using the ISO 14855-1:2012 standard based on the carbon dioxide generated in the process related to the biodegradation process.

To quantitatively evaluate the degree of degradation of starch-based biopolymers, standard methods such as those described in Table 11.1 and other research methods and indices under different environmental conditions are used. However, it is necessary to consider that the process conditions present in industrial waste treatment plants may not be completely reproduced on a laboratory scale. Furthermore, dynamic processes only occur in natural or industrial environments, and experimental tests are often difficult to compare because of the differences in the multitude of matrices that can be used (i.e., soils and biopolymers). Therefore, biopolymers claimed to be biodegradable bioplastics may not degrade in large-scale industrial plants [3].

11.3 Effect of Starch Nanomaterials on the Biodegradability of the Materials

The biodegradability of starch-based nanocomposites can be influenced not only by the environmental conditions but also by the type, nature, concentration, chemical modification, and antimicrobial properties of the nanofiller, biodegradation test methods, and parameters, including temperature, moisture, humidity, pH, quantity, and type of microorganisms, among others [8–10]. Regarding the biopolymer, to be certified as a biodegradable material, as defined by the standards from Table 11.1, at least 90% of its mass should be decomposed into fragments that are less than 2 mm in composting conditions within 6 months.

Biodegradability tests can be performed by analyzing the material weight loss in a specific environment or determining the biogas produced or the carbon dioxide released during the degradation. The former indicates whether, in principle, the microorganisms can attack the substrate, while the latter reflects the metabolic conversion of the substrate in the microbial cell [4, 5].

Starch materials are inherent biodegradables because of their high-water solubility, making them unsuitable for many applications. The water absorption capacity of these materials causes hydrolysis activity, which catalyzes the degradation and breaking of polymer bonds, including glycosidic bonds [12]. During the biodegradation process, the pH of the medium decreases because microorganisms consume the polymeric matrix, producing carbon dioxide and water. The carbon dioxide is solvated, and carbonic acid is formed. Excessive acidification could reduce microorganism activity, limiting materials' biodegradation [13].

This section critically analyzes the effects of incorporating starch-based materials or materials containing SNMts on biodegradability.

11.3.1 Starch-Based Materials

Native starch has been blended with other biopolymers to make materials with improved characteristics. For example, these materials are used to prepare pellets to be processed in a subsequent step to make starch-based films. The biodegradability of these blends has also been tested.

Hubackova et al. [4] prepared blends of pellets of polycaprolactone (PCL) with different types of starch, such as native starch from maize, a genetically modified type of starch from maize, gelatinized maize, and amaranth starch, using glycerol as a plasticizer. Biodegradability tests were performed for each material in a mesophilic anaerobic sludge environment to determine whether the starch type and PCL integration influenced degradation. The different types of starch showed biodegradability varying between 70% (native starch) and 81% (amaranth starch), whereas PCL showed only 2%. Blend pellets of approximately 70:30 PCL/starch showed biodegradability from 5.6% to 10.7% depending on the starch type, the lowest being the native starch. When glycerol was used in the blend, the biodegradability of amaranth starch increased to 31.2%. The low biodegradability values for native starch blends were related to the size (5–20 μm and 1.79 m^2/g) and shape (round and symmetric) of these particles, resulting in a more compact structure with high resistance to anaerobic degradation. The authors concluded that starch grain type, size, and shape significantly affect the resulting biodegradability characteristics and their mechanical and thermal properties.

In another research, Ji et al. [14] prepared starch-sisal fiber composites by adding inorganic (talcum powder, $CaCO_3$) or organic (eggshell powder) materials as fillers, highlighting that the nature of the filler is crucial for the determination of biodegradability. In this case, the eggshell powder, due to its organic matter content, acted as a bridge, favoring hydrolysis and subsequent decomposition into small substances and making degradation easy.

11.3.2 Starch-Based Films

Starch-based films are the most common material reported in the literature and are prepared by several techniques such as extrusion blowing, twin screw extrusion followed by thermo-molding, and casting, depending on the polymer solution characteristics [1, 7, 15–17].

The biodegradability of simple starch-based films has already been widely discussed. Although these discussions mainly focus on the biodegradability rate rather than the effects of the compounds released into the environment, biodegradability does not guarantee that the material is compostable [7].

Several modifications have been proposed in the literature for increasing the moisture resistance of the prepared films. The modifications are oriented to the raw material, i.e., modification of native starch, to prepare blends with other polymers, or to add fillers to the starch matrix. All these approaches primarily seek to increase hydrophobicity, which could cause an evident detriment to the biodegradability rate of the material. However, adding fillers can confer antibacterial and antioxidant characteristics on the starch film [17].

Among the modifications proposed for native starch, crosslinking is the most widely used method for synthesizing modified starch. Zhou et al. [18] prepared modified starch films by inducing the decomposition of benzophenone into free radicals and depriving starch macromolecules of hydrogen under UV irradiation, achieving a film with higher mechanical properties and better water resistance than native starch, which was attributed to the reduction of free hydroxyl groups.

Retrogradation is common in starch-based foods and affects the product's aroma, texture, and nutritional characteristics. This process occurs through the reassociation or recrystallization of polysaccharides in gelatinized starch due to exposure to freeze/thaw cycles or moisture migration [19]. Li et al. [20] prepared compressed films using native maize starch subjected to retrogradation for 0, 8, and 14 days. They observed an increase in the degree of crystallinity with retrogradation time. Enzymatic degradation by fungal α-amylase was evaluated. It was found that films without retrogradation degraded following kinetics of two stages, each first-order, whereas retrograded films followed a simple first-order kinetics. In all cases, the retrograded films presented slower degradation, which was related to the increase in the degree of crystallinity. Regarding the films without retrogradation, these films presented an amorphous structure, which resulted in rapid degradation in the first 90 min, followed by a slower degradation rate of the more resistant portion of the film, similar to the degradation rate of retrograded films.

In another research, Rammak et al. [15] prepared films using native and crosslinked cassava starch with and without adding kaolin, an inorganic material. The physical appearance of these films was analyzed after burial in the soil. The authors found that adding kaolin impedes water penetration into the polymer matrix, which reduces the degradation rate due to the increase in the hydrophobicity of the material.

Ghasemlou et al. [5] used a blend of non-isocyanate polyhydroxyurethanes (PHUs) as a green material along with starch (ST) to prepare ST-PHU films. The authors found that ST-PHU films showed a weight loss of approximately 88% during biodegradation under soil-burial conditions, which was higher than the weight loss obtained for ST films (around 69%). This was explained by the capacity of PHU to act as a water-holding agent, increasing the contact of starch with water. It is worth mentioning that the authors found that the addition of cellulose nanocrystal (CNC) was detrimental to the biodegradation process, which was attributed to the crystallinity of CNC and PHU.

Butylene adipate-co-terephthalate (PBAT) is a hydrophobic biodegradable polymer with similar properties to polyethylene. Blends of starch/PBAT have been proposed to improve properties such as the processability and the brittleness of thermoplastic starch (TPS). However, these blends have limited compatibility between both materials [1, 12]. Pokhrel et al. [1] proposed using citric acid as a compatibilizer. As expected, the PBAT/TPS films with higher starch concentrations (from 5% to 30%) resulted in an increased biodegradation rate. However, Wongphan et al. [12] proposed the modification of TPS to improve its compatibility with PBAT. Native (NS), acetylated (AS), octenyl-succinate (OS), and hydroxypropylated (HS) starch were blended with PBAT for the preparation of films, and their biodegradation rates were analyzed. The authors found that the compatibility between PBAT and modified starch depended on the hydrophobicity of modified starch. At the same time, the authors found that biodegradability depended on starch hydrophilicity (HS~NS>AS>OS).

Zhai et al. [16] prepared starch/PBAT nanocomposite films using nanoclay as a filler by extrusion blowing. Organic montmorillonite modified with octadecyl dimethyl benzyl ammonium chloride was incorporated into the starch matrix film. The authors did not analyze the biodegradation rate or the effects of adding this nanoparticle on the material's biodegradability. However, Gutiérrez et al. [7] prepared corn starch-based films and added a pH-sensitive nanoclay packaged with Jamaica flower (Hibiscus sabdariffa) extract. This nano clay is montmorillonite modified with dimethyl benzyl hydrogenated tallow ammonium. The biodegradability and the ecotoxicity of these films were analyzed, and it was found that this nanoclay system is biodegradable though not compostable, mainly at high concentrations (100 μg/mL), as measured by its effect on lettuce seedlings.

Zoungranan et al. [13] systematically analyzed the influence of humidity, temperature, and enrichment in soil microorganisms during the biodegradation of cassava and corn starch films. The authors found that cassava starch films degraded faster than those based on corn, which is related to the crystalline structure of the starch from each source. Adding Cola cordifolia, a natural ingredient, improved the biodegradability of the materials. Regarding the soil factors analyzed, it was observed that humidity promotes starch biodegradability up to a maximum rate of 15%, above which microbial growth is predominant. The soil enrichment was also limited to a maximum of 20%, explained by the increase in microbial population density. This could affect their mobility and create mutual interactions that reduce or limit their film access. Finally, temperature harms most soil microorganisms, affecting their biodegradability capacity. In addition, the authors compared a normal soil with a sterile one, achieving biodegradation after 14 days of about 40% for cassava-based films in normal soil and less than 5% for the same material in a sterile soil. This highlights the role of microorganisms in the degradation of starch materials.

11.3.3 Starch Nanoparticles

Native starch is widely used to prepare biodegradable starch-based films by mixing with other materials to increase their strength. Native starch is also a raw material for the preparation of starch nanoparticles (SNPs), mainly by the "top-down" approach, in which the SNPs are produced by the breakdown of starch granules [21].

SNPs are highly hydrophilic because of their OH-rich surface [22]. It has been reported that adding starch nanoparticles (SNPs) to films increases the contact angle and decreases the permeability of water vapor, oxygen, and other gasses by increasing the pathway tortuosity [23–27].

SNPs have even been used to reinforce the properties of starch films, taking advantage of the good interaction between the materials [23, 27–29]. Besides, it has been reported that the biodegradability of starch films increases with the increase of SNPs in the matrix, becoming even greater than starch films without SNPs [6, 30, 31]. Kumari et al. [27] also observed improved thermal stability by incorporating SNPs into the starch-film matrix.

Modified SNPs are widely proposed to increase the interaction between SNPs and the polymeric matrix and modify its hydrophilic character [22]. SNPs act as a reinforcing material, increasing tensile strength, and their hydrophilic nature enhances the compatibility of the PBAT/starch blend [25, 30]. Starch NPs have been successfully applied as carriers for drug delivery [32]. The SNPs have also been used in textile applications. Mostafa et al. [33] showed that incorporating 10–50 g/L of SNP into fabric samples allowed us to obtain a reinforced cotton fabric even after 10 washing cycles.

The SNP preparation also influences the final characteristics of the SNPs and their application. Hakke et al. [34] synthesized SNPs using ultrasound-assisted acid hydrolysis. The anionic functional group on the starch nanoparticle (SNP) surface favored the extraction of cupric ions [35]. SNPs were synthesized from barley starch by hydrolysis using sulfuric acid, transforming A+V-type round or disc-shaped native starch into round or irregular sharpened A-type SNPs.

Dong et al. [36] prepared a core-shell starch-based nanoparticle (CSS NP) consisting of a hard starch core and a soft poly(methylacrylate) (PMA) shell for incorporation into a poly(propylene carbonate)/poly(lactic acid) (PPC/PLA) blend. The CCS NPs enhanced the compatibility of the PPC/PLA blend and the mechanical properties of the polymer blends.

Liu et al. [37] investigated the kinetics of the enzymatic hydrolysis of native starch, cooked starch, starch nanocrystals (SNCs, 70–100 nm), and SNPs

(100–300 nm). They found that, as expected, the cooked starch showed a high digestibility rate because the crystalline region was unfolded. SNPs presented the slowest hydrolysis rate, attributed to the compact structure formed during the recrystallization of short glucan chains throughout their preparation and high relative crystallinity (46.32%). However, the SCNs showed even higher crystallinity (62.21%) and a hydrolysis rate similar to that observed for the cooked starch. These results were explained considering that the SNCs have an A-type crystallization characterized by pores and channels contributing to hydrolysis. In contrast, the SNPs exhibit B-type crystallization, which does not present pores. Thus, SNPs are suitable for applications with a low hydrolysis rate. Table 11.2 shows the biodegradability of different starch-based materials performed in different mediums.

Starch has been used to prepare biodegradable films alone or in combination with other materials. The biodegradability of starch films and nanoparticles has been widely tested and accepted. However, biodegradability can change when other materials are added to the starch matrix. In some cases, more biodegradable materials have been obtained by blends with biopolymers or by adding nanoparticles from starch, nanocellulose, or clays. Nevertheless, the material's biodegradability should be analyzed. It is worth mentioning that biodegradability does not guarantee that the material is compostable; therefore, the effects of the compounds released into the environment during the biodegradability process must be studied.

11.4 Effect of Starch Nanomaterials on the Environment

Starch is one of the most abundant natural biopolymers globally, which can be reinforced with SNPs and starch nanocrystals (SNCs) to achieve desirable properties [38, 39]. The difference between SNPs and SNCs was discussed in Chaps. 1, 2 and 3. Despite their advantages, SNMts (SNPs and SNCs) can also harm the environment, which raises concerns about their proper use and disposal. In addition, the potential toxicity of these materials and life cycle assessments (LCA) of starch and starch-based compounds are necessary to ensure their lower environmental impact and sustainability [8, 40, 41].

SNMts are biodegradable and, in theory, degrade rapidly in the environment. Generally, the biodegradation of starch-based nanocomposites has been improved or minimally affected after adding nanofillers and additives with antioxidant and antimicrobial properties.

However, several studies have shown that the degradation of starch SNMts may be slower than expected. Qi et al. [42] found that SNMts can remain in the soil for more than 6 months, negatively impacting soil quality, plants, and wildlife.

During biodegradation, carbon dioxide production, modification of physicochemical and material surface properties, reduction of mechanical properties, and weight loss occur. This is a desirable process favored by microorganisms and environmental factors [1]. However, other compounds could be produced, and their environmental effects are still poorly studied.

Table 11.2 Biodegradability analysis of the different starch-based materials

Environmental type	Material	Weight change (%)	Time	Ecotoxicity	Reference
Soil burial	PBAT film PBAT/TPS 95/5 film PBAT/TPS 80/20 film PBAT/TPS 70/30 film	~5 ~17 ~25 ~27	8 weeks	Ni	[1]
Anaerobic aqueous environment of mesophilic sludge	PCL70/ Meritena Starch pellet PCL70/Amaranth starch pellet PCL61/ Meritena Starch pellet26/Gly13 PCL70/Amaranth starch pellet26/Gly13	6.5 11.9 13.8 30.6	1400 h	Ni	[4]
Vegetable compost (soil)	TPS film TPS+NMnt film TPS+NMnt +JFE film TPS+MntMD +JFE film	100 100 100 100	90 days	All of them presented high cytotoxicity, and they are not compostable	[7]
Soil burial	Cassava starch film only Starch + Cola cordifolia extract film	~40 ~50	14 days	Ni	[13]
Soil burial	Composite foam of starch-sisal fiber Without filler Talcum powder $CaCO_3$ Eggshell powder	71 61 60 67	30 days	Ni	[14]
Soil burial	Native potato starch film Nanocomposite starch film (0.5%)	49 71	1 week	Ni	[31]

Gly glycerol, *JEF* Jamaica flower (*Hibiscus sabdariffa*) extract, *MntMD* montmorillonite modified with dimethyl dihydrogenated tallow ammonium, *NMnt* natural montmorillonite, *PCL* polycaprolactone, *PBAT* polybutylene adipate terephthalate, *TPU* thermoplastic polyurethane film, *TPS* thermoplastic starch, *Ni* not informed

This section reviews starch nanomaterials' potential nanotoxicity and environmental impacts during their degradation. Appropriate handling practices that could be implemented to minimize the environmental and toxic impacts of starch nanomaterials are included.

11.4.1 Potential Nanotoxicity of Starch Nanomaterials

Nanotoxicology is the study of the potentially toxic effects of nanomaterials on living organisms. Although there is limited research on the potential nanotoxicity of starch nanomaterials, some studies have shown that they can cause adverse effects in living organisms.

Regarding the toxicity of starch nanomaterials, there is more information on the nanotoxicity of SNPs. SNPs are particles with at least one dimension smaller than 1000 nm but larger than a single molecule or atom. They are used in various fields, such as food, medicine, and cosmetics. These nanoparticles are derived from starch granules and possess unique physical properties. In addition, they are considered biodegradable, making them suitable for various applications [21].

SNPs have shown potential as safe and effective emulsifiers in various industries, including cosmetics [21]. In the biomedical field, incorporating SNPs into starch materials has been explored to enhance their properties, such as processability and mechanical strength [43]. Furthermore, SNPs have been used in the food industry to encapsulate biologically active food ingredients, demonstrating their versatility and potential in this sector [44].

However, the toxicity of SNPs has been studied in rats, and it was found that they can induce oxidative stress and toxicity [45]. Oxidative stress is an imbalance between free radicals and antioxidants in the body. It can cause damage to organs and tissues and result in various diseases such as cancer, Alzheimer's disease, Parkinson's disease, diabetes, cardiovascular conditions such as high blood pressure, atherosclerosis, stroke, inflammatory disorders, chronic fatigue syndrome, asthma, male infertility, and more [46]. These findings highlight the need for further research to fully understand the potentially toxic effects of SNPs, including their dose-response relationships, and the importance of proper handling and disposal practices to minimize environmental and health impacts. This study contributes to the ongoing effort to ensure the safe and responsible use of nanomaterials in various fields, including food engineering, medicine, and the textile industry [46].

The toxicity of SNMts may vary depending on their size, shape, surface area, and other properties. Further research is needed to understand the potentially toxic effects of these materials and to develop appropriate safety guidelines.

Regarding the proper handling of SNMts, there are some concerns about the toxicity of starch, especially regarding its use in medical and pharmaceutical products. Therefore, it is essential to implement proper handling practices to minimize the potential nanotoxicity of starch nanomaterials. This may include using protective equipment when handling these materials, avoiding inhalation or ingestion, and ensuring proper waste disposal.

Starch is commonly used as a binding and thickening agent in medical and pharmaceutical products. However, its use has been questioned because of possible toxicity. The United States Food and Drug Administration (FDA) presents some recommendations for using starch derivatives as food additives [47], considering the potential risks of inappropriate use for human health. These guidelines could also be applied to the drug delivery systems (DDS). In addition, impurities from solvent systems and other chemicals need to be determined during the starch derivatives preparations [48].

11.4.2 Degradation and Environmental Impacts of Starch Nanomaterials

Nanocomposites derived from renewable resources, specifically starch-based ones, have attracted significant attention because of their environmentally friendly attributes and diverse applications [49, 50]. Starch, a biodegradable and abundant biopolymer, provides an attractive foundation for developing green materials [39, 51].

In general, it is expected that the lower persistence of biodegradable polymers in the environment could reduce the problems related to plastic pollution, for example, the risks related to plastic leakage or microplastic and nanoplastic pollution [52]. Only a few studies have considered the effects of polymer degradation in soil [53].

Most studies do not perform ecotoxicity tests [54]. They focus mainly on evaluating the degree of biodegradation, considering that the degradation of the bioplastic material does not imply the release of toxic compounds. However, accumulation of metabolic intermediates, depletion of oxygen in the soil, or variation in the physical-chemical characteristics of the soil may occur. Another factor to consider is that in a real environment, the results may not be the same as those obtained on a laboratory scale [3].

Physical modification of starch, including methods to improve processing properties, is a major focus of ongoing research [55, 56]. In pursuing a sustainable future, starch-based nanocomposites have been modified via green cross-linkers, enhancing their thermal properties and biodegradation capabilities [57].

The addition of antimicrobials and nanoclay fillers can affect these materials' biodegradation behavior and barrier properties [9, 11]. Despite these advances, these modifications' economic and environmental aspects warrant thorough investigation [40].

As mentioned in Sect. 11.3, Gutiérrez et al. [7] prepared nano clay (montmorillonite modified with dimethyl benzyl hydrogenated tallow ammonium) corn starch-based films with a pH-sensitive Jamaica flower (Hibiscus sabdariffa) extract. The biodegradability and ecotoxicity of these films were analyzed, and it was found that this nanoclay system is biodegradable though not compostable, mainly at high concentrations (100 μg/mL), as measured by its effect on lettuce seedlings. The authors concluded that the cytotoxicity of any nanofiller should be mandatorily evaluated before using it as a reinforcing material of films and that the accumulation of these reinforcement materials, as the modified nanoclays used in that work, could harm the vegetation, giving them limited compostability [7].

SNMts can negatively impact the environment. They can remain in the soil for over a year, negatively affecting soil quality and wildlife [42]. In addition to the possible accumulation of SNMts in the environment, there are concerns about the environmental impacts during the production of these materials. Lavoine et al. [58] found that the production of SNMts can significantly impact greenhouse gas emissions and air and water quality.

LCA has emerged as an important tool for evaluating the environmental impact of starch and starch blends throughout their lifecycle [41, 59]. This holistic approach is essential for understanding the environmental performance of bioplastics and starch plastics [60, 61]. LCA is equally applicable in nanotechnology, providing a sustainability perspective on producing nanocomposites [62].

Using specific types of starch, such as banana, cassava, and maize starch, in nanocomposites has demonstrated potential in various applications, from cosmetics to food packaging [63–66]. The properties of these starches are largely influenced by their structural features, including their amylose content [67, 68]. Manipulating these structural characteristics enables the production of starch-based nanocomposites with enhanced properties and functionalities [69].

Research is also being conducted on the dispersion of multiwalled carbon nanotubes within starch-based films using ionic surfactants and the synthesis of composite materials that include starch and other polymers, such as polystyrene, that demonstrate biodegradability [70, 71]. Furthermore, the addition of fibrous TiO_2 and nano $CaCO_3$ fillers, as well as graphene, to starch-based composites has been shown to improve their structural, mechanical, barrier, and optical characteristics, as well as their biodegradation behavior [72–74].

The research, development, and application of starch-based nanocomposites are progressing rapidly. Their potential to reduce environmental impact and diverse applications promises to shape a more sustainable future [8].

However, proper handling practices help minimize the environmental and potential toxic impacts of SNMts. This may include using starch materials derived from renewable and sustainable sources and implementing recycling systems to reduce the number of materials ending up in landfills, even considering the materials used during processing.

11.4.3 Life Cycle Assessment

LCA is a methodology designed to assess the environmental aspects and potential impacts of a product, process, or service [59, 75]. LCA has been standardized through the ISO 14040 series [8]. The LCA process is typically divided into four stages: principles and framework (ISO 14040:1997), goal and scope definition and inventory analysis (ISO 14041:1998), life cycle impact assessment (ISO 14042: 2003), and interpretation (ISO 14043:2003).

The first stage involves setting the purpose of the study, defining the functional unit (the basis for comparison), and establishing system boundaries. Inventory analysis involves data collection on energy use, material inputs, and environmental releases associated with each product's life cycle stage. The impact assessment stage evaluates potential environmental impacts based on inventory data. Finally, interpretation involves summarizing and discussing the results, assessing the robustness of the study, and making recommendations [40, 60].

LCA can address several categories of environmental impact, such as global warming, ozone depletion, acidification, eutrophication, resource depletion (fossil fuels), ecotoxicity, human toxicity, photo oxidant formation, air smog, agricultural land use, etc., are found for products or processes [59, 60]. This approach has emerged as a valuable tool in various scientific fields, including the study of nanocomposites and biopolymers, to help assess their environmental implications [41, 62].

In the realm of starch-based nanocomposites, LCA presents a systematic approach for evaluating the environmental impacts of these materials throughout their life cycle. Figure 11.1 depicts the simplified framework for the LCA of nanocomposite materials [8].

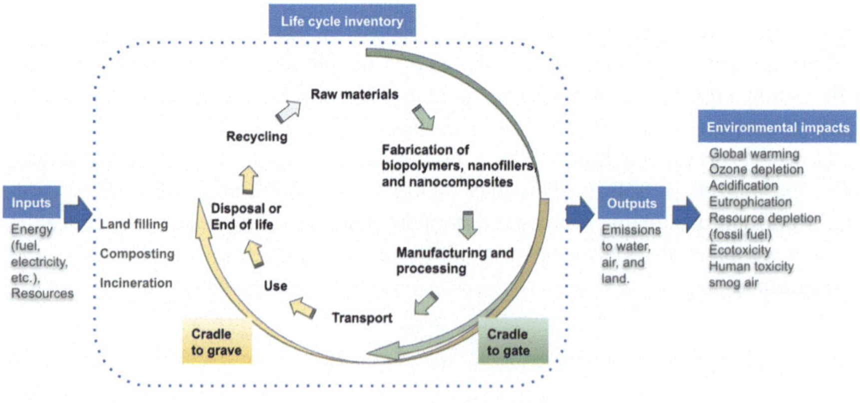

Fig. 11.1 General framework for the LCA of nanocomposite materials. (Adapted with permission from Gamage et al. [8])

Given the burgeoning interest in starch as a sustainable material, it is more important than ever to understand the potential ecological implications of these products from the cradle to the grave [61]. LCA facilitates a comprehensive examination of the environmental benefits and any unexpected impacts that might arise from starch-based nanocomposites' production, utilization, and disposal stages.

Some LCAs have been conducted for blends of starch and other polymers with different compositions. Generally, these studies conclude that starch plastics can reduce energy use and greenhouse gas emissions on a same-weight basis compared to petrochemical plastics.

Guo and Murphy [76] conducted an LCA to evaluate the environmental impact in different applications of mixtures of biopolymers based on starch and PVOH derived from different raw materials (wheat, potato, corn) and compare them with conventional plastics such as high-density polyethylene (HDPE), low-density polyethylene (LDPE) and expanded polystyrene (EPS). Their results suggest no overall environmental advantage for starch, and PVOH blended biopolymers over their petrochemical counterparts across applications. Instead, a case-by-case approach is needed to evaluate the environmental pros and cons based on specific comparisons.

Broeren et al. [61] recently performed an LCA for six different plastic starch compositions. They concluded that starch plastics have been shown to reduce greenhouse gas (GHG) emissions and non-renewable energy use (NREU) but have a greater potential for eutrophication and require greater use of agricultural land compared with the common petrochemical polymers, LDPE, and PP. However, GHG emissions savings are strongly influenced by the plastic composition; therefore, additives can account for up to 40% of GHG emissions from starchy plastics. On the other hand, using recovered starch instead of virgin starch leads to modest reductions in NREU and GHG emissions (<10% in most cases), but up to 60% reductions in eutrophication and agricultural land use.

Regarding the sustainability of extraction of nanofillers' SNCs and organically modified nanoclay montmorillonite (OMMT), global warming and acidification potential indicators of SNC were higher than those of OMMT. However, SNCs have more positive impacts than OMMT, contributing to non-renewable energy and mineral depletion [77].

As the utility of nanocomposites continues to expand, so does the relevance of LCA in their life cycle evaluation. There is an escalating demand for clarity and objectivity in understanding the environmental implications of nanocomposites. Consequently, LCA is anticipated to maintain its crucial role in delivering a rigorous, scientifically-backed evaluation of nanocomposite environmental footprint across their entire life cycle [60].

In addition, the growing global focus on a circular economy underlines the increasing importance of LCA in assessing the feasibility of various recovery and recycling strategies for nanocomposites. For instance, methods to reclaim and reuse materials from end-of-life nanocomposites could considerably reduce their environmental impact. However, these advantages must be carefully counterbalanced by the potential environmental toll of the recovery processes themselves. In this context,

LCA provides valuable insights to inform sustainable decision-making processes [61].

11.5 Conclusions

This chapter reviews degradation standards, the biodegradability of starch-based nanomaterials, and critical analysis of starch-based materials' nanotoxicity and environmental impacts. Starch nanomaterials are promising biomaterials with potential applications in various fields, such as food engineering, medicine, and textiles, because of their unique properties. However, their degradation, potential toxicity, and environmental impacts are emerging concerns. Starch nanomaterials, particularly those reinforced with polysaccharide nanoparticles, have been identified as promising and environmentally friendly alternatives to conventional materials. Although starch is commonly used in various products, including medical and pharmaceutical products, it is important to consider the possible toxicity of starch and its impurities. There is limited research on the potential toxicity of starch nanomaterials, but some studies have shown that they can adversely affect living organisms. The FDA has developed recommendations for using starch derivatives considering the potential risks of inappropriate use for human health. Further studies are needed to fully understand the effects of starch and its safety in different uses and applications. In addition, proper handling practices should be implemented to minimize the risks associated with their use. In addition, LCA provides valuable insights to inform sustainable decision-making processes. Some LCAs have been conducted for blends of starch and other polymers with different compositions. Generally, these studies conclude that starch plastics can reduce energy use and greenhouse gas emissions on the same basis of weight compared to petrochemical plastics. However, further studies are needed to fully understand the effects of starch, its environmental impacts, and safety in different uses and applications.

Acknowledgments This work was (partially) supported by Programa Iberomericano de Ciencia y Tecnología para el Desarrollo (CYTED) (through Red 121RT0108). The authors gratefully acknowledge the Centro Nacional de Alta Tecnología "Dr. Franklin Chang Díaz" (Costa Rica), the Universidad Nacional (Heredia, Costa Rica), and the University of Asunci6n (San Lorenzo, Paraguay) for their support.

Conflicts of Interest The authors declare no conflicts of interest.

References

1. Pokhrel S, Sigdel A, Lach R, Slouf M, Sirc J, Katiyar V, Bhattarai DR, Adhikari R. Starch-based biodegradable film with poly(butylene adipate-co-terephthalate): preparation,

morphology, thermal and biodegradation properties. J Macromol Scie Part A Pure Appl Chem. 2021;58:610–21.
2. Greene J. Chapter 12—Degradation and biodegradation standards for starch-based and other polymeric materials. In: Halley PJ, Avérous L, editors. Starch polymers. Amsterdam: Elsevier; 2014. p. 321–56.
3. Folino A, Pangallo D, Calabrò PS. Assessing bioplastics biodegradability by standard and research methods: current trends and open issues. J Environ Chem Eng. 2023. https://doi.org/10.1016/j.jece.2023.109424.
4. Hubackova J, Dvorackova M, Svoboda P, Mokrejs P, Kupec J, Pozarova I, Alexy P, Bugaj P, Machovsky M, Koutny M. Influence of various starch types on PCL/starch blends anaerobic biodegradation. Polym Test. 2013;32:1011–9.
5. Ghasemlou M, Daver F, Murdoch BJ, Ball AS, Ivanova EP, Adhikari B. Biodegradation of novel bioplastics made of starch, polyhydroxyurethanes and cellulose nanocrystals in soil environment. Sci Total Environ. 2022. https://doi.org/10.1016/j.scitotenv.2021.152684.
6. Nayak SK. Biodegradable PBAT/Starch nanocomposites. Polym PlastTechnol Eng. 2010;49: 1406–18.
7. Gutiérrez TJ, Toro-Márquez LA, Merino D, Mendieta JR. Hydrogen-bonding interactions and compostability of bionanocomposite films prepared from corn starch and nano-fillers with and without added Jamaica flower extract. Food Hydrocoll. 2019;89:283–93.
8. Gamage A, Thiviya P, Mani S, Ponnusamy PG, Manamperi A, Evon P, Merah O, Madhujith T. Environmental properties and applications of biodegradable starch-based nanocomposites. Polymers (Basel). 2022. https://doi.org/10.3390/polym14214578.
9. Mohan TP, Devchand K, Kanny K. Barrier and biodegradable properties of corn starch-derived biopolymer film filled with nanoclay fillers. J Plast Film Sheeting. 2016;33:309–36.
10. Wróblewska-Krepsztul J, Rydzkowski T, Borowski G, Szczypiński M, Klepka T, Thakur VK. Recent progress in biodegradable polymers and nanocomposite-based packaging materials for sustainable environment. Int J Polym Anal Charact. 2018;23:383.
11. Cano AI, Cháfer M, Chiralt A, González-Martínez C. Biodegradation behavior of starch-PVA films as affected by the incorporation of different antimicrobials. Polym Degrad Stab. 2016;132: 11–20.
12. Wongphan P, Panrong T, Harnkarnsujarit N. Effect of different modified starches on physical, morphological, thermomechanical, barrier and biodegradation properties of cassava starch and polybutylene adipate terephthalate blend film. Food Packag Shelf Life. 2022. https://doi.org/10.1016/j.fpsl.2022.100844.
13. Zoungranan Y, Lynda E, Dobi-Brice KK, Tchirioua E, Bakary C, Yannick DD. Influence of natural factors on the biodegradation of simple and composite bioplastics based on cassava starch and corn starch. J Environ Chem Eng. 2020. https://doi.org/10.1016/j.jece.2020.104396.
14. Ji M, Li F, Li J, Li J, Zhang C, Sun K, Guo Z. Enhanced mechanical properties, water resistance, thermal stability, and biodegradation of the starch-sisal fibre composites with various fillers. Mater Des. 2021. https://doi.org/10.1016/j.matdes.2020.109373.
15. Rammak T, Boonsuk P, Kaewtatip K. Mechanical and barrier properties of starch blend films enhanced with kaolin for application in food packaging. Int J Biol Macromol. 2021;192:1013–20.
16. Zhai X, Wang W, Zhang H, Dai Y, Dong H, Hou H. Effects of high starch content on the physicochemical properties of starch/PBAT nanocomposite films prepared by extrusion blowing. Carbohydr Polym. 2020. https://doi.org/10.1016/j.carbpol.2020.116231.
17. Hosseini SN, Pirsa S, Farzi J. Biodegradable nano composite film based on modified starch-albumin/MgO; antibacterial, antioxidant and structural properties. Polym Test. 2021;97: 107182.
18. Zhou Y, Chi Z, Qi X, Wang W, Yu L, Dong Y, Qian C, Fu Y. Degradable photo-crosslinked starch-based films with excellent shape memory property. Int J Biol Macromol. 2021;193: 1685–93.

19. Kong F, Singh RP. Chemical deterioration and physical instability of foods and beverages. In: The stability and shelf life of food. Duxford: Elsevier; 2016. p. 43–76.
20. Li M, Witt T, Xie F, Warren FJ, Halley PJ, Gilbert RG. Biodegradation of starch films: the roles of molecular and crystalline structure. Carbohydr Polym. 2015;122:115–22.
21. Kim HY, Park SS, Lim ST. Preparation, characterization and utilization of starch nanoparticles. Colloids Surf B Biointerfaces. 2015;126:607–20.
22. Angellier H, Molina-Boisseau S, Belgacem MN, Dufresne A. Surface chemical modification of waxy maize starch nanocrystals. Langmuir. 2005;21:2425–33.
23. Ahmad AN, Lim SA, Navaranjan N, Hsu YI, Uyama H. Green sago starch nanoparticles as reinforcing material for green composites. Polymer (Guildf). 2020. https://doi.org/10.1016/j.polymer.2020.122646.
24. Liu Q, Gao L, Qin Y, Ji N, Dai L, Xiong L, Sun Q. Incorporation of oxidized debranched starch/chitosan nanoparticles for enhanced hydrophobicity of corn starch films. Food Packag Shelf Life. 2023. https://doi.org/10.1016/j.fpsl.2023.101032.
25. Marta H, Wijaya C, Sukri N, Cahyana Y, Mohammad M. A comprehensive study on starch nanoparticle potential as a reinforcing material in bioplastic. Polymers (Basel). 2022. https://doi.org/10.3390/polym14224875.
26. Versino F, Ortega F, Monroy Y, Rivero S, López OV, García MA. Sustainable and bio-based food packaging: a review on past and current design innovations. Foods. 2023. https://doi.org/10.3390/foods12051057.
27. Kumari S, Yadav BS, Yadav R. Morphological and thermo-mechanical characterization of sweet potato starch based nanocomposites reinforced with barley starch nanoparticles. J Food Sci Technol. 2022;59:4924–34.
28. Ma X, Jian R, Chang PR, Yu J. Fabrication and characterization of citric acid-modified starch nanoparticles/plasticized-starch composites. Biomacromolecules. 2008;9:3314–20.
29. Piyada K, Waranyou S, Thawien W. Mechanical, thermal and structural properties of rice starch films reinforced with rice starch nanocrystals. Int Food Res J. 2013;20:439–49.
30. González Seligra P, Eloy Moura L, Famá L, Druzian JI, Goyanes S. Influence of incorporation of starch nanoparticles in PBAT/TPS composite films. Polym Int. 2016;65:938–45.
31. Gujral H, Sinhmar A, Nehra M, Nain V, Thory R, Pathera AK, Chavan P. Synthesis, characterization, and utilization of potato starch nanoparticles as a filler in nanocomposite films. Int J Biol Macromol. 2021;186:155–62.
32. El-Naggar ME, El-Rafie MH, El-sheikh MA, El-Feky GS, Hebeish A. Synthesis, characterization, release kinetics and toxicity profile of drug-loaded starch nanoparticles. Int J Biol Macromol. 2015;81:718–29.
33. Mostafa K, Ameen H, Morsy M, El-Ebiassy A, El-Sanabary A, Adel M, Salah A. Production of high-performance textiles via pioneering strengthening approach using starch nanoparticles. J Ind Text. 2020;50:278–92.
34. Hakke VS, Landge VK, Sonawane SH, Babu GUB, Manickam S, Boczkaj G. Cu(II) ions removal from wastewater using starch nanoparticles (SNPs): an eco-sustainable approach. Can J Chem Eng. 2023;101:1815–30.
35. Kumari S, Yadav BS, Yadav RB. Effect of nano-conversion on morphological, rheological and thermal properties of barley starch. J Food Sci Technol. 2022;59:467–77.
36. Dong X, Liu L, Wang Y, Li T, Wu Z, Yuan H, Ma P, Shi D, Chen M, Dong W. The compatibilization of poly (propylene carbonate)/poly (lactic acid) blends in presence of core-shell starch nanoparticles. Carbohydr Polym. 2021. https://doi.org/10.1016/j.carbpol.2020.117321.
37. Liu C, Jiang S, Han Z, Xiong L, Sun Q. In vitro digestion of nanoscale starch particles and evolution of thermal, morphological, and structural characteristics. Food Hydrocoll. 2016;61:344–50.
38. Madhumitha G, Fowsiya J, Mohana Roopan SM, Thakur VK. Recent advances in starch–clay nanocomposites. Int J Polym Anal Charact. 2018;23:331.

39. García NL, Famá L, D'Accorso NB, Goyanes S. Biodegradable starch nanocomposites. In: Thakur VK, Thakur MK, editors. Eco-friendly polymer nanocomposites: processing and properties. New Delhi: Springer India; 2015. p. 17–77.
40. Venkatesh G, Nyflött Å, Bonnerup C, Lestelius M. An economic-environmental analysis of selected barrier-coating materials used in packaging food products: a Swedish case study. Environ Dev Sustain. 2018;20:1483–97.
41. Wani AA, Singh P. Application of life cycle assessment for starch and starch blends. In: Starch-based polymeric materials and nanocomposites : chemistry, processing, and applications. CRC Press: Boca Raton; 2012. p. 373–96.
42. Qi Y, Yang X, Pelaez AM, Huerta Lwanga E, Beriot N, Gertsen H, Garbeva P, Geissen V. Macro- and micro- plastics in soil-plant system: effects of plastic mulch film residues on wheat (Triticum aestivum) growth. Sci Total Environ. 2018;645:1048–56.
43. Zakaria NH, Muhammad N, Abdullah MMAB. Potential of starch nanocomposites for biomedical applications. IOP Conf Ser Mater Sci Eng. 2017. https://doi.org/10.1088/1757-899X/209/1/012087.
44. Rostamabadi H, Falsafi SR, Jafari SM. 15—Nanostructures of starch for encapsulation of food ingredients. In: Jafari SM, editor. Biopolymer nanostructures for food encapsulation purposes. London: Academic; 2019. p. 419–62.
45. Sun Q. Chapter 18—Starch nanoparticles. In: Sjöö M, Nilsson L, editors. Starch in food. 2nd ed. Duxford: Woodhead Publishing; 2018. p. 691–745.
46. What is oxidative stress? Effects on the body and how to reduce. https://www.medicalnewstoday.com/articles/324863. Accessed 4 Dec 2023.
47. Food and Drug Administration D of H and HS Title 21—Food and Drugs. Chapter I -Food and Drug Administration, Department of Health and Human Services. Part 172 Food Additives Permitted for Direct Addition to Food for Human Consumption Subpart A General Provisions.
48. Lemos PVF, Marcelino HR, Cardoso LG, Souza CO de, Druzian JI. Starch chemical modifications applied to drug delivery systems: from fundamentals to FDA-approved raw materials. Int J Biol Macromol. 2021;184:218–34.
49. Ates B, Koytepe S, Ulu A, Gurses C, Thakur VK. Chemistry, structures, and advanced applications of nanocomposites from biorenewable resources. Chem Rev. 2020;120:9304–62.
50. Mohammad F, Arfin T, Bwatanglang IB, Al-lohedan HA. Starch-based nanocomposites: types and industrial applications. In: Sanyang ML, Jawaid M, editors. Bio-based polymers and nanocomposites : preparation, processing, properties & performance. Cham: Springer; 2019. p. 157–81.
51. Arora B, Bhatia R, Attri P. 28—Bionanocomposites: green materials for a sustainable future. In: Hussain CM, Mishra AK, editors. New polymer nanocomposites for environmental remediation. Amsterdam: Elsevier; 2018. p. 699–712.
52. Allouzi MMA, Tang DYY, Chew KW, Rinklebe J, Bolan N, Allouzi SMA, Show PL. Micro (nano) plastic pollution: the ecological influence on soil-plant system and human health. Sci Total Environ. 2021;788:147815.
53. Abe MM, Branciforti MC, Nallin Montagnolli R, Marin Morales MA, Jacobus AP, Brienzo M. Production and assessment of the biodegradation and ecotoxicity of xylan- and starch-based bioplastics. Chemosphere. 2022;287:132290.
54. Arcos-Hernandez MV, Laycock B, Pratt S, Donose BC, Nikolić MAL, Luckman P, Werker A, Lant PA. Biodegradation in a soil environment of activated sludge derived polyhydroxyalkanoate (PHBV). Polym Degrad Stab. 2012;97:2301–12.
55. BeMiller JN. Chapter 5—Physical modification of starch. In: Sjöö M, Nilsson L, editors. Starch in food. 2nd ed. Duxford: Woodhead Publishing; 2018. p. 223–53.
56. Zarski A, Bajer K, Kapuśniak J. Review of the most important methods of improving the processing properties of starch toward non-food applications. Polymers (Basel). 2021. https://doi.org/10.3390/polym13050832.

57. Kumari S, Yadav BS, Yadav RB. Synthesis and modification approaches for starch nanoparticles for their emerging food industrial applications: a review. Food Res Int. 2020. https://doi.org/10.1016/j.foodres.2019.108765.
58. Lavoine N, Desloges I, Dufresne A, Bras J. Microfibrillated cellulose—its barrier properties and applications in cellulosic materials: a review. Carbohydr Polym. 2012;90:735–64.
59. Klöpffer W, Grahl B. Life cycle assessment (LCA): a guide to best practice. Weinheim: Wiley-VCH; 2014. https://doi.org/10.1002/9783527655625.
60. Bishop G, Styles D, Lens PNL. Environmental performance comparison of bioplastics and petrochemical plastics: a review of life cycle assessment (LCA) methodological decisions. Resour Conserv Recycl. 2021;168:105451.
61. Broeren MLM, Kuling L, Worrell E, Shen L. Environmental impact assessment of six starch plastics focusing on wastewater-derived starch and additives. Resour Conserv Recycl. 2017;127:246–55.
62. Joshi S. Can nanotechnology improve the sustainability of biobased products? J Ind Ecol. 2008;12:474–89.
63. Chisenga SM, Workneh TS, Bultosa G, Alimi BA. Progress in research and applications of cassava flour and starch: a review. J Food Sci Technol. 2019;56:2799–813.
64. Zhong Y, Liu L, Qu J, Blennow A, Hansen AR, Wu Y, Guo D, Liu X. Amylose content and specific fine structures affect lamellar structure and digestibility of maize starches. Food Hydrocoll. 2020;108:105994.
65. Thanyapanich N, Jimtaisong A, Rawdkuen S. Functional properties of banana starch (Musa spp.) and its utilization in cosmetics. Molecules. 2021. https://doi.org/10.3390/molecules26123637.
66. Jiménez L, Mena M, Prendiz J, Salas L, Vega-Baudrit J. Polylactic acid (PLA) as a bioplastic and its possible applications in the food industry. Food Sci Nutr. 2019;5:1–6.
67. Govindaraju I, Zhuo G-Y, Chakraborty I, Melanthota SK, Mal SS, Sarmah B, Baruah VJ, Mahato KK, Mazumder N. Investigation of structural and physico-chemical properties of rice starch with varied amylose content: a combined microscopy, spectroscopy, and thermal study. Food Hydrocoll. 2022;122:107093.
68. Pérez S, Baldwin PM, Gallant DJ. Structural features of starch granules I. In: Starch. London: Elsevier Inc; 2009. p. 149–92.
69. Cornejo-Ramírez YI, Martínez-Cruz O, Del Toro-Sánchez CL, Wong-Corral FJ, Borboa-Flores J, Cinco-Moroyoqui FJ. The structural characteristics of starches and their functional properties. CyTA J Food. 2018;16:1003–17.
70. Alves Z, Abreu B, Ferreira NM, Marques EF, Nunes C, Ferreira P. Enhancing the dispersibility of multiwalled carbon nanotubes within starch-based films by the use of ionic surfactants. Carbohydr Polym. 2021;273:118531.
71. Chaudhary AK, Vijayakumar RP. Synthesis of polystyrene/starch/CNT composite and study on its biodegradability. J Polym Res. 2020;27:187.
72. Kochkina NE, Butikova OA. Effect of fibrous TiO2 filler on the structural, mechanical, barrier and optical characteristics of biodegradable maize starch/PVA composite films. Int J Biol Macromol. 2019;139:431–9.
73. Bher A, Unalan IU, Auras R, Rubino M, Schvezov CE. Graphene modifies the biodegradation of poly(lactic acid)-thermoplastic cassava starch reactive blend films. Polym Degrad Stab. 2019;164:187–97.
74. Prusty K, Swain SK. Nano CaCO3 imprinted starch hybrid polyethylhexylacrylate \polyvinylalcohol nanocomposite thin films. Carbohydr Polym. 2016;139:90–8.

75. Kakadellis S, Harris ZM. Don't scrap the waste: the need for broader system boundaries in bioplastic food packaging life-cycle assessment—a critical review. J Clean Prod. 2020;274: 122831.
76. Guo M, Murphy RJ. Is there a generic environmental advantage for starch-PVOH biopolymers over petrochemical polymers? J Polym Environ. 2012;20:976–90.
77. Le Corre D, Angellier-Coussy H. Preparation and application of starch nanoparticles for nanocomposites: a review. React Funct Polym. 2014;85:97–120.

Index

G. Ayala Valencia (ed.), *Starch Nanomaterials and Food Applications*, https://doi.org/10.1007/978-3-031-60086-9

GPSR Compliance

The European Union's (EU) General Product Safety Regulation (GPSR) is a set of rules that requires consumer products to be safe and our obligations to ensure this.

If you have any concerns about our products, you can contact us on ProductSafety@springernature.com

In case Publisher is established outside the EU, the EU authorized representative is:

Springer Nature Customer Service Center GmbH
Europaplatz 3
69115 Heidelberg, Germany

The manufacturer's authorised representative in the EU is Springer Nature Customer Service Centre GmbH, Europaplatz 3, 69115 Heidelberg, Germany. If you have any concerns regarding our products, please contact ProductSafety@springernature.com

Printed and bound by CPI Group (UK) Ltd, Croydon, CR0 4YY
01/10/2026
02225904-0004